Lecture Notes in Computer Science 11832

Wil M. P. van der Aalst · Vladimir Batagelj ·
Dmitry I. Ignatov · Michael Khachay ·
Valentina Kuskova · Andrey Kutuzov ·
Sergei O. Kuznetsov · Irina A. Lomazova ·
Natalia Loukachevitch · Amedeo Napoli ·
Panos M. Pardalos · Marcello Pelillo ·
Andrey V. Savchenko · Elena Tutubalina (Eds.)

Analysis of Images, Social Networks and Texts

8th International Conference, AIST 2019
Kazan, Russia, July 17–19, 2019
Revised Selected Papers

 Springer

Editors
Wil M. P. van der Aalst 🆔
RWTH Aachen University
Aachen, Germany

Dmitry I. Ignatov 🆔
National Research University
Higher School of Economics
Moscow, Russia

Valentina Kuskova 🆔
National Research University
Higher School of Economics
Moscow, Russia

Sergei O. Kuznetsov 🆔
National Research University
Higher School of Economics
Moscow, Russia

Natalia Loukachevitch 🆔
Lomonosov Moscow State University
Moscow, Russia

Panos M. Pardalos 🆔
University of Florida
Gainesville, FL, USA

Andrey V. Savchenko 🆔
National Research University
Higher School of Economics
Nizhny Novgorod, Russia

Vladimir Batagelj 🆔
University of Ljubljana
Ljubljana, Slovenia

Michael Khachay 🆔
Krasovskii Institute of Mathematics
and Mechanics
Yekaterinburg, Russia

Andrey Kutuzov 🆔
University of Oslo
Oslo, Norway

Irina A. Lomazova 🆔
National Research University
Higher School of Economics
Moscow, Russia

Amedeo Napoli 🆔
LORIA
Vandœuvre-lès-Nancy, France

Marcello Pelillo 🆔
Ca Foscari University of Venice
Venice, Italy

Elena Tutubalina 🆔
Kazan Federal University
Kazan, Russia

ISSN 0302-9743   ISSN 1611-3349   (electronic)
Lecture Notes in Computer Science
ISBN 978-3-030-37333-7   ISBN 978-3-030-37334-4   (eBook)
https://doi.org/10.1007/978-3-030-37334-4

LNCS Sublibrary: SL3 – Information Systems and Applications, incl. Internet/Web, and HCI

This Springer imprint is published by the registered company Springer Nature Switzerland AG
The registered company address is: Gewerbestrasse 11, 6330 Cham, Switzerland

Preface

This volume contains the refereed proceedings of the 8th International Conference on Analysis of Images, Social Networks, and Texts (AIST 2019)[1]. The previous conferences (during 2012–2018) attracted a significant number of data scientists – students, researchers, academics, and engineers – working on interdisciplinary data analysis of images, texts, and social networks.

The broad scope of AIST made it an event where researchers from different domains, such as image and text processing, exploiting various data analysis techniques, can meet and exchange ideas. We strongly believe that this may lead to the cross-fertilisation of ideas between researchers relying on modern data analysis machinery.

Therefore, AIST 2019 brought together all kinds of applications of data mining and machine learning techniques. The conference allowed specialists from different fields to meet each other, present their work, and discuss both theoretical and practical aspects of their data analysis problems. Another important aim of the conference was to stimulate scientists and people from industry to benefit from the knowledge exchange and identify possible grounds for fruitful collaboration.

The conference was held during July 17–19, 2019. The conference was organised in Kazan, the capital of the Republic of Tatarstan, Russia, on the campus of Kazan (Volga region) Federal University[2].

This year, the key topics of AIST were grouped into six tracks:

1. General Topics of Data Analysis chaired by Sergei O. Kuznetsov (Higher School of Economics, Russia) and Amedeo Napoli (Loria, France)
2. Natural Language Processing chaired by Natalia Loukachevitch (Lomonosov Moscow State University, Russia), Andrey Kutuzov (University of Oslo, Norway), and Elena Tutubalina (Kazan Federal University, Russia)
3. Social Network Analysis chaired by Vladimir Batagelj (University of Ljubljana, Slovenia) and Valentina Kuskova (Higher School of Economics, Russia)
4. Analysis of Images and Video chaired by Marcello Pelillo (University of Venice, Italy) and Andrey V. Savchenko (Higher School of Economics, Russia)
5. Optimisation Problems on Graphs and Network Structures chaired by Panos M. Pardalos (University of Florida, USA) and Michael Khachay (IMM UB RAS and Ural Federal University, Russia)
6. Analysis of Dynamic Behaviour Through Event Data chaired by Wil M. P. van der Aalst (RWTH Aachen University, Germany) and Irina A. Lomazova (Higher School of Economics, Russia)

---

[1] http://aistconf.org

[2] https://kpfu.ru/eng

The Programme Committee and the reviewers of the conference included 160 well-known experts in data mining and machine learning, natural language processing, image processing, social network analysis, and related areas from leading institutions of 24 countries including Argentina, Australia, Austria, Canada, Czech Republic, Denmark, France, Germany, Greece, India, Iran, Italy, Japan, Lithuania, the Netherlands, Norway, Qatar, Romania, Russia, Slovenia, Spain, Taiwan, Ukraine, and the USA. This year, we received 134 submissions: mostly from Russia but also from Australia, Belarus, Finland, Germany, India, Italy, Norway, Pakistan, Russia, Spain, Sweden, and Vietnam.

Out of 134 submissions, only 27 full papers and 8 short papers were accepted as regular oral papers. Thus, the acceptance rate of this volume was around 24% (not taking into account 21 automatically rejected papers). An invited opinion talk and a tutorial paper are also included in this volume. In order to encourage young practitioners and researchers, we included 36 papers in the companion volume after their poster presentation at the conference. Each submission was reviewed by at least three reviewers, experts in their fields, in order to supply detailed and helpful comments.

The conference featured several invited talks and an industry session dedicated to current trends and challenges.

The invited talks from academia were on Computer Vision and NLP, respectively:

- Ivan Laptev (Inria, Paris; VisionLabs): "Towards Embodied Action Understanding"
- Alexander Panchenko (Skolkovo Institute of Science and Technology, Russia): "Representing Symbolic Linguistic Structures for Neural NLP: Methods and Applications"

The invited industry speakers gave the following talks:

- Elena Voita (Yandex, Russia): "Machine Translation: Analysing Multi-Head Self-Attention"
- Yuri Malkov (Samsung AI Center, Russia): "Learnable Triangulation of Human Pose"
- Oleg Tishutin and Ekaterina Safonova (Iponweb, Russia): "Fraud Detection in Real-Time Bidding".

The program also included a tutorial on high-performance tools for deep models:

- Evgenii Vasilyev (Lobachevski State University of Nizhni Novgorod, Russia), Gleb Gladilov (Intel Corporation, Russia): "Intel® Distribution of OpenVINO™ Toolkit: A Case Study of Semantic Segmentation"

An invited opinion talk on comparison of academic communities formed by the authors of Russian-speaking NLP-oriented conferences was presented by Andrey Kutuzov and Irina Nikishina under the title "Double-Blind Peer-Reviewing and Inclusiveness in Russian NLP Conferences."

We would like to thank the authors for submitting their papers and the members of the Programme Committee for their efforts in providing exhaustive reviews.

According to the programme chairs, and taking into account the reviews and presentation quality, the Best Paper Awards were granted to the following papers:

- Track 1. General Topics of Data Analysis: "Histogram-Based Algorithm for Building Gradient Boosting Ensembles of Piece-Wise Linear Decision Trees" by Alexey Gurianov
- Track 2. Natural Language Processing: "Authorship Attribution in Russian with New High-Performing and Fully Interpretable Morpho-Syntactic Features" by Elena Pimonova, Oleg Durandin, and Alexey Malafeev
- Track 3. Social Network Analysis: "Analysis of Students Educational Interests Using Social Networks Data" by Evgeny Komotskiy, Tatiana Oreshkina, Liubov Zabokritskaya, Marina Medvedeva, Andrey Sozykin, and Nikolai Khlebnikov
- Track 4. Analysis of Images and Video: "Data Augmentation with GAN: Improving Chest X-rays Pathologies Prediction on Class-Imbalanced Cases" by Tatiana Malygina, Elena Ericheva, and Ivan Drokin
- Track 5. Optimisation Problems on Graphs and Network Structures: "Efficient PTAS for the Euclidean Capacitated Vehicle Routing Problem with Non-Uniform Non-Splittable Demand" by Michael Khachay and Yuri Ogorodnikov
- Track 6. Analysis of Dynamic Behaviour Through Event Data: "Method to Improve Workflow Net Decomposition for Process Model Repair" by Semyon Tikhonov and Alexey Mitsyuk

We would also like to express our special gratitude to all the invited speakers and industry representatives.

We deeply thank all the partners and sponsors. Especially, the hosting university, our main sponsor and the co-organiser this year, the National Research University Higher School of Economics, as well as Springer, who sponsored the Best Paper Awards.

Our special thanks go to Springer for their help, starting from the first conference call to the final version of the proceedings. Last but not least, we are grateful to Airat Khasianov and Valery Solovyev from the Higher Institute of Information Technology and Intelligent Systems of KFU, and all the organisers, especially to Yuri Dedenev, and the volunteers, whose endless energy saved us at the most critical stages of the conference preparation.

Here, we would like to mention that the Russian word "aist" is more than just a simple abbreviation (in Cyrillic) – it means "a stork". Since it is a wonderful free bird, a symbol of happiness and peace, this stork gave us the inspiration to organise the AIST conference series. So we believe that this young and rapidly growing conference will likewise bring inspiration to data scientists around the world!

October 2019

Wil M. P. van der Aalst
Vladimir Batagelj
Dmitry I. Ignatov
Michael Khachay
Valentina Kuskova
Andrey Kutuzov
Sergei O. Kuznetsov
Irina A. Lomazova
Natalia Loukachevitch
Amedeo Napoli
Panos M. Pardalos
Marcello Pelillo
Andrey V. Savchenko
Elena Tutubalina

Organisation

Programme Committee Chairs

Wil M. P. van der Aalst	RWTH Aachen University, Germany
Vladimir Batagelj	University of Ljubljana, Slovenia
Michael Khachay	Krasovskii Institute of Mathematics and Mechanics of Russian Academy of Sciences and Ural Federal University, Russia
Valentina Kuskova	National Research University Higher School of Economics, Russia
Andrey Kutuzov	University of Oslo, Norway
Sergei O. Kuznetsov	National Research University Higher School of Economics, Russia
Amedeo Napoli	Loria – CNRS, University of Lorraine, Inria, France
Irina A. Lomazova	National Research University Higher School of Economics, Russia
Natalia Loukachevitch	Computing Centre of Lomonosov Moscow State University, Russia
Panos M. Pardalos	University of Florida, USA
Marcello Pelillo	University of Venice, Italy
Andrey V. Savchenko	National Research University Higher School of Economics, Russia
Elena Tutubalina	Kazan Federal University, Russia

Proceedings Chair

Dmitry I. Ignatov	National Research University Higher School of Economics, Russia

Steering Committee

Dmitry I. Ignatov	National Research University Higher School of Economics, Russia
Michael Khachay	Krasovskii Institute of Mathematics and Mechanics of Russian Academy of Sciences and Ural Federal University, Russia
Alexander Panchenko	University of Hamburg, Germany, and Université catholique de Louvain, Belgium
Andrey V. Savchenko	National Research University Higher School of Economics, Russia
Rostislav Yavorskiy	National Research University Higher School of Economics, Russia

Programme Committee

Anton Alekseev	St. Petersburg Department of Steklov Institute of Mathematics of the Russian Academy of Sciences, Russia
Ilseyar Alimova	Kazan Federal University, Russia
Vladimir Arlazarov	Smart Engines Ltd and Federal Research Centre "Computer Science and Control" of the Russian Academy of Sciences, Russia
Aleksey Artamonov	Neuromation, Russia
Ekaterina Artemova	National Research University Higher School of Economics, Russia
Jaume Baixeries	Universitat Politècnica de Catalunya, Spain
Amir Bakarov	National Research University Higher School of Economics, Russia
Vladimir Batagelj	University of Ljubljana, Slovenia
Laurent Beaudou	National Research University Higher School of Economics, Russia
Malay Bhattacharyya	Indian Statistical Institute, India
Chris Biemann	University of Hamburg, Germany
Elena Bolshakova	Moscow State Lomonosov University, Russia
Andrea Burattin	Technical University of Denmark, Denmark
Evgeny Burnaev	Skolkovo Institute of Science and Technology, Russia
Aleksey Buzmakov	National Research University Higher School of Economics, Russia
Ignacio Cassol	Universidad Austral, Argentina
Artem Chernodub	Ukrainian Catholic University and Grammarly, Ukraine
Mikhail Chernoskutov	Krasovskii Institute of Mathematics and Mechanics of the Ural Branch of the Russian Academy of Sciences and Ural Federal University, Russia
Alexey Chernyavskiy	Samsung R&D Institute, Russia
Massimiliano de Leoni	University of Padua, Italy
Ambra Demontis	University of Cagliari, Italy
Boris Dobrov	Moscow State Lomonosov University, Russia
Sofia Dokuka	National Research University Higher School of Economics, Russia
Alexander Drozd	Tokyo Institute of Technology, Japan
Shiv Ram Dubey	Indian Institute of Information Technology, India
Svyatoslav Elizarov	Alterra AI, USA
Victor Fedoseev	Samara National Research University, Russia
Elena Filatova	City University of New York, USA
Goran Glavaš	University of Mannheim, Germany
Ivan Gostev	National Research University Higher School of Economics, Russia
Natalia Grabar	Université Lille, France

Artem Grachev	Samsung R&D Institute and National Research University Higher School of Economics, Russia
Dmitry Granovsky	Yandex, Russia
Dmitry Gubanov	Trapeznikov Institute of Control Sciences of the Russian Academy of Sciences, Russia
Dmitry I. Ignatov	National Research University Higher School of Economics, Russia
Dmitry Ilvovsky	National Research University Higher School of Economics, Russia
Max Ionov	Goethe University Frankfurt, Germany, and Lomonosov Moscow State University, Russia
Vladimir Ivanov	Innopolis University, Russia
Anna Kalenkova	National Research University Higher School of Economics, Russia
Ilia Karpov	National Research University Higher School of Economics, Russia
Nikolay Karpov	Sberbank and National Research University Higher School of Economics, Russia
Egor Kashkin	Vinogradov Russian Language Institute of the Russian Academy of Sciences, Russia
Yury Kashnitsky	Koninklijke KPN N.V., The Netherlands
Mehdi Kaytoue	Infologic and LIRIS – CNRS, France
Alexander Kazakov	Matrosov Institute for System Dynamics and Control Theory, Siberian Branch of the Russian Academy of Sciences, Russia
Alexander Kelmanov	Sobolev Institute of Mathematics, Siberian Branch of the Russian Academy of Sciences, Russia
Attila Kertesz-Farkas	National Research University Higher School of Economics, Russia
Mikhail Khachay	Krasovsky Institute of Mathematics and Mechanics, Russia
Alexander Kharlamov	Institute of Higher Nervous Activity and Neurophysiology of the Russian Academy of Sciences, Russia
Javad Khodadoust	Payame Noor University, Iran
Donghyun Kim	Kennesaw State University, USA
Denis Kirjanov	National Research University Higher School of Economics, Russia
Sergei Koltcov	National Research University Higher School of Economics, Russia
Jan Konecny	Palacký University Olomouc, Czech Republic
Anton Konushin	Lomonosov Moscow State University and National Research University Higher School of Economics, Russia
Andrey Kopylov	Tula State University, Russia
Mikhail Korobov	ScrapingHub Inc., Ireland

Evgeny Kotelnikov	Vyatka State University, Russia
Ilias Kotsireas	Maplesoft and Wilfrid Laurier University, Canada
Boris Kovalenko	National Research University Higher School of Economics, Russia
Fedor Krasnov	Gazprom Neft, Russia
Ekaterina Krekhovets	National Research University Higher School of Economics, Russia
Tomas Krilavicius	Vytautas Magnus University, Lithuania
Anvar Kurmukov	Kharkevich Institute for Information Transmission Problems of the Russian Academy of Sciences, Russia
Valentina Kuskova	National Research University Higher School of Economics, Russia
Valentina Kustikova	Lobachevsky State University of Nizhny Novgorod, Russia
Andrey Kutuzov	University of Oslo, Norway
Andrey Kuznetsov	Samara National Research University, Russia
Sergei O. Kuznetsov	National Research University Higher School of Economics, Russia
Florence Le Ber	Université de Strasbourg, France
Alexander Lepskiy	National Research University Higher School of Economics, Russia
Bertrand M. T. Lin	National Chiao Tung University, Taiwan
Benjamin Lind	Anglo-American School of St. Petersburg, Russia
Irina A. Lomazova	National Research University Higher School of Economics, Russia
Konstantin Lopukhin	Scrapinghub, Ireland
Anastasiya Lopukhina	National Research University Higher School of Economics, Russia
Natalia Loukachevitch	Lomonosov Moscow State University, Russia
Ilya Makarov	National Research University Higher School of Economics, Russia
Tatiana Makhalova	National Research University Higher School of Economics, Russia, and Loria – Inria, France
Olga Maksimenkova	National Research University Higher School of Economics, Russia
Alexey Malafeev	National Research University Higher School of Economics, Russia
Yury Malkov	Samsung AI Center, Russia
Valentin Malykh	Vk.ru and Moscow Institute of Physics and Technology, Russia
Nizar Messai	Université François Rabelais Tours, France
Tristan Miller	Austrian Research Institute for Artificial Intelligence, Austria
Olga Mitrofanova	St. Petersburg State University, Russia

Alexey A. Mitsyuk	National Research University Higher School of Economics, Russia
Evgeny Myasnikov	Samara State University, Russia
Amedeo Napoli	Loria – CNRS, University of Lorraine, Inria, France
Long Nguyen The	Irkutsk State Technical University, Russia
Huong Nguyen Thu	Irkutsk State Technical University, Russia
Kirill Nikolaev	National Research University Higher School of Economics, Russia
Damien Nouvel	Inalco University, France
Dimitri Nowicki	Glushkov Institute of Cybernetics of the National Academy of Sciences, Ukraine
Evgeniy M. Ozhegov	National Research University Higher School of Economics, Russia
Alina Ozhegova	National Research University Higher School of Economics, Russia
Alexander Panchenko	University of Hamburg, Germany and Skolkovo Institute of Science and Technology, Russia
Panos M. Pardalos	University of Florida, USA
Marcello Pelillo	University of Venice, Italy
Georgios Petasis	National Centre for Scientific Research Demokritos, Greece
Anna Petrovicheva	Xperience AI, Russia
Alex Petunin	Ural Federal University, Russia
Stefan Pickl	Bundeswehr University Munich, Germany
Vladimir Pleshko	RCO LLC, Russia
Aleksandr I. Panov	Institute for Systems Analysis of the Russian Academy of Sciences, Russia
Maxim Panov	Skolkovo Institute of Science and Technology, Russia
Mikhail Posypkin	Dorodnicyn Computing Centre of the Russian Academy of Sciences, Russia
Anna Potapenko	National Research University Higher School of Economics, Russia
Surya Prasath	Cincinnati Children's Hospital Medical Center, USA
Andrea Prati	University of Parma, Italy
Artem Pyatkin	Novosibirsk State University and Sobolev Institute of Mathematics of the Siberian Branch of the Russian Academy of Sciences, Russia
Irina Radchenko	ITMO University, Russia
Delhibabu Radhakrishnan	Kazan Federal University and Innopolis, Russia
Vinit Ravishankar	University of Oslo, Norway
Evgeniy Riabenko	Facebook, UK
Anna Rogers	University of Massachusetts Lowell, USA
Alexey Romanov	University of Massachusetts Lowell, USA
Yuliya Rubtsova	Ershov Institute of Informatics Systems, Siberian Branch of the Russian Academy of Sciences, Russia
Alexey Ruchay	Chelyabinsk State University, Russia

Additional Reviewers

Saba Anwar
Vladimir Bashkin
Anna Berger
Vladimir Berikov
Rémi Cardon
Sofia Dokuka
Dmitrii Egurnov
Elizaveta Goncharova
Ivan Gostev
Artem Grachev

Natalia Korepanova
Anna Muratova
Victor Nedel'Ko
Alexander Plyasunov
Vitaly Romanov
Alexander Semenov
Grigory Serebryakov
Andrei Shcherbakov
Oleg Slavin
Kirill Struminskiy

Organising Committee

Andrey Novikov (Head of Organisation)	National Research University Higher School of Economics, Russia
Elena Tutubalina (Local Organising Chair)	Kazan Federal University, Russia
Yury Dedenev (Venue Organisation)	Kazan Federal University, Russia
Anna Ukhanaeva (Communications and Website Support)	National Research University Higher School of Economics, Russia
Valeria Andrianova (Travel Support)	National Research University Higher School of Economics, Russia
Ilseyar Alimova	Kazan Federal University, Russia
Anna Kalenkova	National Research University Higher School of Economics, Russia
Ayrat Khasyanov	Kazan Federal University, Russia
Zulfat Miftahutdinov	Kazan Federal University, Russia
Valery Solovyev	Kazan Federal University, Russia

Volunteer

Anita Kurova   Kazan Federal University, Russia

Sponsors

National Research University Higher School of Economics, Russia
Springer

Invited Talks

Towards Embodied Action Understanding

Ivan Laptev[1,2]

[1] Inria, Paris
[2] VisionLabs, the Netherlands
ivan.laptev@inria.fr

Abstract. Computer vision has come a long way towards automatic labelling of objects, scenes and human actions in visual data. While this recent progress already powers applications such as visual search and autonomous driving, visual scene understanding remains an open challenge beyond specific applications. In this talk I outline limitations of human-defined labels and argue for the task-driven approach to scene understanding. Towards this goal I describe our recent efforts on learning visual models from narrated instructional videos. I present methods for automatic discovery of actions and object states associated with specific tasks such as changing a car tire or making coffee. Along these efforts, I describe a state-of-the-art method for text-based video search using our recent dataset with automatically collected 100M narrated videos. Finally, I present our work on visual scene understanding for real robots where we learn agents to discover sequences of actions for completing particular tasks.

Keywords: Action understanding · Text-based video search · Visual scene understanding · Narrated videos

Representing Symbolic Linguistic Structures for Neural NLP: Methods and Applications

Alexander Panchenko

Skolkovo Institute of Science and Technology, Russia
A.Panchenko@skoltech.ru

Abstract. In this talk, I will speak of several papers, which will be presented at ACL 2019: some of them on the main conference, some at the workshops and related events. The central topic of many of them will be how one can make use of symbolic linguistic structures, such as knowledge bases and graphs of taxonomic relations in the era of neural NLP models. It is not obvious to directly encode graph structures due to their sparseness in a neural network (and almost any lexical resource could be considered as a form of a multi-label weighted graph). On the other hand, hundreds of man-years were spent to manually encode some linguistic information into these resources and it may be a big miss to not use them. However, to date, most of the neural NLP models rely on word and character embeddings which are derived from text only, potentially limiting their performance.

Keywords: Natural language processing · Symbolic linguistic structures · Word embeddings · Knowledge bases · Taxonomic relations

Invited Industry Talks

Machine Translation: Analysing Multi-head Self-attention

Elena Voita

Yandex, Moscow, Russia
`lena-voita@yandex-team.ru`

Abstract. Attention is an integral part of the state of the art architectures for NLP. At a high-level, attention mechanism enables a neural network to "focus" on relevant parts of its input more than the irrelevant parts. If an attention mechanism is designed in a way that it is able to focus on different input aspects simultaneously, it is called "multi-head attention". Multi-head attention is a key component of the Transformer, a state-of-the-art architecture for neural machine translation. Multi-head attention was shown to make more efficient use of the model's capacity, but its importance for translation and roles of individual "heads" are not clear.

In this talk, I briefly describe standard attention in sequence to sequence models, as well as the Transformer architecture with multi-head self-attention. Then, we evaluate the contribution made by individual attention heads to the overall performance of the Transformer and analyse the roles played by them. I show that the most important and confident heads play consistent and often linguistically-interpretable roles. When pruning heads using a method based on stochastic gates and a differentiable relaxation of the L_0 penalty, we observe that specialised heads are last to be pruned. Our novel pruning method removes the vast majority of heads without seriously affecting performance.

The talk is based on our recent work [1].

Keywords: Natural language processing · Text annotation · Text classification · Sequence labelling

Reference

1. Voita, E., Talbot, D., Moiseev, F., Sennrich, R., Titov, I.: Analyzing multi-head self-attention: specialized heads do the heavy lifting, the rest can be pruned. ACL (1), 5797–5808 (2019)

Learnable Triangulation of Human Pose

Yury Malkov

Samsung AI Center, Moscow, Russia
goran@informatik.uni-mannheim.de

Abstract. We present two novel solutions for multi-view 3D human pose estimation based on new learnable triangulation methods that combine 3D information from multiple 2D views. Crucially, both approaches are end-to-end differentiable, which allows us to directly optimise the target metric. We demonstrate transferability of the solutions across datasets and significantly improve the multi-view state of the art on the Human3.6M dataset.

Keywords: Natural language processing · Word embeddings · Lexico-semantic relations

Fraud Detection in Real Time Bidding

Oleg Tishutin and Ekaterina Safonova

IPONWEB, Moscow, Russia
goran@informatik.uni-mannheim.de

Abstract. In the talk, we address the types of fraud in RTB (Real Time Bidding) ecosystem (bots, ad stacking, spoof sites). Then we discuss what kind of fraud can be resolved by means of various approaches including machine learning, e.g. modified bid clustering for good traffic (human) and bad (bot). We also discuss which clustering method is better, which way of learning (supervised/unsupervised) is suitable, how feature selection may help in terms of fighting fraud. As for the technical part, we discuss the impact of different parameters (e.g., size of learning sample, number of Google Cloud Engine machines needed) and possible ways of computational optimisation.

Keywords: Real time bidding · Fraud detection · Web advertising · Machine learning

Contents

Natural Language Processing

Optimization Problems on Graphs and Network Structures

Analysis of Dynamic Behavior Through Event Data

Invited Opinion Talk

Double-Blind Peer-Reviewing and Inclusiveness in Russian NLP Conferences

Andrey Kutuzov[1]([✉])(iD) and Irina Nikishina[2,3](iD)

[1] University of Oslo, Oslo, Norway
andreku@ifi.uio.no
[2] Skolkovo Institute of Science and Technology (Skoltech), Moscow, Russia
[3] National Research University Higher School of Economics, Moscow, Russia
irina.nikishina@mail.ru

Abstract. Double-blind peer reviewing has been proved to be pretty effective and fair way of academic work selection. However, to the best of our knowledge, nobody has yet analysed the effects caused by its introduction at the Russian NLP conferences. We investigate how the double-blind peer reviewing influences gender and location (according to authors' affiliations) biases and whether it makes two of the conferences under analysis more inclusive. The results show that gender distribution has become more equal for the Dialogue conference, but did not change for the AIST conference. The authors' location distribution (roughly divided into 'central' and 'not central') has become more equal for AIST, but, interestingly, less equal for Dialogue.

1 Setting the Question

Double-blind peer-reviewing means that the authors of the submitted papers do not know the names of the reviewers, and the reviewers do not know the names of the authors.

Peer review originates from the publishing process of Philosophical Transactions journal in the middle of the eighteenth century: its reviewing policy implied sending manuscripts to experts before publishing [3]. By the middle of twentieth century, peer reviewing has become the widely acknowledged standard for all top-tier international journals and conferences. Despite the long history, first papers devoted to single-blind and double blind review comparison date back only to the 1980s [4].

In comparison to the double-blind system, other setups where reviewers know the names of the authors have an obvious shortcoming: human subjectivity. In other words, the reviewers' decisions are inevitably biased (consciously or unconsciously). For instance, according to [5], papers by well-known authors are accepted 1.5 times more often, by well-known companies—2 times more often, with a female first author—20% less often. Double-blind peer reviewing successfully tackles the problem, significantly alleviating the bias [1]. Apparently,

W. M. P. van der Aalst et al. (Eds.) AIST 2019, LNCS 11832, pp. 3–8, 2019.
https://doi.org/10.1007/978-3-030-37334-4_1

it does not solve all the existing issues, but it does make the scientific program more diverse and the conference itself more inclusive.

The main conferences in Computational linguistics and Natural Language Processing in Russia also try to keep up with that trend: AIST[1] switched completely to double-blind reviewing starting from 2017, Dialogue[2] did the same in 2019.

In this paper we set to find out whether the 'double-blind turn' influences the most widespread biases about gender and place of origin of the authors of the accepted papers. In particular, our research questions are:

1. Did the ratio of female authors in the accepted papers increased after the introduction of double-blind reviewing?
2. Did the number of 'non-centrally located' authors increased after the introduction of double-blind reviewing?

2 Inclusiveness in Russian NLP Conferences

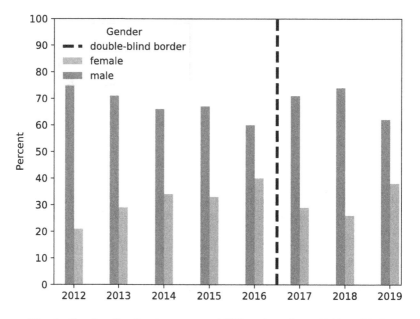

Fig. 1. Gender distribution among AIST authors from 2012 to 2019

In order to measure inclusiveness, we use data from the RusNLP project[3] [2] for the annual AIST (years 2012–2019) and Dialogue (years 2000–2019) conferences. Gender and geography metadata was annotated manually, based on

---

[1] https://aistconf.org/.
[2] http://www.dialog-21.ru/en/.
[3] https://nlp.rusvectores.org.

authors' names and affiliations. Unfortunately, the RusNLP database does not contain information about the order of the authors, so we counted all of them equally. While annotating the 'location' or 'city' attribute, we used the following notation: 'centre' stands for Moscow, Saint-Petersburg and authors from outside Russia, while 'province' stands for all other regions and cities in Russia. We then calculated the percentage of male-female and central-provincial authors for each year and each venue.

In order to measure the difference in these percentages before and after the introduction of double-blind reviewing for AIST, we applied the Welch T-test. For Dialogue, we have only one data point with the double-blind reviewing (year 2019), thus the T-test is ill-defined, and we simply checked whether the absolute difference between the percentages exceeds the standard deviation of the respective values for the years before the double-blind introduction (2000–2018).

2.1 Gender Distribution

Starting with AIST, we first calculated the average percentage of its female authors before and after double-blind introduction (31 before and 31 after). Obviously, there is no statistically significant difference here, according to the Welch T-test: $statistic = -0.08$, $P = 0.94$. The yearly percentages are visualised in Fig. 1. In this and all the following plots, the dashed vertical line denotes the year after which the venue switched to the double-blind process.

The picture is different for the gender distribution of the Dialogue authors. Before the double-blind peer review, on average 57% of authors were males and 43% were females. This changed to 45% and 55% respectively in 2019 (see Fig. 2). Thus, the ratio of female authors increased by 12 points. Naively comparing this value to the standard deviation of the yearly female percentages before the introduction of double blind reviewing (it is 5 across 18 years), we observe that the increase value exceeds the standard deviation more than two times. From this, we conclude that the difference is significant, and the number of female authors has indeed increased.

2.2 City/Location Distribution

For AIST, the average yearly percentage of 'central' authors before double-blind reviewing was 79%, but after introducing it in 2017, this value fell to 56%. The Welch T-test confirms that the difference is statistically significant: $statistic = 2.48$, $P = 0.048$. 'Provincial' authors indeed benefited from the double-blind process. This can also be clearly seen in Fig. 3, which additionally shows the geographical location of the conference itself in each respective year ('E-burg' stands for Ekaterinburg). Interestingly, the introduction of double-blind reviewing has significantly decreased the ratio of 'central' authors, even though at the same time the conference itself moved to Moscow (years 2017 and 2018). This additionally confirms the significance of the discovered trend.

For Dialogue, there were on average 89% of 'central' authors before and 98% after the introduction of double-blind reviewing. Thus, the percentage

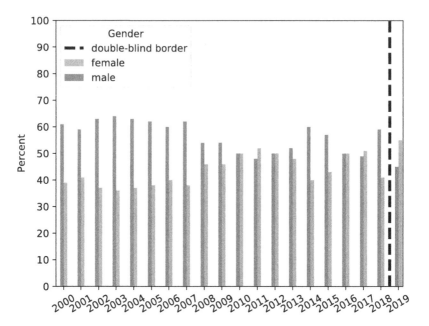

Fig. 2. Gender distribution among Dialogue authors from 2000 to 2019

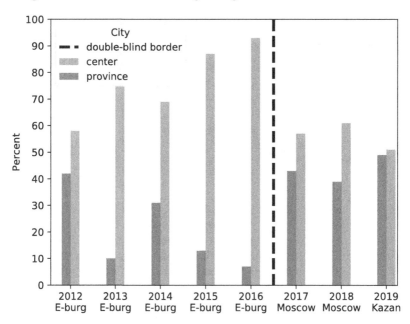

Fig. 3. City distribution among AIST authors from 2012 to 2019

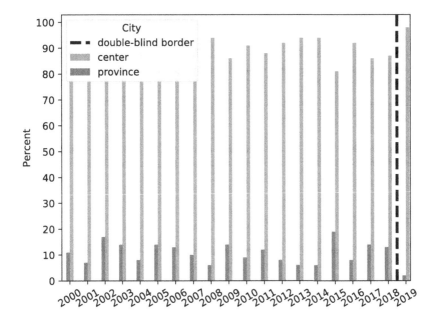

Fig. 4. City distribution among Dialogue authors from 2000 to 2019

of 'provincial' authors has actually *decreased* from 11% to 2% (see Fig. 4). This absolute difference of 9 is much higher than the standard deviation from the years 2000–2018 (4), so it seems to be significant. We believe this is caused by a random fluctuation: in Dialogue-2019, there is actually only *one* accepted paper authored by persons not located in Moscow, Saint-Petersburg or outside of Russia. This single paper is responsible for all the 2% of 'provincial' authors that we observe. It is of course difficult to do any conclusions on such an anecdotal evidence. More observations are certainly needed in the years to come.

3 Results

There are several significant changes occurring after introducing double-blind reviewing for both AIST and Dialogue conferences. First of all, there are no changes found in gender distribution of authors for AIST, which we guess means it has been egalitarian enough in this respect from the very start. At the same time, for Dialogue, we observe a significant increase in the ratio of female authors.

In case of location distribution, the results are rather controversial: there is a significant increase in the ratio of 'provincial' authors for AIST, but for Dialogue, there is a significant *decrease*. However, since the Dialogue statistics after the introduction of double-blind reviewing is based on a single observation, it is not fully reliable. We certainly have to wait until the year 2020 to see the forthcoming trends.

Another limitation of this pilot research is that obviously other factors may influence the distribution changes, e.g. conference location, or topical popularity. We have yet to find out how to exclude the influence of such extra factors. Finally, in the future we plan to expand our analysis by including the data from the AINL conference[4].

And of course, we welcome everyone to submit to all of the conferences mentioned above.

Acknowledgements. The article was prepared within the framework of the HSE University Basic Research Program and funded by the Russian Academic Excellence Project '5-100'.

References

1. Budden, A.E., Tregenza, T., Aarssen, L.W., Koricheva, J., Leimu, R., Lortie, C.J.: Double-blind review favours increased representation of female authors. Trends Ecol. Evol. **23**(1), 4–6 (2008)
2. Nikishina, I., Bakarov, A., Kutuzov, A.: RusNLP: semantic search engine for Russian NLP conference papers. In: van der Aalst, W.M.P., et al. (eds.) AIST 2018. LNCS, vol. 11179, pp. 111–120. Springer, Cham (2018). https://doi.org/10.1007/978-3-030-11027-7_11
3. Spier, R.: The history of the peer-review process. Trends Biotechnol. **20**(8), 357–358 (2002)
4. Surwillo, W.W.: Anonymous reviewing and the peer-review process. Am. Psychol. **41**(2), 218 (1986)
5. Tomkins, A., Zhang, M., Heavlin, W.D.: Single versus double blind reviewing at WSDM 2017. arXiv preprint arXiv:1702.00502 (2017)

---

[4] https://ainlconf.ru/.

Tutorial

Intel Distribution of OpenVINO Toolkit: A Case Study of Semantic Segmentation

Valentina Kustikova[1(✉)], Evgeny Vasiliev[1], Alexander Khvatov[1],
Pavel Kumbrasiev[1], Ivan Vikhrev[1], Konstantin Utkin[1], Anton Dudchenko[1],
and Gleb Gladilov[2]

[1] Lobachevsky State University of Nizhni Novgorod, 606950 Nizhny Novgorod,
Russia
valentina.kustikova@itmm.unn.ru, eugene.unn@gmail.com,
khvatov.alexander@gmail.com, pavel.kumbrasev@gmail.com,
laind4471@gmail.com, megarungle@gmail.com, da394372@gmail.com
[2] Intel Corporation, 603024 Nizhny Novgorod, Russia
gleb.gladilov@intel.com

Abstract. We provide an overview of the Intel Distribution of Open-
VINO toolkit. The application of the OpenVINO toolkit is represented
on the case study of semantic segmentation of on-road images. We provide a step-by-step tutorial for the problem solution based on the Dila-
tion10 model, trained on the Cityscapes dataset. The Inference Engine
component of the OpenVINO toolkit is used to implement inference of
deep model. We focus on synchronous inference mode. Comparison of
on-road semantic segmentation models supported by the OpenVINO
toolkit is provided. Performance analysis of the Dilation10 inference
implemented using various deep learning tools is carried out.

Keywords: Intel Distribution of OpenVINO toolkit · Deep learning ·
Inference · Semantic segmentation

1 Introduction

The importance of computer vision is increasingly growing. The classical computer vision problems (detection, classification, recognition, etc.) are widely used
in video surveillance and video analytics systems. Commercial companies special-
izing in the development of such systems attempt to find a framework that allows
them to quickly deploy a final software product. Intel Distribution of OpenVINO
toolkit [1] focuses on the developing cross-platform solutions of computer vision
problems based on image processing, machine learning, and deep learning. The
toolkit provides a wide range of algorithms optimized to achieve maximum performance on the Intel hardware (Intel CPUs, Intel Processor Graphics, Intel
Movidius Neural Compute Stick, Intel FPGAs, Intel Gaussian Mixture Models).
Heterogeneous execution of algorithms on various Intel accelerators using the
same application programming interface is also supported.

The goal of this article is to provide an overview of the OpenVINO toolkit
and represent a case study of its application on the semantic segmentation of

© Springer Nature Switzerland AG 2019
W. M. P. van der Aalst et al. (Eds.) AIST 2019, LNCS 11832, pp. 11–23, 2019.
https://doi.org/10.1007/978-3-030-37334-4_2

on-road images. The paper is structured as follows. Section 2 states the problem. Section 3 provides an overview of existing deep models for semantic segmentation. Section 4 describes the Dilation10 model. In Sect. 5, the toolkit's components are emphasized. Section 6 is a step-by-step tutorial for the inference implementation using the Inference Engine component of the OpenVINO toolkit. In Sect. 7, we compare deep networks for semantic segmentation of on-road images and provide performance analysis of various inference implementations. Section 8 concludes the paper.

2   Problem Statement

The problem of semantic segmentation involves determining the class of objects to which each pixel of the image belongs. Supposed the input image resolution is $w \times h$, where w is the image width and h is the image height, c is a number of channels. The output is a map of pixel confidences for each of the available object classes. This map is represented by a three-dimensional tensor of the shape $w \times h \times k$, where k is a number of object classes. The channel number i of the constructed tensor corresponds to the confidence of belonging pixels to the class number i. The channel number i of the map for the pixel (x, y) corresponds to the class of this pixel if the maximum confidence over all classes is achieved.

3   Related Work

Deep learning is one of the most effective approaches to solving semantic segmentation problem. A typical network topology for semantic segmentation is the ***encoder-decoder architecture***. This architecture assumes that image processing is divided into two stages, corresponding to the two parts of the neural network. ***Encoder*** provides a high-level representation of the original image. ***Decoder*** creates a map of confidences based on a compressed representation and the subsequent construction of a segmented image. DeconvNet [2], SegNet [3] and FCN [4] are typical examples of such models. Usually, encoders are based on the existing convolutional networks such as VGG-16 [17]. Encoder is complemented by a decoder that is represented by a set of ***deconvolutional layers*** arranged mirrorwise relative to the original network. In most networks, the transition to a compressed representation is carried out together with a reduction of the feature map dimension. So, small objects in the image are inevitably lost. To reduce this effect, approaches based on taking into account features from different scales are used (ParseNet [5], UNet [6]). The authors of the ENet [7] and ICNet [8] models offer compact encoder-decoder architectures, allowing to obtain comparable quality on the well-known datasets and demonstrate higher performance.

The second group of models assumes a simple ***interpolation*** [9,10] or ***dilated convolutions*** [11,12] are used to restore the image resolution. In [27] the ideas of encoder-decoder architecture and dilated convolutions are combined.

The third group of deep models involves the use of graph probabilistic algorithms such as ***Markov random fields*** (MRF) and ***conditional random*** *fields* (CRF). In [3,9,13–15] convolutional neural networks are combined with graph probabilistic models. Convolutional networks allow to extract image features and perform coarse segmentation, while graph models allow to refine the obtained segmentation. These models demonstrate high quality segmentation, but tend to work slowly [16].

4   Dilation10

Further, the Dilation10 model for semantic segmentation of on-road scenes is used. This section briefly describes the overall structure of Dilation10 [27]. First, we consider the concept of *dilated convolution*. Let $F : \mathbb{Z}^2 \to \mathbb{R}$ is a discrete function, $\Omega_r = [-r,r] \cap \mathbb{Z}^2$, $k : \Omega_r \to \mathbb{R}$ is a discrete filter of size $(2r+1)^2$. Then l-*dilated convolution* is as follows.

$$(F *_l k)(p) = \sum_{s+lt=p} F(s)k(t).$$

Dilated convolution supports exponentially expanding receptive field without losing resolution or coverage.

The idea of *a context module* based on a sequence of dilated convolutions is proposed in [27]. This module increases the performance of dense prediction architectures by aggregating multi-scale contextual information. Dilation10 is based on VGG-16 [17] and contains the context module consisting of 10 convolutional layers. The Dilation10 architecture is as follows. It contains all convolutional layers preceding the last max-pooling layer of VGG-16. Fully-connected layers are replaced by a sequence of dilated convolutions and dropouts followed by the context module (Table 1). The rectified linear unit (ReLU) follows after each convolutional layer excepting the last one. The final layer of the network is deconvolutional one. Note that Dilation10 is a fully convolutional network.

Table 1. The structure of the Dilation10 context module

Layer	1	2	3	4	5	6	7	8	9	10
Kernel size	3×3	3×3	3×3	3×3	3×3	3×3	3×3	3×3	3×3	1×1
l-dilated convolution	1	1	2	4	8	16	32	64	1	1

Dilation10 was trained on the Cityscapes dataset [18] using Caffe [20]. This dataset contains 2975 training images, image resolution is 2048×1024 pixels, and the number of object classes is 19 (road, sidewalk, building, wall, fence, pole, light, sign, vegetation, terrain, sky, person, rider, car, truck, bus, train, motorcycle, bicycle). The input shape of the network is $b \times 3 \times 1396 \times 1396$, where b is a batch size, the output shape is $b \times 19 \times 1024 \times 1024$.

Dilation10 outperformed all prior models (FCNs, ParseNet and other) on the Cityscapes dataset as reported in [19]. The *intersection over union* (IoU) score is used for qualitative comparison of segmentation models. Here we represent the

IoU score for the major object classes: road – 97.6, building – 89.9, traffic signs – 65.2, person – 78.9, car – 93.3, vegetation – 91.8. Mean IoU over all 19 classes is 67.1. High quality of semantic segmentation of on-road scenes for the key object classes proves the perspectives of practical use of Dilation10.

5   Intel Distribution of OpenVINO Toolkit

5.1   Components

The OpenVINO toolkit [1] consists of the several major parts.

1. **Deep learning for computer vision**. This part includes the Deep Learning Deployment Toolkit (DLDT) [29] to effectively execute pretrained deep neural network models using a high-level application programming interface.
2. **Traditional computer vision**. The development of computer vision applications based on OpenCV [24] or OpenVX API [25] is supported.
3. **Additional packages** for Intel FPGAs, Intel Movidius Neural Compute Stick, Intel GMM-GNA and media encode/decode functions.

Further, we consider the DLDT component in detail, DLDT is used in the tutorial.

5.2   Deep Learning Deployment Toolkit

DLDT consists of the following components.

1. **Model Optimizer**. This component allows you to convert pretrained models from the Caffe [20], TensorFlow [21], MXNet [22] library formats, and from the Open Neural Network Exchange Format (ONNX) [23] to the optimized intermediate representation of the OpenVINO toolkit. Model Optimizer provides platform-independent optimizations, in particular, the merging of layers and their replacement.
2. **Inference Engine**. This is a C++ library that provides APIs in C++ and Python for deep models' inference on various target Intel's platforms. Inference Engine allows to use models in the format generated by Model Optimizer. Inference is a feed forward of the network for a given set of images. Inference optimization is provided through the construction and analysis of a computational graph, effective planning of calculations and compression of the deep model. The inference execution on various devices is implemented through *plugins*. There are *synchronous* (or latency) and *asynchronous* (or throughput) execution modes. The first mode supposes the next inference request is executed after the completion of the previous one. The synchronous mode is used when we need to minimize the execution time of a single request. The second mode assumes constructing a queue of inference requests, several requests can be executed in parallel. This mode is useful while minimizing the overall execution time of the request set.

3. **Open Model Zoo**. This component contains a large number of trained deep models that can be useful for the projects where the OpenVINO toolkit is used. There are networks to solve a variety of problems (object detection, text detection, action recognition, etc.). The developers also provide Model Downloader, which is able to download well-known models, as well as models trained by Intel employees. It is free available on GitHub [26].

4. **Samples**. The OpenVINO toolkit contains samples to quickly study the toolkit. Examples are developed in the C++ and Python languages.

The scheme of using DLDT in the tutorial is shown below (Fig. 1). It involves the following steps.

1. Download a deep model using **Model Downloader**.
2. Convert the model to the intermediate representation of the OpenVINO toolkit by calling **Model Optimizer**.
3. Load input data and infer the model using the **Inference Engine** component, receive the model output for the subsequent interpretation.

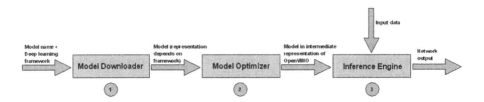

Fig. 1. Using DLDT implemented in the tutorial

6 Tutorial

6.1 Development Steps

The application of OpenVINO is represented on the case study of semantic segmentation of on-road scenes using the Dilation10 model described above. We use DLDT for inference implementation and OpenCV for image processing. Implementation is provided in Python 3. The development is structured as follows.

1. Install the OpenVINO toolkit.
2. Learn the Python API of Inference Engine.
3. Prepare the Dilation10 model (it is called "dilation" in Model Downloader).
4. Implement inference of deep models for semantic segmentation.
5. Execute inference implementation on a device.

Let us consider each of these steps in detail.

6.2   Install the Intel Distribution of OpenVINO Toolkit

The OpenVINO toolkit is an open source and cross-platform tool with official support of the following operating system: Windows 10 and well-known Linux distributions (Ubuntu 16.04 LTS 64-bit, CentOS 7.6 64-bit and Yocto Project Poky Jethro v2.0.3 64-bit). However, the toolkit can be executed on the previous versions of Windows; for other Linux distributions, you can build the source code of the toolkit by yourself. The installation guide is represented in the documentation [28]. This tutorial is developed on Windows, the default settings are used. Further, we suppose the installation directory of the toolkit is stored in the environment variable `%OPENVINO_DIR%`.

6.3   Python API of the Inference Engine Component

Python API of Inference Engine provides a simplified interface for the functionality of Inference Engine. It allows working with deep models, load and configure plugins, and execute model inference. Here we give a brief description of main classes and methods of the `openvino.inference_engine` [30] package, used further during the inference implementation.

IENetwork. The `IENetwork` class contains information about the neural network, which is loaded from the intermediate representation of Inference Engine.

The class constructor provides reading the model (`.xml`) and the trained weights (`.bin`). The constructor has the following prototype.

```
__init__(model: str, weights: str)
```

It takes the file names containing the intermediate representation of the model (`.xml`) and the trained model weights (`.bin`).

IEPlugin. The `IEPlugin` class is the main interface of the plugin, in which the model is loaded to execute inference on a specific device.

The class constructor takes the name of the target device (`CPU`, `GPU`, `FPGA`, `MYRIAD`, or `HETERO`) and the directory containing the available plugins.

```
__init__(device: str, plugin_dirs=None)
```

The model is loaded into the plugin using the `load(...)` method.

```
load(network: IENetwork, num_requests: int=1, config=None)
```

This method takes the object `network` of the loaded model, the number of inference requests `num_requests`, a dictionary of keys and their values `config` of the plugin configuration. The output is the `ExecutableNetwork` object.

ExecutableNetwork. The `ExecutableNetwork` class is used to represent the executable network loaded into the plugin and executed on the device.

The basic class methods provide the deep model inference in synchronous and asynchronous modes. Further, the synchronous inference mode is implemented, so we consider the corresponding method.

```
infer(inputs=None)
```

The method input is a dictionary that maps input layer names to `numpy.ndarray` objects of proper shape with input data for the layer. The output is a dictionary that maps output layer names to `numpy.ndarray` with output data of the layer.

6.4   Prepare the Dilation10 Model

Preparing the Dilation10 model for solving the task involves loading the trained model in the Caffe format using Model Downloader. To load this model, go to the directory containing the proper tool and execute the script `downloader.py` with the parameters specified below.

```
cd %OPENVINO_DIR%\deployment_tools\model_downloader
python downloader.py --name dilation
```

This script downloads files `dilation.caffemodel` and `dilation.prototxt` to the directory `semantic_segmentation\dilation\cityscapes\caffe`. Let us denote this directory as `%DILATION_DIR%`.

The following use of the model by Inference Engine requires its conversion to the intermediate format. Conversion is carried out by moving to the directory containing the proper OpenVINO component and calling it with the correct parameters. The example below converts the Dilation10 model (parameters `--input_model` and `--input_proto`), trained using Caffe (`--framework`). The trained weights are 32-bit real numbers (`--data_type FP32`). The model input is a 4-dimensional tensor (`--input_shape [1,3,1396,1396]`), and the mean pixel intensity in the training dataset is $[72.39, 82.91, 73.16]$ (`--input data --mean_values data[72.39,82.91,73.16]`). The model output is interpreted as the probability that a pixel belongs to a certain object class (`--output prob`).

```
cd %OPENVINO_DIR%\deployment_tools\model_optimizer
python mo.py --framework caffe --data_type FP32 --input_shape [1,3,1396,1396] --input data
    --mean_values data[72.39,82.91,73.16] --output prob \
    --input_model %DILATION_DIR%\dilation.caffemodel \
    --input_proto %DILATION_DIR%\dilation.prototxt
```

Completing the model conversion, there are three files located in the directory `%OPENVINO_DIR%\deployment_tools\model_optimizer`.

– `dilation.xml` is the model description in the intermediate representation.
– `dilation.bin` is a file containing trained parameters of the model in the intermediate representation.
– `dilation.mapping` is a file containing the correspondence between the layers of the initial model description and the intermediate representation.

6.5   Inference Implementation

The source code of the application that provides the deep model inference will be placed in the file `semseg_inference.py`. Assumed this file is stored in a directory whose value is written to the `%CURRENT_DIR%` environment variable. The inference implementation is divided into several functions:

– `build_argparser(...)` is the function to parse command line arguments;
– `prepare_model(...)` is the function of loading the model and checking the presence of the implementation of all layers;
– `convert_image(...)` is the function of preparing images fed to the network;
– `infer_sync(...)` is the inference function;
– `segmentation_output(...)` is the function for the model output processing;
– `main(...)` is the main function which sequentially calls the previous ones.

The application receives the following input parameters.

1. Description (parameter `--model`) and weights (`--weights`) of the model.
2. Image file name or directory containing images (`--input`).
3. Batch size of images processed in a single network pass (`--batch_size`).
4. Library containing the implementation of model layers that are not available in the standard CPU plugin (`--cpu_extension`, it is not required for the dilation model).
5. Directory containing available plugins (`--plugin_dir`).
6. Device for inference execution (`--device`). The device parameter takes the values `CPU`, `GPU`, `FPGA`, or `MYRIAD`.
7. Color map (`--color_map`) that corresponds to the set of object classes. Each line contains the object class color in the BGR format.

We offer to implement *the function of parsing command line arguments* using the `argparse` package.

To load and prepare a deep model for the inference we use the `openvino.inference_engine` module and its member classes `IENetwork` and `IEPlugin`. The function of preparing the model involves creating plugin and model objects using constructors. In addition, this function checks the availability of the implementation of all layers in the CPU plugin (use the `device` property and the `get_supported_layers(...)` method of the plugin object).

To prepare input data for the model you should create a list of processed images, based on the corresponding application parameter, and implement the `get_image_list(...)` function. If a directory is a command line argument, then we walk through the directory contents and create a list of full file names. Then it is needed to read each image, resize it to the size of a neural network input and rearrange dimensions using OpenCV. We recommend to call the following functions and methods of the image object: `cv2.imread(...)`, `cv2.resize(...)` and `transpose(...)`. To get the network input shape you have to take the `inputs` property of the network object. The described sequence is implemented as a function `convert_image(...)`.

The next step is to implement *the inference function*. The input parameters of this function are a list of images, the network executed on the device, and the model description. We infer a deep model using the `infer(...)` method of the executable network object. The described algorithm is implemented in the `infer_sync(...)` function.

The model output is a 4-dimensional tensor, the first dimension corresponds to the number of processed images, the other two to the height and width of the images, and the fourth one to the number of object classes. Let us implement *the function of converting the network output to a segmented image*. The function inputs are the network output and a color map in which the class index is associated with a color in the BGR format (3 unsigned char numbers). Each pixel intensity corresponds to the color of the class to which this pixel belongs. We construct the segmented image using OpenCV and save the final image to the current directory. The described algorithm is implemented in the `segmentation_output(...)` function.

The resolution of the segmented images corresponds to the output dimension of the last layer of the model. We implement *the function to resize the segmented image to the size of the original image*. The input parameters are two lists of the original and segmented images. We perform the following actions: read both images, take the width and height of the original one, scale the segmented image to the size of the original one, save the resized image. The described algorithm is implemented in the `size2origin(...)` function.

The main function combines the developed functionality. It contains sequential calls of the functions for parsing command line arguments, preparing the model and input images, loading the model into the plugin and creating the executable model, executing inference and processing the model output. Further, we release memory used for storing network and plugin. To trace actions we recommend to import `logging` package.

The developed application is free available on GitHub [31]. We provide the developed script, the IPython notebook, and the test image.

6.6   Inference Execution

The developed script is executed via the command line. An example of the command line is represented below. The current directory (`%CURRENT_DIR%`) corresponds to the location of the executed script `semseg_inference.py`. Input parameters of the script are the intermediate representation (`dilation.xml`) and the trained weights (`dilation.bin`) of the model, the image (`image.png`) for semantic segmentation, the color map (`color_map.txt`) for the Dilation10 model, plugin name (`CPU`), and the batch size supplied to the network (`-b 1`).

```
cd %CURRENT_DIR%
python semseg_inference.py -m %OPENVINO_DIR%\deployment_tools\model_optimizer\dilation.xml \
    -w %OPENVINO_DIR%\deployment_tools\model_optimizer\dilation.bin \
    -i image.png --color_map color_map.txt -d CPU -b 1
```

If the command line arguments are entered correctly, a log is displayed on the screen, it reflects all stages of the application execution (IPython notebook [31]).

7   Experiments

7.1   Infrastructure

The following hardware and software are used to verify the inference implementation and to analyse inference performance.

- **CPU**: Intel® Core™ i5-8600K CPU@3.60 GHz.
- **GPU**: Intel® UHD Graphics 630, Coffee Lake GT2.
- **RAM**: 32 GB.
- **OS**: Ubuntu 16.04.
- **Software**: Python 3.5.2, GCC 5.4.0, Intel Distribution of OpenVINO toolkit 2018 R5 (OpenCV 4.0.1), Intel Optimization for Caffe v1.1.6 (from Anaconda Cloud), Caffe v1.0 (from Anaconda Cloud).

7.2 Models

There are several models for semantic segmentation of on-road images provided by Model Downloader.

1. **dilation**. This is the Dilation10 model [27] described above.
2. **semantic-segmentation-adas-0001**. This model is trained on the Mapillary Vistas Dataset [32] by Intel engineers. It provides multiclass (road, sidewalk, building, wall, fence, pole, traffic light, traffic sign, vegetation, terrain, sky, person, rider, car, truck, bus, train, motorcycle, bicycle, ego-vehicle) segmentation based on ICNet [8]. The input shape is $1 \times 3 \times 1024 \times 2048$.
3. **semantic-segmentation-adas-0001-fp16**. This model is the same as previous one, ***-fp16** refers to using half-precision floating point format (16-bit).
4. **road-segmentation-adas-0001**. The specified model is trained on the Mighty dataset [33] and provides multiclass (background, road, curbs, marks) segmentation based on the ENet model [7], using depthwise convolutions and without ELU (Exponential Linear Unit) operations and without concatenation. The input tensor size is $1 \times 3 \times 512 \times 896$. There are corresponding ***-fp16** and ***-int8** models supported half-precision floating point format and **int8** precision for network weights and operations.

Further, during the qualitative comparison, the first three models (**dilation**, ***-adas-0001** and ***-adas-0001-fp16**) are used since these models are trained to segment similar object classes. The developed inference implementation is used for these models, the difference is in the arguments.

7.3 Test Data

Further, we represent qualitative analysis on three types of images.

- **Image #1** from the Cityscapes dataset [18], the resolution is 1024×2048.
- **Image #2** from the Mapillary Vistas Dataset [32], 2448×3264.
- **Image #3** is the third-party image obtained from the DVR, 1080×1920.

7.4 Qualitative Analysis

Results of the semantic segmentation for the test images are represented below (Fig. 2). It is obvious that the **dilation** model does not detect small objects; this is most clearly seen at the edges of the segmented images (Image #1

Fig. 2. Semantic segmentation of on-road images

and #3). The effect occurs because of scaling the original image to the network input and the network output to the size of the original image. It can be eliminated, since the model is fully convolutional one. Support of arbitrary input sizes for such models is currently tested in the OpenVINO toolkit. The ***-adas-0001[-*]** models provide more accurate segmentation because of the topology is based on ICNet [8], and the network input is a pyramid of original image scales. Note, the use of half-precision (***-adas-0001-fp16**) does not lead to a loss of segmentation accuracy on all test images. Comparison of the third-party image segmentations shows that the **dilation** model better segments the building and sidewalk classes, while ***-adas-0001[-*]** detect car bonnet captured by the DVR. In general, the considered models demonstrate comparable results.

7.5 Performance Analysis

The Inference Engine component is a tool for efficiently inference of deep models on the Intel hardware. In this regard, we represent the experiments on the Intel CPU and Intel Processor Graphics of the test computer. Several framework configurations are used to compare inference performance. Inference Engine is executed on the CPU using the MKLDNN plugin in multi-threading mode. OpenCV is executed on the CPU using the dnn module, dnn is able to use

Inference Engine as backend. Intel Distribution of Caffe and Caffe are executed on the CPU.

During a single experiment, the time of a single network pass on a batch consisting of one image is measured. Each experiment is repeated 100 times; *the average time of a single network pass* is calculated based on the obtained times. The obtained performance results (Table 2) demonstrate that the inference implementation based on Inference Engine is approximately 3 times faster than Caffe-based ones. The dnn module of the OpenCV library with the Inference Engine backend shows the similar average time of a single network pass as the direct execution of Inference Engine. Note that the OpenCV inference implementation is faster than Caffe-based implementations.

Table 2. Average time of a single network pass (in seconds)

Model	Inference Engine	OpenCV			Intel Optimization for Caffe	Caffe
	CPU (MKLDNN)	CPU (dnn)	CPU (MKLDNN)		CPU	CPU
Dilation	**9.347**	17.54	**9.219**		30.218	27.919

8   Conclusion

An overview of the OpenVINO toolkit has been represented. The practical application of the OpenVINO toolkit has been demonstrated on the semantic segmentation of on-road images. We have described the problem solution using the Dilation10 model and provided inference performance analysis. The developed tutorial is a complete guide for implementing the deep models' inference based on the Python API of Inference Engine.

Acknowledgements. The research was supported by the Intel Corporation. The authors thank company's employees for their help and attention to the research.

References

1. Intel Distribution of OpenVINO toolkit. https://software.intel.com/en-us/openvino-toolkit
2. Noh, H., et al.: Learning deconvolution network for semantic segmentation. In: Proceedings of the IEEE ICCV, pp. 1520–1528 (2015)
3. Badrinarayanan, V., et al.: SegNet: a deep convolutional encoder-decoder architecture for image segmentation. arXiv preprint arXiv:1511.00561 (2015)
4. Long, J., et al.: Fully convolutional networks for semantic segmentation. In: Proceedings of the IEEE Conference on CVPR, pp. 3431–3440 (2015)
5. Liu, W., Rabinovich, A., Berg, A.C.: ParseNet: looking wider to see better. arXiv preprint arXiv:1506.04579 (2015)
6. Ronneberger, O., Fischer, P., Brox, T.: U-Net: convolutional networks for biomedical image segmentation. arXiv preprint arXiv:1505.04597 (2015)

7. Paszke, A., et al.: ENet: a deep neural network architecture for real-time semantic segmentation. arXiv preprint arXiv:1606.02147 (2016)
8. Zhao, H., et al.: ICNet for real-time semantic segmentation on high-resolution images. arXiv preprint arXiv:1704.08545 (2017)
9. Chen, L.C., et al.: Semantic image segmentation with deep convolutional nets and fully connected CRFs. arXiv preprint arXiv:1412.7062 (2014)
10. Chen, L.C., et al.: Attention to scale: scale-aware semantic image segmentation. In: Proceedings of the IEEE Conference on CVPR, pp. 3640–3649 (2016)
11. Lin, G., et al.: Efficient piecewise training of deep structured models for semantic segmentation. In: Proceedings of the IEEE Conference on CVPR, pp. 3194–3203 (2016)
12. Wu, Z., Shen, C., Hengel, A.: Wider or deeper: revisiting the ResNet model for visual recognition. arXiv preprint arXiv:1611.10080 (2016)
13. Zheng, S., et al.: Conditional random fields as recurrent neural networks. In: Proceedings of the IEEE International Conference on Computer Vision, pp. 1529–1537 (2015)
14. Cogswell, M., et al.: Combining the best of graphical models and ConvNets for semantic segmentation. arXiv preprint arXiv:1412.4313 (2014)
15. Liu, Z., et al.: Deep learning Markov random field for semantic segmentation. arXiv preprint arXiv:1606.07230 (2016)
16. Sidnev, A., et al.: Semantic segmentation review (2018). https://software.intel.com/en-us/download/semantic-segmentation-review
17. Simonyan, K., Zisserman, A.: Very deep convolutional networks for large-scale image recognition. arXiv preprint arXiv:1409.1556 (2015)
18. The Cityscapes Dataset. https://www.cityscapes-dataset.com
19. Cordts, M., et al.: The cityscapes dataset for semantic urban scene understanding. In: Proceedings of the IEEE Conference on CVPR (2016)
20. Caffe. http://caffe.berkeleyvision.org
21. TensorFlow. https://www.tensorflow.org
22. MXNet. https://mxnet.incubator.apache.org
23. ONNX. Open Neural Network Exchange Format. https://onnx.ai
24. OpenCV. http://opencv.org
25. OpenVX. https://www.khronos.org/openvx
26. Open Model Zoo repository. https://github.com/opencv/open_model_zoo
27. Yu, F., Koltun, V.: Multi-scale context aggregation by dilated convolutions. arXiv preprint arXiv:1511.07122 (2016)
28. Intel Distribution of OpenVINO toolkit. Documentation. https://docs.openvino toolkit.org
29. Deep Learning Deployment Toolkit. https://github.com/opencv/dldt
30. Overview of Inference Engine Python API. https://docs.openvinotoolkit.org/2019_R1/_inference_engine_ie_bridges_python_docs_api_overview.html
31. Solving the problem of semantic segmentation using the Inference Engine of Intel Distribution of OpenVINO toolkit. https://github.com/itlab-vision/tutorials/tree/solution/aist2019-openvino-semseg
32. Mapillary Vistas Dataset. https://www.mapillary.com/dataset/vistas
33. Mighty AI. https://mighty.ai/blog/tag/open-dataset

General Topics of Data Analysis

Experimental Analysis of Approaches to Multidimensional Conditional Density Estimation

Anna Berger[1] and Sergey Guda[1,2]

[1] Institute of Mathematics, Mechanics, and Computer Science, Southern Federal University, Milchakova 8a, 344090 Rostov-on-Don, Russia
anna.ig.berger@gmail.com, gudasergey@gmail.com
[2] The Smart Materials Research Center, Southern Federal University, Sladkova Street 174/28, 344090 Rostov-on-Don, Russia

Abstract. Recently several original methods for conditional density estimation (CDE) have been developed. The abundance of information comprised by the full conditional density of target variables is great when compared to the regression or quantile regression estimates. Still, there are only few independent experimental investigations of these methods, especially concerning a multidimensional target variable, and this paper aims to address this issue. We consider several approaches such as kernel density estimation, reduction to binary classification, Naïve Bayes, Bayesian Network, "varying coefficient" approach, random forests and Approximate Bayesian Computation applied to a conditional density estimation problem. We examine these methods when applying to various datasets together with the dependency of the methods' performance on different parameters including the number of irrelevant covariates, smoothness, and flatness of the distribution. Considered datasets include artificial models with required properties and with the known exact value of CDE evaluation measure and a real-world dataset arisen from the problem of structure recognition by XANES spectra, which is reduced to a regression task with a complex multimodal probability distribution of the target variable. The special attention is paid to the computation of the evaluation measure as the methods based on the direct optimization of the loss employ its imprecise but fast approximation which results in the poor prediction quality for datasets with a small target variance.

Keywords: Conditional density estimation · Multidimensional regression · Experimental analysis

1 Introduction

Conditional density estimation of a random variable $z \in \mathbb{R}$ can be considered a generalization of a regression problem. Standard regression returns point estimation of a target variable z. It minimizes the standard deviation leading to a

This work was supported by the Russian Foundation for Basic Research, project 18-02-40029.

prediction of the target expectation. In case of a multimodal or skewed distribution, the expectation value has a smaller probability than the mode—the most probable value. But a researcher expects namely the latter to obtain as a result of the regression algorithm as it is implied by the word "expectation".

To overcome the limitation of the point regression estimation, in many applications such as time series analysis, the quantile regression is employed. For the given probability, it determines the minimal interval having the form $(q, +\infty)$ and containing target value [1]. This approach makes it possible to predict the interval containing the target variable for the given confidence level. Another way of generalization is to consider regression function $X \to Z$ as a manifold in ambient space $X \times Z$. It enables the construction of prediction regions with a given confidence level [3, 9, 10].

Still, the distribution of the target variable can happen to be too complex to be well estimated via quantile regression: for instance, if multimodality or a significant skew of the response is present in the data. In this case, the standard and quantile regression may be insufficient for the proper data analysis and solving the problem, while the estimator of full conditional density provides a more comprehensive accounting of the target variable.

Several recent works utilize the CDE of the full probability distribution in various application domains and, by doing so, achieve substantial improvements. The approach proves itself especially in settings with complicated sources of errors which are widespread in physics in general [2]—and in Cosmology [14], in particular.

The other possible scope of application for CDE is multitask learning. If the objective function can be decomposed into several autonomous parts (e.g. error squares for different target coordinates), then the multitask problem splits itself into an independent problem for each target component. However, estimation of some non-decomposable target, such as the mode of the target variable, is a different matter. In this case, one cannot optimize the components separately as the combination of univariate target component modes is, generally speaking, not the mode of the multivariate target. The mode regression was developed for univariate case (see [8, 15]). Nevertheless, to the best of the authors' knowledge, currently, there are no methods for multi-target mode regression, except using CDE. While namely these methods, for example, are needed to solve the problem of predicting the molecular structure by a given XANES spectrum (see [4] Section 3.5.2). Due to the independence of the XANES spectrum on symmetry transformations of molecule geometry and some geometry parameters, the molecule geometry probability distribution has multiple modes, which are hard to predict with ordinary regression techniques.

As of today, there is a lack of comparison of different CDE methods in literature. This paper aims to fill this gap and provides the experimental overview and comparison of these methods when applying to the data from various distributions. We demonstrate the soundness of such methods as kernel density estimation, reducing to binary classification, Naïve Bayes, Bayesian Network, "varying coefficient" approach, random forests, Approximate Bayesian Computation and study the dependency of the methods performance quality on volatility degree when x changes, correlation between z components, the number of irrelevant covariates, flatness of the underlying distribution.

For measuring quality performance CDE loss is employed as the most commonly used measure for this type of problem. It implies calculating several multivariate integrals, which are not always computed precisely especially in the algorithms, which directly optimize the CDE loss. That motivates us to study the error of computing the CDE loss by different methods separately.

There are several techniques, which can be adopted for conditional density estimation. Kernel density estimation is one of the simplest of them though it heavily suffers from the curse of dimensionality in the multivariate setting. Another view on a CDE problem can be reformulating it in terms of a binary classification problem and further utilizing existing powerful binary classifiers. The assumption of conditional independence of response variables results in the Naïve Bayes approach, and, as in many cases it is not fulfilled, but the relationship among the target variables is known, the Bayesian Network as well can be built to model the outcome. A different view of the problem is presented by a group of "varying coefficient" approaches which include FlexCode [7] and RFCDE [13] and exploit expanding the conditional density function into orthogonal series. The combination of CDE and Approximate Bayesian Computation (ABC) incorporates the advantages of both approaches and leads to better density estimates [6].

The remainder of the paper is structured as follows. We discuss the existing approaches to CDE in more detail in Sect. 2. Section 3 addresses the chosen performance measure and its approximations employed in several methods. In Sect. 4 the datasets under consideration are described. In Sect. 5 we present the results and the reflections on our findings. We conclude the research and present some suggestions for future work in Sect. 6.

2   CDE Approaches

Assume we observe the finite sample of data $\{(\boldsymbol{x}_i, \boldsymbol{z}_i)\}_{i=1}^n$ with multidimensional covariates $\boldsymbol{x}_i \in \mathbb{R}^m$ and a multidimensional response $\boldsymbol{z}_i \in \mathbb{R}^\ell$. The goal of conditional density estimator methods is to reconstruct the full conditional probability density function $p(\boldsymbol{z}|\boldsymbol{x})$ as, in general, it provides us with a more comprehensive understanding of the underlying probability distribution than point estimations of conditional mean and variance. The following paragraphs give a brief description of the methods which can be employed for conditional density estimation.

Kernel Density Estimation. The first approach to this problem is estimating $p(\boldsymbol{x}, \boldsymbol{z})$ and $p(\boldsymbol{x})$ separately and then blending them as $p(\boldsymbol{z}|\boldsymbol{x}) = \frac{p(\boldsymbol{x},\boldsymbol{z})}{p(\boldsymbol{z})}$. The estimation of a joint probability function $p(z, x)$ and a marginal probability function $p(\boldsymbol{x})$ can be performed with kernel density estimators (KDE).

Binary Classification Approach. One more approach to conditional density estimation is based on reformulating the problem of density estimation as a binary classification problem. To accomplish it, we assign class $c = 1$ to all points of the given dataset, sample new points from a uniform distribution in a bound rectangle of our data sample and then treat the latter as the elements of the class $c = 2$. Applying then any existing classifier such as logistic regression or decision trees, we calculate probabilities $p(c|\boldsymbol{x}, \boldsymbol{z})$ and $p(c|\boldsymbol{x})$.

The number of sampled points is the same as in the given data sample to keep the dataset balanced. This approach allows us to estimate again $p(\boldsymbol{x}, \boldsymbol{z})$ and $p(\boldsymbol{x})$ separately and, after that, compute $p(\boldsymbol{z}|\boldsymbol{x})$:

$$p(\boldsymbol{x}, \boldsymbol{z}) = \frac{1}{V_{xz}} \frac{p(0|\boldsymbol{x}, \boldsymbol{z})}{p(1|\boldsymbol{x}, \boldsymbol{z})}, \quad p(\boldsymbol{x}) = \frac{1}{V_x} \frac{p(0|\boldsymbol{x})}{p(1|\boldsymbol{x})}, \quad p(\boldsymbol{z}|\boldsymbol{x}) = \frac{p(\boldsymbol{x}, \boldsymbol{z})}{p(\boldsymbol{z})},$$

where V_{xz} and V_x are the volumes of bound rectangles of $\{(\boldsymbol{x}_i, \boldsymbol{z}_i)\}_{i=1}^n$ and $\{\boldsymbol{x}_i\}_{i=1}^n$ samples correspondingly.

Naïve Bayes Approach. Naïve Bayes approach can be applied under the assumption of conditional independence of $\boldsymbol{z}$ coordinates—response variables $z^{(i)}$. It enables application of univariate methods to the problems with multidimensional $\boldsymbol{z}$.

Bayesian Network Approach. The extension of the previous estimator is the approach based on a Bayesian network built on the response variables $z^{(i)}$. The parent-child dependencies of response variables $z^{(i)}$ should be defined at the training step and should represented via a directed acyclic graph. Then, the conditional density estimation is performed as follows:

$$p(\boldsymbol{z}|\boldsymbol{x}) = p(z^{(1)}, z^{(2)}, \dots, z^{(\ell)}|\boldsymbol{x}) = \prod_{i=1}^{\ell} p(z^{(i)}|\mathrm{Parents}(z^{(i)}), \boldsymbol{x}).$$

FlexCode approach proposed in the paper [7] involves expanding the conditional density function into a series in which each coefficient can be estimated directly via regression if the chosen basis is orthonormal. This approximation reduces the multidimensional conditional density estimation problem to point estimation problem which is more straightforward to fulfill.

RFCDE: Random Forests for Conditional Density Estimation. The paper [13] further develops the idea of a series expansion and focuses on building ensembles of regression trees suggesting the method called Random Forests for Density Estimation (RFCDE). Its main contribution is the way of choosing the partition splits in the nodes of the trees: instead of minimizing traditional mean-squared loss, they optimize the loss specific to CDE (which will be discussed in Sect. 3). Several simplifications allow them to keep the optimization process computationally feasible.

NNKCDE: Approximate Bayesian Computation Method (ABC-CDE method). The last paper to consider in the scope of this research is [6], which suggests an efficient methodology for non-parametric conditional density estimation for the problems with inaccessible or intractable likelihood and available though limited data simulations. It aims for estimating the posterior density with the means of Approximate Bayesian Computation (ABC). The suggested framework combines the advantages of both ABC and CDE approaches and proposes not only the way of estimating the posterior density upon observing high-dimensional data but also the way of comparing the performance of ABC and other related methods and choosing the optimal summary statistics.

During the first step of the algorithm, the training set is constructed via a simple ABC rejection sampling algorithm for choosing the sample points x close to the given x_0 according to some pre-defined distance function $d(x, x_0)$ and δ such as $d(x, x_0) < \delta$. This training set is further employed for building the conditional density estimator. For this purpose the authors of the paper [6] adopt the FlexCode estimator from their previous paper [7], nevertheless mentioning the flexibility of the estimator choice. All the advantages mentioned above of the FlexCode estimator apply to this problem as well.

3   CDE Measure

One of the most straightforward approaches to assessing the quality of the obtained density function is evaluating some of its point estimates such as median, mean or any appropriate raw or central n-th moments of the distribution. The problem is that in many cases, such as multimodality, asymmetry and heteroscedastic noise, the point estimates do not fully describe the underlying data structure and therefore, cannot be considered representative for estimation quality assessment.

In order to quantify the distribution in a more comprehensive way, we need to measure the difference between the estimation and the true values in all data points of the sample. The most commonly used metric for estimating the discrepancy between exact $p(z|x)$ and approximate $\hat{p}(z|x)$ is the mean integrated square error (MISE):

$$\text{MISE} = \mathbb{E}_x \left(\int (\hat{p}(z|x) - p(z|x))^2 dz \right).$$

As it is hard to calculate, the reduced measure is used:

$$L(p, \hat{p}) = \iint \hat{p}^2(z|x) dF(x) dz - 2 \iint \hat{p}(z|x) p(z|x) dF(x) dz$$
$$= \iint \hat{p}^2(z|x) dF(x) dz - 2 \iint \hat{p}(z|x) dF(x, z) \tag{1}$$

where $F(x)$ is a marginal cumulative distribution function. $L(p, \hat{p})$ differs from MISE by a constant, which doesn't depend on the estimator $\hat{p}$. MISE generalizes the mean squared error and controls the overall MSE of the entire density function. It is closely related to the L_2 error of estimating a function.

If the "varying coefficient" approach [7] is employed, then the MISE measure can be rewritten in a form more convenient for optimization:

$$\widehat{L}(p,\hat{p}) = \sum_{i=1}^{I} \frac{1}{n}\sum_{k=1}^{n}\hat{\beta}_i^2(\boldsymbol{x}_k) - 2\frac{1}{n}\sum_{k=1}^{n}\hat{p}(\boldsymbol{z}_k|\boldsymbol{x}_k). \tag{2}$$

This is also the loss employed in [13] for defining the optimal split in tree nodes. Its properties allow the execution to be performed in a parallel manner which makes the optimizational process relatively fast.

For our experiments, we employ two approaches to computing the MISE: straightforward (1) which involves computing the multiple integral with rectangle rule in the bounded rectangle of the training sample and the approximate one (2) suggested in [7]. We also examine these MISE losses by comparing them with the best possible loss computed for the datasets with the pre-defined distribution and study the errors gained.

4   Datasets

We evaluate the approaches discussed in Sect. 2 on several datasets to reveal the strengths and weaknesses of the methods and to assess the problems they could help to overcome. They include artificial datasets with required properties and with the known exact value of CDE evaluation measure and practical significant multimodal dataset [11, 12].

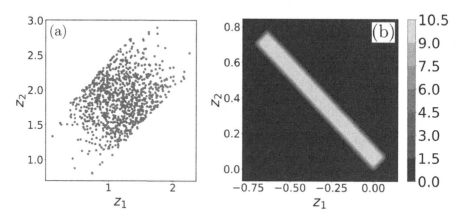

Fig. 1. (a) SURD ($n_x = 10$, $n_y = 1$, $n_z = 2$, $\boldsymbol{d} = [0.1, 1]$, $\sigma = 0.01$, $\boldsymbol{\alpha} = [0.3, 0.3]$, $\varphi = \frac{\pi}{4}$), (b) density of the distribution given in (a) at $\boldsymbol{x} = \boldsymbol{0}$.

The artificial datasets are generated from one general scheme of sampling by fixing different parameters. We consider the smoothed uniform rotated distribution SURD (n_x, n_y, n_z, $\boldsymbol{d}$, σ, $\boldsymbol{\alpha}$, φ):

$$x^{(i)}, \; y^{(j)} \sim \text{Uniform}(0,1), \; i = 1..n_x, \; j = 1..n_y,$$

$$\widetilde{z}^{(k)} \sim \text{Uniform}(0, d^{(k)}) + \text{Normal}(0, \sigma), k = 1..n_z,$$

$$z = Q\widetilde{z} + \alpha \sum_{i=1}^{n_x} x_i. \tag{3}$$

Here $x^{(i)}$ are relevant covariates, $y^{(j)}$ are irrelevant covariates, Q stands for the orthogonal rotation matrix which for every pair of coordinates $(2k-1, 2k)$, $k = 1..\frac{n_z}{2}$ rotates the point by the degree of rotation φ. The variance of the distribution along axes is determined by the vector d while the α parameter explicitly controls the dependence of z on the values of x. The σ parameter regulates the smoothness of the distribution. Figure 1 shows the example of sampling according to the described scheme together with the density of the distribution at the fixed point $x = 0$.

The artificial univariate multimodal dataset Fork (n_x, n_y, m, σ) (considered in the RFCDE paper [13]) is generated by the following scheme to verify our findings from the multivariate case: $x^{(i)}, \; y^{(j)} \sim \text{Uniform}(0,1), \; i = 1..n_x, \; j = 1..n_y, \; s \sim \text{Multinomial}\left(1, \frac{1}{m}, ..., \frac{1}{m}\right), \; v = (v_1, ...v_m), \; v_i = -1 + \frac{2(i-1)}{m-1}, \; k = \langle s, v \rangle, \; z \sim \text{Normal}\left(k \sum_{i=1}^{n_x} x^{(i)}, \sigma\right)$. Here $x^{(i)}$ are relevant covariates, $y^{(j)}$ are irrelevant covariates, s is an unobserved one-hot binary vector covariate which introduces multimodality in the conditional densities, $m \geq 1$ is the parameter controlling the number of peaks in the multimodal distribution. Angle brackets $\langle \cdot, \cdot \rangle$ in the equation for k denote the dot product.

The practically significant multimodal dataset containing examples for the PyFitIt software [11,12] was built by calculating XANES spectra for various geometry modifications of $[Fe(terpy)_2]^{2+}$ complex by FDMNES [5]. Then the spectra were smoothed to get the same shape as the experimental one. The argument x here is a XANES spectrum (dimension $= 86$), the target variable z is a 6-dimensional vector of geometry parameters. The dataset was constructed in such a way that the partial probability distribution $p(z)$ of the target z is uniform in a rectangle $[-0.3, 0.5]^6$. To check the results, we use the dataset Feterpy_combined with reduced number of geometry parameters: 3-dimensional z. Feterpy dataset contains 729 spectra, Feterpy_combined—500.

5   Experiments

During our experiments we determine the relative error of the method as

$$\frac{|L_{cur} - L_{best}|}{|L_{best}|}, \tag{4}$$

where L is defined in (1). All KDE estimators are run with a Gaussian kernel and a bandwidth h set to 0.4. The base classifier for the Binary Classification Approach is the LGBM classifier from the LightGBM framework that uses tree-based learning algorithms with $n\_estimators = 200$ and $learning\_rate = 0.005$.

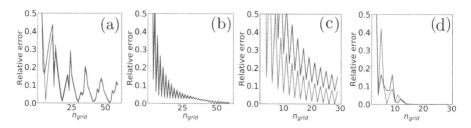

Fig. 2. Relative errors for straightforward (red) and approximate (blue) approaches for computing the MISE loss for (a) SURD ($n_x = 2$, $n_y = 1$, $n_z = 2$, $d = [0.1, 1]$, $\sigma = 0$, $\alpha = [0, 0]$, $\varphi = \frac{\pi}{4}$), (b) SURD ($n_x = 2$, $n_y = 1$, $n_z = 2$, $d = [1, 1]$, $\sigma = 0$, $\alpha = [0, 0]$, $\varphi = \frac{\pi}{4}$), (c) SURD ($n_x = 2$, $n_y = 1$, $n_z = 3$, $d = [1, 1, 1]$, $\sigma = 0$, $\alpha = [0, 0, 0]$, $\varphi = \frac{\pi}{4}$), (d) Fork ($n_x = 10$, $n_y = 1$, $m = 2$, $\sigma = 1$). (Color figure online)

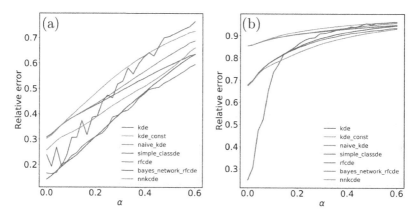

Fig. 3. Relative errors for CDE methods depending on $\alpha = [\alpha, \alpha]$, $\alpha \in (0, 0.6)$ for (a) SURD ($n_x = 10$, $n_y = 0$, $n_z = 2$, $d = [1, 1]$, $\sigma = 0$, $\varphi = \frac{\pi}{4}$) and (b) same as (a) except $d = [0.1, 1]$.

The base one-dimensional estimator for the Naïve Bayes approach is set to KDE with the same parameters. The RFCDE model is employed with the following set of parameters: $n\_trees = 1000$, $mtry = 4$, $n\_basis = 15$, $node\_size = 20$. The base estimator for Bayesian Network is the RFCDE model with the above mentioned parameters. We utilize the NNKCDE model and set its parameters to $k = 10$, $bandwidth = 0.2$. Any other parameters in the algorithms under consideration are set to default.

We start our experiments by assessing the errors obtained by two approaches to calculating the MISE measure discussed in Sect. 3. The parameter under investigation is the number of grid points taken along each axis to compute the integrals in either straightforward (1) or approximate (2) manner. Figure 2 displays that the approximate approach needs a more frequent grid to achieve the competitive relative error rate in comparison with more computationally expensive straightforward numerical integration especially for $n_z = 3$. Another interesting

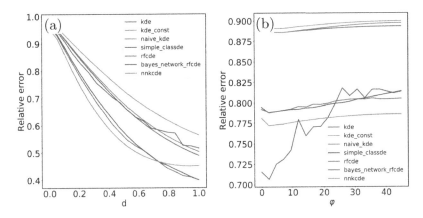

Fig. 4. Relative errors for CDE methods depending on (a) $d = [d, 1]$, $d \in (0, 1)$ for SURD ($n_x = 10$, $n_y = 0$, $n_z = 2$, $\alpha = [0.3, 0.3]$, $\sigma = 0$, $\varphi = 0$) and (b) $\varphi \in \left(0, \frac{\pi}{4}\right)$ for SURD ($n_x = 10$, $n_y = 0$, $n_z = 2$, $\alpha = [0.1, 0.1]$, $d = [0.1, 1]$, $\sigma = 0$).

observation is the fact that the data points flatness along one of the axes leads to the worse performance of both approaches which can be explained by the inability of both numerical integration and employed cosine basis to calculate integral for the flattened data. It can be also noted that increasing the dimensionality of the target variable n_z heavily influences the relative error rate for the thinner grid. In the univariate case, two approaches to estimating the MISE loss gained by the estimator appear to exhibit better results as it is shown in Fig. 2d.

Henceforth, in our experiments, we employ the straightforward approach for evaluating the MISE loss as it requires fewer points of the grid to accomplish more true-to-life values obtained by the loss of the algorithm.

The next parameter to study is α which regulates the dependence of z on the values of x. For simplicity we set $\alpha = [\alpha, \alpha]$. We evaluate all conditional density estimators discussed in Sect. 2 except FlexCode which implementation provided by the authors of the paper [7] does not support multidimensional datasets. According to Fig. 3, all concerned CDE algorithms indicate the similar tendency: the stronger the dependence between z and x, the more difficult is to provide the correct estimations for the fixed x. Flatness of data along one of the axes even stronger obstructs the correct conditional density estimation as can be seen when analyzing Fig. 3a against Fig. 3b.

Furthermore, we examine the dependency on data flatness along one of the axes separately as it proved to be critical for the quality of estimation in the previous experiments. Without loss of generality, we vary the first component of d. Figure 4a confirms the idea expressed in the previous paragraphs: the closer the variances along axes to each other, the better the algorithms model the underlying distribution. Figure 4a indicates as well that the family of orthogonal series approaches shows the best performance among concerned methods though still suffering from an imbalanced variance of data along different axes.

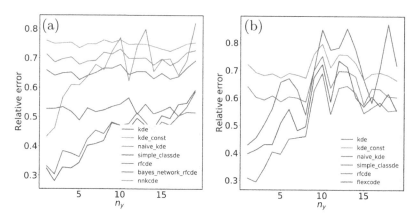

Fig. 5. Relative errors for CDE methods depending on $n_y \in [1, 20]$ for (a) SURD ($n_x = 1$, $n_z = 2$, $\alpha = [2, 2]$, $d = [1, 1]$, $\sigma = 0$, $\varphi = 0$) (b) Fork ($n_x = 1$, $m = 2$, $\sigma = 0.1$).

Another interesting parameter to discuss is φ which determines the dependency between target variables $z^{(i)}$ by controlling the degree of the data points rotation Fig. 4b. One can observe that the Binary Classification Approach approximation becomes noticeably inaccurate with amplification of the angle φ. The estimation appears to be imprecise as the volume of the bounded rectangles V_x and V_{xz} can be sufficiently larger than the area in which the true density values have fallen. This fact leads to the approximation error becoming larger and explains the deteriorated performance of this method with an increased φ.

Finally, we evaluate the influence of the number of irrelevant covariates n_y on the performance of the CDE algorithms. Figure 5a indicates that while RFCDE, NNKCDE and Bayesian Network approach (employing RFCDE one-dimensional estimator internally) show the lowest relative error among all the methods, their performance deteriorates with growing n_y. A similar tendency can be observed for the family of orthogonal series approaches in the univariate case in Fig. 5b.

Practically significant datasets have the drawback: the exact value of the CDE evaluation measure is not known. So, the relative error (4) can't be calculated and we have to be content with the value (1). We calculate the measure values for considered algorithms for the dataset with 6-dimensional z and for the dataset with combined geometry parameters (see Feterpy_combined in [11]), which has 3-dimensional z. The results are collected in Table 1. The NNKCDE

Table 1. Values of CDE evaluation measure (1) for Feterpy datasets

	kde	kde_const	naive_kde	simple classde	rfcde	bayes_network rfcde	nnkcde
Feterpy	−0.6	−0.4	−0.6	3.9	-	−4.5	−7.7
Feterpy combined	−1.4	−0.8	−1.3	−1.8	−5.5	−4.2	−8.7

algorithm outperforms the others both by quality and speed. RFCDE required more than 80 GB of memory for Feterpy dataset and didn't finish calculation.

6   Conclusion

This research aims to investigate several approaches to conditional density estimation since point estimations or quantile regression do not suffice, for example, if a multimodal or skewed distribution is considered. Specifically, we perform the comparative study of several methods employing inherently different techniques: Kernel Density Estimators, Binary Classification Approach, Bayesian Networks, the family of orthogonal series approaches and one of the most recent models—NNKCDE which implements the ABC-rejection scheme. We provide an overview of these methods and varying the parameter of two synthetic datasets (for multivariate and univariate cases) and practically significant multimodal dataset, we demonstrate the strengths and the weaknesses of the algorithms under consideration. Not only concern our experiments the algorithms themselves, but they also tackle the problem of the quality of the loss computation performed with two approaches.

There are multiple directions in which this research may progress. During the experiments, we did not manage to examine datasets with a high dimensional target variable z. The problem is that the straightforward (1) way of calculating MISE measure is computationally exhaustive while the second (2) produces inaccurate estimation. So we have to either consider another measure or invent new algorithms of the MISE optimization suitable for a high-dimensional target.

References

1. Angrist, J.D., Pischke, J.S.: Mostly Harmless Econometrics: An Empiricist's Companion. Princeton University Press, Princeton (2008)
2. Bohm, G., Zech, G.: Introduction to Statistics and Data Analysis for Physicists, vol. 1. Desy, Hamburg (2010)
3. Burnaev, E., Nazarov, I.: Conformalized kernel ridge regression. In: 2016 15th IEEE International Conference on Machine Learning and Applications (ICMLA), pp. 45–52. IEEE (2016). https://doi.org/10.1109/ICMLA.2016.0017
4. Guda, A.A., Guda, S.A., Lomachenko, K.A., et al.: Quantitative structural determination of active sites from in situ and operando XANES spectra: from standard AB initio simulations to chemometric and machine learning approaches. Catal. Today (2018). https://doi.org/10.1016/j.cattod.2018.10.071
5. Guda, S.A., Guda, A.A., Soldatov, M.A., et al.: Optimized finite difference method for the full-potential XANES simulations: application to molecular adsorption geometries in MOFs and metal-ligand intersystem crossing transients. J. Chem. Theory Comput. **11**(9), 4512–4521 (2015). https://doi.org/10.1021/acs.jctc.5b00327
6. Izbicki, R., Lee, A.B., Pospisil, T.: ABC-CDE: toward approximate Bayesian computation with complex high-dimensional data and limited simulations. J. Comput. Graph. Stat. 1–20 (2019). https://doi.org/10.1080/10618600.2018.1546594

7. Izbicki, R., Lee, A.B., et al.: Converting high-dimensional regression to high-dimensional conditional density estimation. Electron. J. Stat. **11**(2), 2800–2831 (2017). https://doi.org/10.1214/17-EJS1302

8. Kemp, G.C., Silva, J.S.: Regression towards the mode. J. Econ. **170**(1), 92–101 (2012). https://doi.org/10.1016/j.jeconom.2012.03.002

9. Kuleshov, A.P., Bernstein, A., Burnaev, E.: Conformal prediction in manifold learning. In: 7th Symposium on Conformal and Probabilistic Prediction and Applications, COPA 2018, Maastricht, The Netherlands, 11–13 June 2018, pp. 234–253 (2018)

10. Kuleshov, A.P., Bernstein, A., Burnaev, E.: Kernel regression on manifold valued data. In: 5th IEEE International Conference on Data Science and Advanced Analytics, DSAA 2018, Turin, Italy, 1–3 October 2018, pp. 120–129 (2018). https://doi.org/10.1109/DSAA.2018.00022

11. Martini, A., Guda, S.A., Guda, A.A., et al.: PyFitit: the software for quantitative analysis of XANES spectra using machine-learning algorithms. Mendeley Data (2019). https://doi.org/10.17632/dwrb56xrx6.1

12. Martini, A., Guda, S.A., Guda, A.A., et al.: PyFitit: the software for quantitative analysis of XANES spectra using machine-learning algorithms. Comput. Phys. Commun. (2019, to appear)

13. Pospisil, T., Lee, A.B.: RFCDE: random forests for conditional density estimation. arXiv preprint arXiv:1804.05753 (2018)

14. Rau, M.M., Seitz, S., Brimioulle, F., et al.: Accurate photometric redshift probability density estimation-method comparison and application. Mon. Not. R. Astron. Soc. **452**(4), 3710–3725 (2015). https://doi.org/10.1093/mnras/stv1567

15. Yao, W., Li, L.: A new regression model: modal linear regression. Scand. J. Stat. **41**(3), 656–671 (2014). https://doi.org/10.1111/sjos.12054

Histogram-Based Algorithm for Building Gradient Boosting Ensembles of Piecewise Linear Decision Trees

Aleksei Guryanov$^{(\boxtimes)}$ (iD)

Lomonosov Moscow State University, Leninskie Gory 1, 119991 Moscow, Russia
guryanov93@gmail.com

Abstract. One of the most popular machine learning algorithms is gradient boosting over decision trees. This algorithm achieves high quality out of the box combined with comparably low training and inference time. However, modern machine learning applications require machine learning algorithms, that can achieve better quality in less inference time, which leads to an exploration of grading boosting algorithms over other forms of base learners. One of such advanced base learners is a piecewise linear tree, which has linear functions as predictions in leaves. This paper introduces an efficient histogram-based algorithm for building gradient boosting ensembles of such trees. The algorithm was compared with modern gradient boosting libraries on publicly available datasets and achieved better quality with a decrease in ensemble size and inference time. It was proven, that algorithm is independent of a linear transformation of individual features.

Keywords: Machine learning · Ensembles · Gradient boosting · Piecewise linear decision trees

1 Introduction

Gradient boosting over decision trees is one of the most popular machine learning algorithms. It was proposed in [1]. It possesses several useful properties that allow it to be used efficiently in many real-world scenarios:

- achieves great quality on a large variety of datasets out of the box;
- has built-in feature selection and regularizations;
- allows straightforward management of model complexity through tree depth and number of trees;
- has several efficient implementations;
- algorithmic complexity of training and inference time can be estimated based on properties of the dataset and the hyperparameters.

However, modern applications of machine learning demand modifications to machine learning algorithms to achieve better quality in less inference time.

W. M. P. van der Aalst et al. (Eds.) AIST 2019, LNCS 11832, pp. 39–50, 2019.
https://doi.org/10.1007/978-3-030-37334-4_4

Catboost [2], for example, employs to that purpose decision trees, that have the same split rule on each level, which allows for significant inference time reduction through SIMD-accelerated computations. One other type of base learner is a piecewise linear decision tree [3,4], that has the same split rules as plain decision trees, but predicts a linear function in every leaf. In this paper, an efficient and algorithmically-sound algorithm for building gradient boosting ensembles of piecewise linear trees is proposed. The proposed algorithm is implemented using Python and Numba [5] for JIT-compilation of computationally-intensive parts, and the implementations show improvement in quality over current popular gradient boosting libraries as shown on publicly available datasets.

The paper is organized as follows. First, a brief overview of second-order gradient boosting, which is a basis of several popular gradient boosting libraries [6,7], is given. Then the concept of piecewise linear decision trees is explained, with the explanation of certain modifications, exclusive to this paper. Then several key formulas are derived, which allow the training of piecewise linear decision trees as described in the previous section. Then an algorithm for building such trees as part of gradient boosting ensemble is explained. The next section shows that it is possible to prove specific properties of the suggested algorithm, such as indifference to a linear transformation of individual features. The next section describes experimental research of the suggested algorithm in comparison to several other gradient boosting libraries. Then a brief overview of several other works devoted to piecewise linear decision trees is given. The last chapter details the conclusion of the work.

2   Second-Order Gradient Boosting

Given a dataset of samples $D = (x_i, y_i)$ with m features, where $|D| = n$ is the number of samples, and $x_i \in R^m$, gradient boosting algorithm trains a sequence of size T of decision trees $\{t_k\}_1^T$ to minimize global loss function L. The final output of gradient boosting is the summation of these trees $\hat{y} = \sum_{k=1}^{T} t_k(x_i)$. Let $l : R^2 \to R$ be the loss function for a single sample. Then the global loss function is usually represented as a summation of individual losses over training dataset D and a regularization term $\Omega(t_k)$ to prevent overfitting:

$$L = \sum_{i=1}^{n} l(\hat{y}_i, y_i) + \sum_{k=1}^{T} \Omega(t_k). \tag{1}$$

Let $\hat{y}_i^{(k)}$ be the predicted value of x_i after iteration k. At iteration $k+1$ a new tree t_{k+1} is trained to minimize the following loss

$$L^{(k+1)} = \sum_{i=1}^{n} l(\hat{y}_i^{(k+1)}, y_i) + \sum_{k=1}^{T} \Omega(t_k)$$

$$= \sum_{i=1}^{n} l(\hat{y}_i^{(k)} + t_k(x_i), y_i) + \sum_{k=1}^{T} \Omega(t_k) \tag{2}$$

With second-order approximation, and assuming all predictors from 1 to k fixed, the loss at step k can be represented as

$$L^{(k+1)} \approx C + \Omega(t_k) + \sum_{i=1}^{n} (g_i t_k(x_i) + \frac{1}{2} h_i t_k(x_i)^2), \tag{3}$$

where C is a constant value independent of the tree, $g_i = \frac{\delta l(\hat{y}_i, y_i)}{\delta \hat{y}_i}(\hat{y}_i^{(k)})$, $h_i = \frac{\delta^2 l(\hat{y}_i, y_i)}{\delta^2 \hat{y}_i}(\hat{y}_i^{(k)})$ – gradients and hessians of the loss with respect to the algorithms prediction in the point, equal to the prediction of the ensemble at step k.

Given the fixed structure of a single decision tree, we can rewrite its loss as a sum of losses over each existing leaf:

$$L^{(k+1)} = \sum_{leaf \in t_k} L_{leaf}^{(k+1)} + \Omega(t_k)$$

$$L_{leaf}^{(k+1)} = \sum_{i \in leaf} (g_i t_k(x_i) + \frac{1}{2} h_i (t_k(x_i))^2). \tag{4}$$

Greedy algorithms for building gradient boosting ensembles over decision trees can use Eq. 4 as an information gain for node splitting criterion. Gradient boosting algorithm can be used to minimize any twice differentiable loss function, which includes cross-entropy for binary and multiclass classification and mean squared error for regression tasks.

3 Piecewise Linear Decision Trees

Piecewise linear tree is a decision tree, which has a linear function as a predictor in each leaf. This paper suggests building such decision trees using additive linear functions in each node over the same feature this specific node has split over. Every internal node in such tree has FeatureIdx, Threshold to define the split rule, links to LeftChild and RightChild nodes, as well as LeftCoefficient and RightCoeffient to define linear function over feature in the split rule. Every leaf node in such tree has corresponding LeafValue. Algorithm to compute prediction of such a tree for an object $x \in R^m$ is described in Algorithm 1.

3.1 Gradient Boosting over Piecewise Linear Decision Trees

Assuming fixed tree structure, we can specify the accumulated linear part of the prediction of this tree in every object i as f_i. It is possible to add a linear component over feature j to a leaf in the tree, and prediction of this modified tree structure in the leaf can be rewritten as $t_k^*(x_i) = f_i + wx_i^j + b, i \in leaf$, where x_i^j is the value of feature j of object x_i. The regularization loss, given two regularization coefficients λ_w and λ_b for linear and constant prediction components respectively, is represented as

$$\Omega(t_k) = \sum_{w \in t_k.LinearCoefficients} \frac{1}{2} \lambda_w \cdot w^2 + \sum_{b \in t_k.LeafValues} \frac{1}{2} \lambda_b \cdot b^2. \tag{5}$$

Algorithm 1. Prediction of piecewise linear decision tree

Input: Root, instance of root node
Input: x, array[1..m] of object features
Node ← Root
Prediction ← 0.0
while *Node is not leaf* **do**
 FeatureIdx ← SplitFeatureIdx of Node
 Threshold ← SplitFeatureThreshold of Node
 Value ← x[FeatureIdx]
 if *Value ≤ Threshold* **then**
  LinearCoefficient ← LeftCoefficient of Node
  Node ← LeftChild of Node
 else
  LinearCoefficient ← RightCoefficient of Node
  Node ← RightChild of Node
 Prediction ← Prediction + Value · LinearCoefficient
Prediction ← Prediction + LeafValue of Node
Result: Prediction

Loss from Eq. 4 can be represented as a quadratic polynomial with respect to w and b:

$$L_{\text{leaf}j} = \frac{1}{2}w^2 \sum_{i \in \text{leaf}} (h_i(x_i^j)^2 + \lambda_w) + \frac{1}{2}b^2 \sum_{i \in \text{leaf}} (h_i + \lambda_b)$$
$$+ w \sum_{i \in \text{leaf}} (g_i x_i^j + h_i f_i x_i^j) + b \sum_{i \in \text{leaf}} (g_i + h_i f_i) \tag{6}$$
$$+ wb \sum_{i \in \text{leaf}} h_i x_i^j + \sum_{i \in \text{leaf}} (g_i f_i + \frac{1}{2}h_i f_i^2).$$

It is possible to introduce certain abbreviations of several accumulated values:

$$\text{GHF}_{\text{N}} = \sum_{i \in N} g_i + h_i f_i, \quad \text{GXHFX}_{\text{N}}^{\text{j}} = \sum_{i \in N} g_i x_i^j + h_i f_i x_i^j$$
$$\text{H}_{\text{N}} = \sum_{i \in N} h_i, \quad \text{HX}_{\text{N}}^{\text{j}} = \sum_{i \in N} h_i x_i^j, \quad \text{HXX}_{\text{N}}^{\text{j}} = \sum_{i \in N} h_i (x_i^j)^2. \tag{7}$$

With these abbreviations it is possible to rewrite the original quadratic polynomial from Eq. 6 as

$$L_{\text{leaf}}^j = \frac{1}{2}w^2(\text{HXX}_{\text{N}}^{\text{j}} + \lambda_w) + \frac{1}{2}b^2(\text{H}_{\text{N}} + \lambda_b)$$
$$+ w \cdot \text{GXHFX}_{\text{N}}^{\text{j}} + b \cdot \text{GHF}_{\text{N}} \tag{8}$$
$$+ wb \cdot \text{HX}_{\text{N}}^{\text{j}} + \sum (g_i f_i + \frac{1}{2}h_i f_i^2).$$

Minimization of such quadratic polynomial with non-degenerate values of w and b requires a positive-definite matrix of second derivatives. After omission of regularization terms to ensure computational stability, Sylvester's criterion gives us the following requirements:

$$\mathtt{HXX_N^j} > 0$$
$$\delta = \mathtt{HXX_N^j} \cdot \mathtt{H_N} - (\mathtt{HX_N^j})^2 > 0 \tag{9}$$

Assuming these requirements are fulfilled, we can compute optimal value of linear coefficient w, constant b and minimal value of loss:

$$w_{\mathrm{Linear}}^j(N) = \frac{(\mathtt{GHF_N}) \cdot (\mathtt{HX_N^j}) - (\mathtt{GXHFX_N^j}) \cdot (\mathtt{H_N} + \lambda_b)}{(\mathtt{HXX_N^j} + \lambda_w) \cdot (\mathtt{H_N} + \lambda_b) - (\mathtt{HX_N^j})^2}$$

$$b_{\mathrm{Linear}}^j(N) = \frac{(\mathtt{GXHFX_N^j}) \cdot (\mathtt{HX_N^j}) - (\mathtt{GHF_N}) \cdot (\mathtt{HXX_N^j} + \lambda_w)}{(\mathtt{HXX_N^j} + \lambda_w) \cdot (\mathtt{H_N} + \lambda_b) - (\mathtt{HX_N^j})^2}$$

$$\mathrm{Loss}_{\mathrm{Linear}}(N, j) = \frac{Q^j(N)}{(\mathtt{HXX_N^j} + \lambda_w) \cdot (\mathtt{H_N} + \lambda_b) - (\mathtt{HX_N^j})^2} \tag{10}$$
$$+ \sum_{i \in N} (g_i f_i + \frac{1}{2} h_i f_i^2)$$

$$Q^j(N) = (\mathtt{GXHFX_N^j})^2 (\mathtt{H_N} + \lambda_b) + (\mathtt{GHF_N})^2 (\mathtt{HXX_n^j} + \lambda_w)$$
$$- 2 \cdot (\mathtt{GXHFX_N^j})(\mathtt{GHF_N})(\mathtt{HX_N^j})$$

It is also essential to specify the value of coefficient b and the value of optimal loss in the case, where we assign zero as the value of linear coefficient w.

$$b_{\mathrm{Constant}}(N) = -\frac{\mathtt{GHF_N}}{\mathtt{H_N} + \lambda_b}$$
$$\mathrm{Loss}_{\mathrm{Constant}}(N) = -\frac{1}{2}\frac{(\mathtt{GHF_N})^2}{\mathtt{H_N} + \lambda_b} + \sum_{i \in N} (g_i f_i + \frac{1}{2} h_i f_i^2) \tag{11}$$

If we are considering split of parent node P into two nodes L and R over feature j, then we can use as an information gain for splitting the node the difference between the loss of finalizing the node P with a constant value and the sum of two losses of finalizing nodes L and R as linear nodes.

$$\mathrm{Gain}^j(P, L, R) = \mathrm{Loss}_{\mathrm{Constant}}(P) - \mathrm{Loss}_{\mathrm{Linear}}(L, j) - \mathrm{Loss}_{\mathrm{Linear}}(R, j) \tag{12}$$

Since elements in parent node P are separated into two child nodes L and R, the terms $\sum(g_i f_i + \frac{1}{2} h_i f_i^2)$ in $Loss_{Linear}$ and $Loss_{Constant}$ are reduced. This allows to simplify the computation of $Gain$:

$$\mathrm{Gain}^j(P, L, R) = \mathrm{Loss}_{\mathrm{Constant}}^*(P) - \mathrm{Loss}_{\mathrm{Linear}}^*(L, j) - \mathrm{Loss}_{\mathrm{Linear}}^*(R, j)$$

$$\mathrm{Loss}_{\mathrm{Linear}}^*(N, j) = \frac{Q^j(N)}{(\mathtt{HXX_N^j} + \lambda_w) \cdot (\mathtt{H_N} + \lambda_b) - (\mathtt{HX_N^j})^2} \tag{13}$$

$$\mathrm{Loss}_{\mathrm{Constant}}^*(N) = -\frac{1}{2}\frac{(\mathtt{GHF_N})^2}{\mathtt{H_N} + \lambda_b}$$

3.2   Algorithm

Algorithm for building a piecewise linear decision tree mostly repeats the algorithm for building a plain decision tree. The significant difference is the requirement to update values of accumulated predictions after every split based on the deduced linear coefficient of splitted nodes to represent correct accumulated linear prediction and modified procedure for finding optimal split and linear function in each new node.

Algorithm 2. Building of piecewise linear decision tree

    Input: X, array[1..n,1..m] matrix of features
    Input: gradients: array[1..n] of gradients of boosting step
    Input: hessians: array[1..n] of hessians of boosting step
    AccumulatedPredictions: array[1..n] ← [0...0]
    Root: TreeNode
    NodesToSplit: array[TreeNode] ← [Root]
    while *not StopCondition* **do**
       Node, SplitInfo ← FindBestSplit(NodesToSplit)
       LeftNode, RightNode ← SplitNode(Node, SplitInfo)
       Remove Node from NodesToSplit
       Add LeftNode, RightNode to NodesToSplit
       UpdateAccumulatedPredictions(LeftNode, RightNode, SplitInfo)
    Result: Decision tree with root in Root

Since changes to a node do not change optimal split for other leaf nodes in the tree, it is possible to compute the best splitting only for new nodes in the tree. Stopping condition usually consists of restrictions on a tree structure, such as a limit of tree depth or a number of leaves in the tree. Sylvester's criterion is described in Eq. 9. It is possible that for a given node there would be no split with a positive gain. That means that current node should not be split, it is removed from the pool of available to split nodes and finalized into a leaf.

Algorithm 3. Finding best split for a node

    Input: P, parent node to split
    featureIdx ← 1
    while *featureIdx ≤ m* **do**
       while *Available splits for featureIdx exist* **do**
          SplitCandidate ← available split for featureIdx
          Remove SplitCandidate from available splits
          L, R ← split parent node P by SplitCandidate
          if *Sylvester's criterion is met for L and R* **then**
             gain = Gain(P, L, R)
          else
             gain = 0
          score ← max(gain, score)
       Increment featureIdx
    Result: featureIdx, SplitCandidate, score with max score

Training data is separated into bins for every feature based on the percentiles of this feature before the gradient boosting algorithm, as suggested in [7]. Only splits between the bins are considered. Such optimization reduces the computational complexity of the algorithm and allows the usage of efficient histogram-based computations. It is possible to efficiently update accumulated values in Eq. 7 using histograms and sorting of available splits by feature value to ensure linear complexity of Algorithm 3 regarding the number of objects n in the training set.

3.3   Properties

It is possible to prove that, under certain conditions, the suggested algorithm is stable with regard to linear transformations of features. More specifically, predictions of the suggested algorithm on the training set will not change after applying the nondegenerate linear transformation to some or all features of the data set and retraining the model.

Theorem 1. *Assuming regularization terms λ_w and λ_b are equal to zero, predictions of gradient boosting over piecewise linear decision trees on training data do not change after non degenerate linear transformation of individual feature $x_i^{j^*} = \alpha \cdot x_i^j + \beta, \alpha \neq 0$, retraining of the ensemble and application of the ensemble on the training data.*

Proposition 1. *Given a node N with fixed values of feature j, gradients g_i, hessians h_i and accumulated predictions f_i in objects of node N, and feature x^j, for other feature $x^{j^*} = \alpha \cdot x^j + \beta, \alpha \neq 0$, the following equations are correct:*

$$\text{HXX}_N^{j^*} \cdot \text{H}_N - (\text{HX}_N^{j^*})^2 = \alpha^2 * (\text{HXX}_N^j \cdot \text{H}_N - (\text{HX}_N^j)^2)$$
$$Q^{j^*}(N) = \alpha^2 \cdot Q^{j^*}(N)$$
$$\text{Loss}^*_{\text{Linear}}(N, j^*) = \text{Loss}^*_{\text{Linear}}(N, j) \tag{14}$$
$$\text{Gain}^{j^*}(P, L, R) = \text{Gain}^j(P, L, R)$$
$$w^{j^*}_{\text{Linear}}(N) = \frac{1}{\alpha} w^j_{\text{Linear}}(N).$$

The equations above mean, that for fixed gradients, hessians, accumulated predictions and tree structure, linear transformation of feature does not change information gain of splits for this feature. This means that Algorithm 2 chooses the same splitting rule over all features as it would if there was no linear transformation of a specific feature. Linear coefficient for the feature changed with linear transformation would be reduced by a factor of $/alpha$, which means that accumulated prediction will be different by a constant $f_i^* = f_i + c, c = \frac{\alpha}{\beta}$.

Proposition 2. *Given a node N with fixed values of feature j, gradients g_i, hessians h_i and accumulated predictions f_i in objects of node N, and feature x^j, changes to f_i by a constant to be equal to $f_i^* = f_i + c$ keep the following equations correct:*

$$w_{\text{Linear}}^{j}(N, f_i^*) = w_{\text{Linear}}^{j}(N, f_i)$$
$$b_{\text{Linear}}^{j}(N, f_i^*) = b_{\text{Linear}}^{j}(N, f_i) - c \qquad (15)$$
$$\text{Gain}^{j}(P, L, R, f_i^*) = \text{Gain}^{j}(P, L, R, f_i).$$

Since final prediction of the algorithm on objects of training set can be represented as $t_k(x_i) = f_i + b_{\text{Node}_{x_i}}$, where f_i are values of accumulated predictions after the last split is finalized, it is possible to affirm, that final prediction of an algorithm with a linearly transformed feature is not changed.

4   Experiments

The suggested algorithm, in future named Piecewise Linear, is compared based on training time, inference time and maximum achieved quality with 3 other popular gradient boosting libraries: Xgboost [6], Lightgbm [7] and CatBoost [2]. Hyperparameters of gradient boosting decision trees are listed in Table 1. Experimental evaluation is conducted on three publicly available datasets from UCI machine learning repository: HIGGS[1] and HEPMASS[2] for binary classification and YEAR[3] for regression. Dataset HIGGS consists of 11000000 objects with 28 real-valued features, dataset HEPMASS consists of 7000000 objects with 28 real-valued features, dataset YEAR consists of 515345 objects with 90 real-valued features. 5% of each dataset were separated and used as a test set for quality measurement; other 95% was used as a training set. Training and testing sets were given as an input to all four machine learning algorithms without any preprocessing. The target variable for YEAR dataset was normalized to have zero mean. Details of the experimental setup are specified in Table 1. Details of the gradient boosting libraries setup are specified in Table 2. For all datasets and for all four libraries three values were computed: final quality of the ensemble on test dataset measured in ROC-AUC for binary classification tasks and root mean squared error (RMSE) for regression task, time to train full ensemble and time to apply ensemble on the whole dataset. These results are specified in Tables 3, 4 and 5. Quality of the gradient boosting ensembles as a function of the number of trees used in prediction (also frequently called staged quality) is represented in plots in Fig. 1.

Experimental results show, that implementation of the suggested algorithm can achieve better quality than other popular gradient boosting libraries in the

Table 1. Experimental setup

Operating system	CPU	Memory
Ubuntu 16.04.1	Xeon(R) CPU E5-2630 v4 x2	DDR4 256 GB

---

[1] https://archive.ics.uci.edu/ml/datasets/HIGGS.
[2] https://archive.ics.uci.edu/ml/datasets/HEPMASS.
[3] https://archive.ics.uci.edu/ml/datasets/YearPredictionMSD.

Table 2. Gradient boosting library setup

Library	Version	Parameter	Value
Piecewise linear		λ_w	1.0
		λ_b	1.0
		Learning rate	0.1
		Max depth	5
		Num iteration	250
LightGBM	2.2.3	λ	1.0
		Learning rate	0.1
		Max depth	5
		Num iteration	250
Xgboost	0.90	λ	1.0
		Learning rate	0.1
		Max depth	5
		Num iteration	250
CatBoost	0.15.2	λ	1.0
		Learning rate	0.1
		Max depth	5
		Num iteration	250

same number of iterations. Training time for the suggested algorithm is 2–4 times slower on average for classification tasks and 4–20 times slower for regression tasks than the other gradient boosting libraries. Slower training times can be attributed to many potential areas of optimization in the implementation of suggested algorithm. The potential for the optimization is great for regression tasks, since mean squared error loss has constant hessian, and this property can be used to dramatically reduce the amount of computations and the amount of data to be processed during computations of gradient and hessian based statistics. Despite the addition of linear components in the prediction of Piecewise Linear decision trees, inference time for this algorithm does not increase significantly. The only outlier with respect to inference time is Catboost, which uses a special form of decision trees, optimized for inference. This means, that compared to gradient boosting algorithms based on classic decision trees, suggested algorithm can achieve better quality without increasing inference time, which is crucial in high-load applications of machine learning.

5   Related Work

Several other works have proposed algorithms for building gradient boosting ensembles of piecewise linear decision trees. In [8] an algorithm is proposed, that for every possible split in the node computes optimal linear predictor in left and

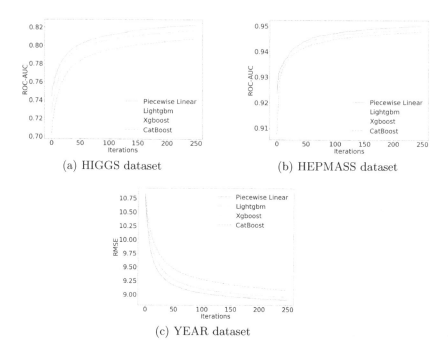

(a) HIGGS dataset   (b) HEPMASS dataset

(c) YEAR dataset

Fig. 1. Staged quality of gradient boosting libraries

right child nodes and the value of gain associated with optimal linear predictors. However such algorithm is not computationally tractable for any significantly large dataset, and it was acknowledged as such. One of the proposals to make algorithm computationally tractable was to use plain splits with constant value on first several levels of decision trees to avoid using heavy computations in nodes, where several elements are high. In [9] an efficient algorithm for building piecewise linear decision trees was proposed. The main differences with previous work include using only features that the tree was split over as a linear prediction in leaf and assuming that objects inside every bin, created over feature j have the same value, equal to the average value of feature j in the bin. These and other modifications reduce the training time of piecewise linear decision trees. However, an algorithm based on the substitution of feature values with a constant inside a bin makes the solution to a problem of loss minimization from Eq. 4 incorrect. More research is required to assess all effects of such substitution on the performance of the machine learning model.

Table 3. Experimental results, HIGGS dataset

Library	ROC-AUC	Training time	Inference time
Piecewise linear	**0.8221**	5 min 56 s	16.5 s
Lightgbm	0.8161	1 min 19 s	16 s
Xgboost	0.8161	2 min 34 s	12.5 s
CatBoost	0.7902	3 m 56 s	1.38 s

Table 4. Experimental results, HEPMASS dataset

Library	ROC-AUC	Training time	Inference time
Piecewise linear	**0.9500**	4 min 7 s	9.34 s
Lightgbm	0.9486	1 min 5 s	10.9 s
Xgboost	0.9485	1 min 42 s	8.13 s
CatBoost	0.9477	2 m 46 s	616 ms

Table 5. Experimental results, YEAR dataset

Library	RMSE	Training time	Inference time
Piecewise linear	**8.901**	1 min 56 s	857 ms
Lightgbm	8.955	5.78 s	1.67 s
Xgboost	8.9485	30.8 s	1.39 s
CatBoost	9.08	18 s	41 ms

6 Conclusion

An efficient algorithm for building boosting ensembles of piecewise linear decision trees was suggested in this paper. It was proved, that ensembles built with suggested algorithm under certain constraints are invariant of nondegenerate linear transformations of features. The algorithm was implemented, and its performance was measured on publicly available datasets. It was shown, that gradient boosting ensembles built with suggested algorithms show better quality with lesser number of iterations, with a certain increase in training time of one decision tree, and a negligible increase in inference time of one decision tree.

References

1. Friedman, J.H.: Stochastic gradient boosting. Comput. Stat. Data Anal. **38**(4), 367–378 (2002). https://doi.org/10.1016/S0167-9473(01)00065-2
2. Dorogush, A.V., Ershov, V., Gulin, A.: CatBoost: gradient boosting with categorical features support. arXiv preprint arXiv:1810.11363 (2018)
3. Gama, J.: Functional trees. Mach. Learn. **55**(3), 219–250 (2004). https://doi.org/10.1023/B:MACH.0000027782.67192.13

4. Chaudhuri, P., et al.: Piecewise-polynomial regression trees. Statistica Sinica **4**, 143–167 (1994)
5. Lam, S.K., Pitrou, A., Seibert, S.: Numba: a LLVM-based python JIT compiler. In: Proceedings of the Second Workshop on the LLVM Compiler Infrastructure in HPC. ACM (2015). https://doi.org/10.1145/2833157.2833162
6. Chen, T., Guestrin, C.: XGBoost: a scalable tree boosting system. In: Proceedings of the 22nd ACM SIGKDD International Conference on Knowledge Discovery and Data Mining. ACM (2016). https://doi.org/10.1145/2939672.2939785
7. Ke, G., et al.: LightGBM: a highly efficient gradient boosting decision tree. In: Advances in Neural Information Processing Systems (2017)
8. de Vito, L.: LinXGBoost: extension of XGBoost to generalized local linear models. arXiv preprint arXiv:1710.03634 (2017)
9. Shi, Y., Li, J., Li, Z.: Gradient boosting with piece-wise linear regression trees. arXiv preprint arXiv:1802.05640 (2018). APA

Deep Reinforcement Learning Methods in Match-3 Game

Ildar Kamaldinov and Ilya Makarov$^{(\boxtimes)}$

National Research University Higher School of Economics, Moscow, Russia
irkamaldinov@edu.hse.ru, iamakarov@hse.ru

Abstract. A large number of methods are being developed in the deep reinforcement learning area recently, but the scope of their application is limited. The number of environments does not always allow for a comprehensive assessment of a new agent training algorithm. The main purpose of this article is to present another environment for Match-3 game that could be expanded, which would have a connection with the real business. The results for the most popular deep reinforcement learning algorithms are presented as a baseline.

Keywords: Deep reinforcement learning · Match-3 video game · Game environment · Deep learning

1 Introduction

In this paper, we study Match-3 tile-matching game, in which manipulating tiles leads to their disappearance according to a matching criterion. In many such games, it requires to gather certain number of identical tiles in adjacent manner, for e.g., place three tiles in a row or in a column. Match-3 games are very popular last years as top SmartPhone applications, especially for road trip time filling. These games have very simple and easy-to-learn gameplay and short sessions. Example of the game board in the classic Match-3 game "Bejeweled" is shown in Fig. 1. Usually, such a game must not require great skill to play such a game, but then developers need a proper level-design with fast deploy of new levels while being sure that they meet difficulty constraints for players. Besides visual and novel developing counterparts, the difficulty is the key variable in players retention and conversion to payers.

In fact, the complexity of the level should be known in advance, so that testing departments could evaluate it. Usual test time for one level playing by one human is several days. Computer science approach for this problem is to use simulation [10] in order to evaluate level complexity from self-playing agents and measure their performance wile learning to play the current level. Taking into

I. Makarov—The work was supported by the Russian Science Foundation under grant 17-11-01294 and performed at National Research University Higher School of Economics, Russia.

account the described disadvantages of special departments and simulations, the solution may be using reinforcement learning (RL).

From a reinforcement learning point of view, Match-3 game can be described as a model-based environment if the agent can choose among available moves, and model-free if the agent can not choose.

Firstly, deep reinforcement algorithms using deep learning showed outstanding results on Atari 2600. For the first time, the results surpassing human ones were achieved with Deep Q-learning [8]. The algorithm learns by deep reinforcement learning to maximize its score given only the pixels and game score as the input. Then Atari 2600 became the key way to evaluate deep RL algorithms.

Given the success of deep RL in Atari 2600 and the challenges that business faces, using these methods for Match-3 can help because, first, playing a large number of sessions can take less time, and, second, it is possible to get more human-like results when using imitation learning.

Our main contribution is a new open-source environment with *gym* [1] interface which is easy to use and extend. It is the first free implementation for the Match-3 game in python for reinforcement research purposes. Previously, game developers ran experiments with closed source environments without reproducibility.

Throughout this work, we used a classic configuration of Match-3 game board from "Bejeweled" or a custom set of 30 levels. We also provide information on popular algorithms performance in model-free setting and their comparison with random action policy and each other.

Section 2 describes key reinforcement learning concepts and algorithms used in the research, Double Dueling DQN, Asynchronous Actor-Critic Agents and Proximal Policy Optimization. It also includes a description of applying RL concepts to Match-3 games. Section 3 describes the developed environment and testing scenarios for a game bot.

Fig. 1. Bejeweled game board

2 Deep Reinforcement Learning Overview

2.1 General Overview

Reinforcement learning paradigm is based on interaction between an agent and the environment. An agent interact with the environment, changing its state and receiving reward for the current action.

In fully observable setting, each step the agent observes state (s_t) of the environment and can choose any action (a_t) from action space leading to a new state of the environment (s_{t+1}) and receiving some reward (r_t).

To choose the action for a given state, the agent uses a policy, which can be deterministic or stochastic. In deep reinforcement learning we deal with parameterized policies:

$$a_t = \mu_\theta(s_t) \tag{1}$$
$$a_t = \pi_\theta(\cdot|s_t) \tag{2}$$

The goal of the agent is to maximize cumulative reward over a trajectory $(s_0, a_0, s_1, a_1, ...)$. Usually, some finite part of discounted return is taken as approximation of the sum over an infinite number of steps with each reward discounted by factor γ:

$$R(\tau) = \sum_{t=0}^{\infty} \gamma^t r_t. \tag{3}$$

The discount rate determines the present value of future rewards: a reward received t time steps in the future is worth only γ^t times what it would be worth if it were received immediately. Mathematically, if $\gamma < 1$, the infinite sum has a finite value as long as the reward sequence is bounded [16].

2.2 Reinforcement Learning Algorithms

The first fork in RL algorithms is the following question: whether the agent has access to a model of the environment (or transition model). If the agent can predict a next state of the environment and appropriate reward it is the case of model-based reinforcement learning when the agent can plan its actions. In most cases model of the environment is hidden and the agent can not explicitly use it for predicting future states, therefore the agent should learn from interacting with it. This is the case of model-free RL.

The model-free RL consists of two main approaches: Q-learning and policy optimization. By definition Q-function is an expected return for the given state, an arbitrary action, and policy:

$$Q^\pi(s, a) = E_{\tau \sim \pi}[R(\tau)|s_0 = s, a_0 = a] \tag{4}$$

in deep RL applications Q-function is parameterized by θ, a key benefit of the algorithm is that it can be performed off-policy, i.e. it can use previous data with any policy. This fact allows using collected data more efficiently. An example of Q-learning is DQN algorithm [8], which outplayed the best players of a bunch of Atari games.

Policy optimization algorithms try to represent parameterized policy $\pi_\theta(a|s)$ by doing gradient ascent on some objective. In contrast to Q-learning, they are called on-policy models meaning that updating parameters should be applied by using data collected with recent policy. Since we should collect data for each policy, these algorithms are less sample efficient.

Further, we will describe some algorithms used for the Match-3 game.

Deep Q-Learning. Deep Q-network (DQN) algorithm optimizes an objective typically based on Bellman equations for the optimal value functions:

$$V^* = \max_a \mathrm{E}_{s' \sim P}[r(s,a) + \gamma V^*(s')] \tag{5}$$

$$Q^*(s,a) = \mathrm{E}_{s' \sim P}[r(s,a) + \gamma \max_{a'} Q^*(s',a')], \tag{6}$$

where $s' \sim P$ is shorthand for $s' \sim P(\cdot|s,a)$, indicating that the next state s' is sampled from the environment's transition rules. DQN uses parameterization on Q-function and applies gradient descent. For taking maximum over possible actions we store old version of Q-network (called *target network*) and for updating parameters authors exploited an additional source of data, i.e. previous transitions (called *experience replay* in the original paper).

Double DQN. Because of upward bias in estimating maximum over actions [17], another model proposes to use a target network for getting maximum value for the current state, but the action is taken from the current Q-function. In that case, loss function on the transition will be the following:

$$L_i(\theta_i) = \mathrm{E}_{s,a,s',r}(r + \gamma Q(s' \arg\max_{a'} Q(s',a';\theta);\theta_i^-) - Q(s,a;\theta_i))^2 \tag{7}$$

where θ_i^- is parameters of a target network on iteration i.

Prioritized Experience Replay. For performing gradient descent on 7, we select transitions from replay buffer randomly, but if rewards are rare, transitions in the buffer are almost useless, experience replay buffer [12] store errors of last received Q-value on transitions and sample them based on errors.

Dueling DQN. Q-function inputs state and action, but it seems natural to represent Q-functions by two parts, first one is a value for given state and second is an advantage from acting on the current state [18]:

$$Q(s,a;\theta,\alpha,\beta) = V(s;\theta,\beta) + \left(A(s,a;\theta,\alpha) - \frac{1}{|\mathcal{A}|} \sum_{a'} A(s,a';\theta,\alpha) \right), \tag{8}$$

where θ denotes parameters of convolutional layers, α and β are parameters of two streams of fully-connected layers, advantage function is $A(\cdot)$ and $|\mathcal{A}|$ denotes action space size.

Now we will describe on-policy algorithms, starting with the key algorithm Vanilla Policy Gradient, because it is the basis of modern methods.

Vanilla Policy Gradient. Maximization of the expected return

$$J(\pi_\theta) = E_{\tau \sim \pi_\theta} R(\tau) \tag{9}$$

by the following gradient

$$\hat{g} = \frac{1}{|\mathcal{D}|} \sum_{\tau \in \mathcal{D}} \sum_{t=0}^{T} \nabla_\theta \log \pi_\theta(a_t|s_t) R(\tau). \tag{10}$$

where $\mathcal{D}$ is set of trajectories and $|\mathcal{D}|$ denotes the number of trajectories in D.

Asynchronous Advantage Actor-Critic. A3C algorithm [7] maintains a policy network and an estimation of the value function $V(s_t; \theta_V)$. The update on policy performed by the algorithm that can be seen as

$$\nabla_\theta \log \pi(a_t|s_t; \theta) A(s_t, a_t; \theta, \theta_V), \tag{11}$$

where $A(s_t, a_t; \theta, \theta_V)$ is an estimate of the advantage function given by

$$\sum_{t=0}^{k-1} \gamma^i r_{t+i} + \gamma^k V(s_{t+k}; \theta_V) - V(s_t; \theta_V). \tag{12}$$

Agents act in parallel and accumulate updates for improving training stability.

Proximal Policy Optimization. In PPO with the clipped objective, we want to maximize the ratio of the probability under the new and old policies $(\mathcal{R}_t)$ multiplied by advantage (A_t). Clipping is necessary to prevent abrupt policy changes. The final objective is the following [14]:

$$L^{CLIP}(\theta) = E_t[\min(r_t(\theta) A_t, clip(\mathcal{R}_t(\theta), 1 - \varepsilon, 1 + \varepsilon) A_t)], \tag{13}$$

where ε is a hyperparameter and *clip* denotes clipping which removes the incentive for moving $\mathcal{R}_t$ outside of the interval $[1 - \varepsilon, 1 + \varepsilon]$.

This objective implements a way to do a Trust Region update [13] which is compatible with Stochastic Gradient Descent and simplifies the algorithm by removing the KL penalty and need to make adaptive updates.

2.3   Match-3 Case

Let us define core Deep RL concepts for Match-3 game case.

State. The current state in Match-3 is a game board with shapes on it (see Fig. 2). Size of the board and the number of available shapes are parameters of the game level affecting its difficulty.

We represent by colour movable shapes, while black shapes are immovable thus increasing level difficulty. For this study, we allow to try to move black shapes but such move will lead no effect on board thus increasing first successful step metric, so agent will learn to not to touch such shapes.

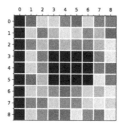

Fig. 2. Match-3 state

Action Space. The gameplay allows to swap two neighboring shapes on the board according to common row or column, diagonals are not included for neighboring proximity. Match-3 environment can be classified as model-based if the agent knows possible moves which lead to deleting shapes, and as model-free if not, that difference appears as two possible action spaces.

Model-Based Action Space. Since the environment is simple it is possible to find successful moves by brute force and delegate choice one of them to an agent.

Model-Free Action Space. Most of the actions will not change the state because only successful moves are allowed, therefore in model-free setting agent should find proper actions and learn their positions based on encoder of board space state.

Trajectory. A trajectory is considered as a sequence of changes between game board states after shape movements, and board update after removing shapes form the board on successful actions. In general, Match-3 game environment is generated randomly from previous state and removed shapes, excluding one dead-end case when there are no admissible moves and the whole board is shuffled from existing position in order to continue playing.

Reward. For the state-action pair reward is the number of deleted shapes, including those that may be removed after filling opened positions by randomly chosen shapes. For a model-free setting, we aim to training agent searching for admissible moves, so that discounted reward will be much lower of it tries actions not leading to shape removal.

Return. Real mobile or web applications consist of finite time (number of action) games with different additional goals, for e.g., collect elements of particular shapes in fixed number of steps, but core gameplay remains the same. For our study, we formulate the goal of an agent as to maximize the sum of reward collected after 100 steps.

2.4   Related Work

One of the most relevant works in Match-3 agents were published by King company researchers [4,10,11]. Mostly, they focus on predicting average human success rate (AHSR) from metrics of an agent while we try to learn playing Match-3 using DRL and only then evaluate game level from the agent metrics.

In [10], Monte-Carlo Tree Search with Upper Confidence Bound selection strategy was suggested to best handcrafted heuristics and human game testers [5]. The main drawback is time required to simulate enough game situations.

In [11], authors used deep neural network (DNN) to predict next move from gameplay data of MCTS-bot, compute AHSR and thus test level in real-time. In [4], instead of MCTS-bot, actions of real users were used as an input for DNN providing best results in AHSR prediction in accuracy and prediction time. The key drawback of this approach is data collection requiring human playing.

3   Environment and Scenario

We developed a new Match-3 environment with gym interface [1] well known among RL researchers[1].

3.1   Scenarios

Scenario 1. Bejeweled Board. Default board has seven shapes, it is square-shaped with the length of border equal to 8, the minimum length of matching is 3. That configuration is identical to "Bejeweled Classic". This board is simple even number of states is equal to 7^{64}.

Scenario 2. Board With Immovable Shapes. More popular board with shapes which can not be moved. The number of states is less than in the first scenario, but patterns of moving are more difficult. For this scenario, we developed 30 levels.

For testing policies in the second scenario, we apply two approaches. First, using the same board, and second, all rotations except initial board state. It allows as to examine agent in different situations.

3.2   Policy Tests

Below, we enumerate possible scenarios for testing policy, which acts in the described environment. One for evaluating the ability to find available action and second for collected reward taking into account only successful moves.

Cumulative Reward. We use mean and median cumulative reward on episodes.

---

[1] Detailed implementation can be found at https://github.com/kamildar/gym-match3.

First Correct Move Test. By sorting the algorithm's outputs we can find the first successful move to evaluate the ability to search for moves. This test is necessary because the previous does not take into account unavailable moves which do not change the state on the board.

Length of the test episode is set to 100. The number of iterations for the tests is set to 1000.

4   Proposed Approach

Let us describe input state representation and architecture of the encoder network for DRL agent.

4.1   Game State

We use similarity of Match-3 game board to other board games like chess, shogi and 'go', so we adapted input for the network from AlphaZero [15]. Each of N shapes on the board are decomposed into separate level transforming two-dimensional board into $W \times H \times N$ tensor. For the second scenario with immovable shapes the first dimension of the state-tensor reserved for immovable shapes mask. We also tried reshaping of state-tensor to one-dimensional vector, but it worked worse.

4.2   Architecture

Deep neural network architecture consists of two main parts: an encoder network and dense layers specified for algorithms (actor and critic parts).

Encoder. An encoder architecture includes only one depthwise separable convolution [3] with 3×3 filter and ReLU activation. This network uses the same convolution filter for each layer of the input tensor. By this way we exploited the nature of Match-3 games, there is no difference between shapes except immovable. For all algorithms feature extractor was trained from scratch. We also tested regular convolution with 3×3 filter and ReLU activation. During tests, we found that there is no strong dependence on the encoder architecture.

DQN, PPO, A3C. Intermediate layers specific to each algorithm have ReLU as activation, size of hidden layers is equal to 128. Detailed implementation can be found in RLlib [6].

5   Results

Only Asynchronous Advantage Actor-Critic showed good performance, dueling double DQN and PPO showed result worse than random policy.

5.1   First Scenario

The algorithms were trained for 4 h on 16 CPUs, A3C continued to improve after 4 h. The plot of the dynamics of the average cumulative reward per episode in the first testing scenario without immovable shapes is in Fig. 3. PPO performs as random policy, DQN degraded during training and only A3C gave results better than random (see Fig. 1 and Table 1).

Table 1. Algorithms performance for Match-3 game in the 1st scenario

	Rank			Reward		
	Mean	Median	Std	Mean	Median	Std
Random	13.38	9.5	13.64	4.27	3.0	2.37
A3C	12.01	8.0	12.34	4.30	3.0	2.55
DQN	14.44	11.0	13.64	4.23	3.0	2.44
PPO	14.62	11.0	13.02	4.47	3.0	2.71

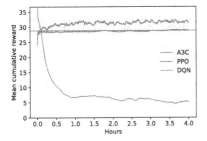

Fig. 3. Dynamics of training in the 1st scenario

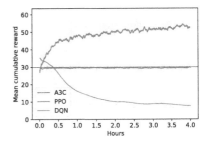

Fig. 4. Dynamics of training in the 2nd scenario

5.2   Second Scenario

The algorithms were trained for 4 h in the second scenario on 16 CPUs. The plot of the dynamics of the average cumulative reward per episode in the second testing scenario without immovable shapes is in Fig. 4.

For the second scenario, we tested policies on rotated board as an examination of generalization. Results for both cases are shown in the Table 2.

A3C was successfully trained in both cases, but it was the best in the first successful move test, while it was worse than random in the average reward of successful moves. It is clear that the improvement in the rank test is due to a decrease in the average reward per successful move since in model-free setting searching moves is more valuable. Testing agent on the rotated levels showed A3C-trained agent has learned to play and can do so even at unfamiliar levels.

5.3 Individual Levels

In this subsection, we will describe some particular levels among default the set.

Table 2. Algorithms performance for Match-3 game in the 2nd scenario

	Rank			Reward		
	Mean	Median	Std	Mean	Median	Std
Base levels						
Random	14.92	11	15.41	5.33	3	4.24
DQN	14.05	9	15.44	5.08	3	3.78
PPO	16.93	13	15.58	5.2	3	3.75
A3C	7.5	4	9.61	5.28	3	3.66
Rotated levels						
Random	15.55	10	16.37	3.88	3	5.23
DQN	14.68	9	16.00	3.84	3	5.14
PPO	16.42	12	15.16	4.15	3	5.29
A3C	7.76	4	10.63	3.77	3	5.38

In the Fig. 5, level on the left is the padded Bejeweled board, it is padded since the biggest board among default levels has 9×9 size, then Bejeweled is 8×8 only, another difference is the greater action space, agent can try to move padded figures, but it will not have an impact on the board. We tested A3C on that level only and it gave results better than agent trained on Bejeweled board only (12.01 vs. 11.04 for the mean rank test).

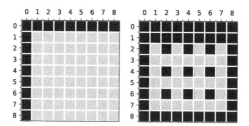

Fig. 5. Padded Bejeweled and 6th levels

The level on the right is the worst for random policy because an effective action space is very small and a lot of moves will not lead any changes on the board, at the same time A3C outperform random policy the most. The mean rank of the A3C is 6.33, random policy's result is 28.54.

6   Discussion

We present a useful approach for encoder architecture producing feature maps that better reflect the behavior of the real human who needs to visually assess the board and then choose the right move without hints from the game. For model-based environment, one does not need to use an encoder (MCTS with UCB1 [2]) or use encoder that evaluates already known moves as it happened when using DQN for Go [9].

This is important since that is a connection between scientific purposes of the environment and their connection to real business. It is a big challenge to test a new level in Match-3 game before its worldwide release. The problem of automatic testing is human-like playing. An approach based on "playtesting" departments in big game companies is time-consuming. Another one is Monte-Carlo simulations, which is much faster, but results can be far away from real-human metrics. An approach based on reinforcement learning in the model-free setting can be more appropriate in that sense.

Despite choosing model-free setting as close to human-playing scenario, a trained agent does not necessarily require to guess instantly correct moves and may require additional steps to make a good guess.

7   Conclusion and Future Work

This article presented the open source environment to study deep reinforcement learning in Match-3 game scenarios. The article described the key concepts of deep RL, basic algorithms and Match-3 game in terms of reinforcement learning. It explained the value of such an environment and the task of predicting the complexity of the level in a Match-3 game. The first experiments have been carried out in the model-free setting, because it may allow obtaining metrics that are more similar to human behavior. Baselines are obtained using state-of-the-art deep RL algorithms.

Future work could focus on a model-based approach and explore agents' ability to plan moves at new (rotated levels). It is also possible to improve the environment to make it more similar to popular projects: goal-oriented levels, more mechanics and more levels.

References

1. Brockman, G., et al.: OpenAI Gym (2016)
2. Browne, C., et al.: A survey of Monte Carlo tree search methods. IEEE Trans. Comput. Intell. AI Games 4(1), 1–43 (2012). https://doi.org/10.1109/TCIAIG. 2012.2186810.   http://www.incompleteideas.net/609dropbox/otherreadingsandre sources/MCTS-survey.pdf
3. Chollet, F.: Xception: deep learning with depthwise separable convolutions. In: Proceedings of the IEEE Conference on Computer Vision and Pattern Recognition, pp. 1251–1258. IEEE (2017)

4. Eisen, P., Gudmundsson, S.F., Dowling, J.: Simulating human game play for level difficulty estimation with convolutional neural networks. Technical report, KTH, School of Information and Communication Technology (ICT) (2017). http://kth. diva-portal.org/smash/get/diva2:1149021/FULLTEXT01.pdf

5. Kocsis, L., Szepesvári, C.: Bandit based Monte-Carlo planning. In: Fürnkranz, J., Scheffer, T., Spiliopoulou, M. (eds.) ECML 2006. LNCS (LNAI), vol. 4212, pp. 282–293. Springer, Heidelberg (2006). https://doi.org/10.1007/11871842_29

6. Liang, E., et al.: RLlib: abstractions for distributed reinforcement learning. arXiv preprint arXiv:1712.09381, December 2017. http://arxiv.org/abs/1712.09381

7. Mnih, V., et al.: Asynchronous methods for deep reinforcement learning. In: International Conference on Machine Learning, pp. 1928–1937 (2016)

8. Mnih, V., et al.: Playing Atari with deep reinforcement learning. CoRR (2013). http://arxiv.org/abs/1312.5602

9. Mnih, V., et al.: Human-level control through deep reinforcement learning. Nature **518**(7540), 529–533 (2015). https://doi.org/10.1038/nature14236. http://www.nature.com/articles/nature14236

10. Poromaa, E.R.: Crushing candy crush: predicting human success rate in a mobile game using Monte-Carlo tree search. Technical report, KTH, School of Computer Science and Communication (CSC) (2017). http://kth.diva-portal.org/smash/get/diva2:1093469/FULLTEXT01.pdf

11. Purmonen, S.: Predicting game level difficulty using deep neural networks. Technical report, KTH, School of Computer Science and Communication (CSC) (2017). http://kth.diva-portal.org/smash/get/diva2:1154062/FULLTEXT01.pdf

12. Schaul, T., Quan, J., Antonoglou, I., Silver, D.: Prioritized experience replay (2015)

13. Schulman, J., Levine, S., Abbeel, P., Jordan, M., Moritz, P.: Trust region policy optimization. In: International Conference on Machine Learning, pp. 1889–1897 (2015)

14. Schulman, J., Wolski, F., Dhariwal, P., Radford, A., Klimov, O.: Proximal policy optimization algorithms, July 2017. http://arxiv.org/abs/1707.06347

15. Silver, D., et al.: Mastering Chess and Shogi by self-play with a general reinforcement learning algorithm, December 2017. http://arxiv.org/abs/1712.01815

16. Sutton, R.S., Barto, A.G., et al.: Introduction to Reinforcement Learning, vol. 135. MIT Press, Cambridge (1998)

17. Van Hasselt, H., Guez, A., Silver, D.: Deep reinforcement learning with double q-learning. In: Thirtieth AAAI Conference on Artificial Intelligence (2016)

18. Wang, Z., Schaul, T., Hessel, M., van Hasselt, H., Lanctot, M., de Freitas, N.: Dueling network architectures for deep reinforcement learning, November 2015. http://arxiv.org/abs/1511.06581

Distance in Geographic
and Characteristics Space
for Real Estate Price Prediction

Evgeniy M. Ozhegov$^{(\boxtimes)}$, Alina Ozhegova , and Ekaterina Mitrokhina

Research Group for Applied Markets and Enterprises Studies,
National Research University Higher School of Economics, Perm, Russia
tos600@gmail.com, arbuzanakova@gmail.com, mitkaty83@gmail.com

Abstract. The common approach to predict the price of residential property is the hedonic price model and its extension to the case of spatial autoregression. The hedonic approach models the dependence between the price and internal characteristics of an apartment, house characteristics and external characteristics. To account for the unobserved quality of the surrounding environment price model includes factors of spatial price correlation, where the distance is usually measured as the distance in geographic space. Determining the price the seller focuses not only on the observed and unobserved factors of the apartment, house and its environment but also on the prices of similar marketed objects which can be selected both by geographic proximity and by characteristics similarity. In this paper, we use ensemble clustering approach to measure objects proximity and test that the proximity of objects in the characteristics space along with spatial correlation explains the significant variation in prices that in turn leads to an improvement of predictive ability of the model.

Keywords: Real estate price · Spatial models · Distance measures

1 Introduction

Choice of a place to live can be challenging because each one comes with different features and amenities. Therefore apartment characteristics are considered by people before purchasing a flat and consequently are reflected in the property prices. The relation between real estate value and characteristics of apartments is widely studied. A hedonic price model is extensively used in real estate research. Employing a hedonic price model reveals that internal property characteristics are crucial for determining real estate value. Authors often include total or living square footage, kitchen area, floor level and the number of rooms. Some researchers pay special attention to external apartment factors. It is obvious when a buyer thinks about buying an apartment, she does not just about consider how fancy kitchen is or what is the number of bedrooms and bathrooms. She enters the community and surroundings also [3]. Buyers often search for

© Springer Nature Switzerland AG 2019
W. M. P. van der Aalst et al. (Eds.) AIST 2019, LNCS 11832, pp. 63–72, 2019.
https://doi.org/10.1007/978-3-030-37334-4_6

location that deliver a higher quality of life, therefore, the consumer's attitude towards the surrounding environment is important also [13]. Environmental factors can largely influence the value of a property.

Potential buyers are willing to pay more for apartments located near green spaces. Security is another important factor when buying an apartment [8,14]. The willingness of individuals to pay for sure depends on local crime risk also [4,11]. Certainly, the prices of apartments are also affected by closeness to urban transit services, railway stations, and airports. Properties of higher market prices, compared with their cheaper counterparts, tend to benefit more from spatial proximity to the bus stop locations [7,15]. While transport hubs can reduce an apartment price due to created externalities [1]. Additionally, positive or negative effect on a real estate price can be caused by the location of shopping malls [16].

However, in the hedonic price model setting, one can take advantage of geocoded data. Such data contain information on location of marketed objects and allow to control on spatial autocorrelation of prices, ignoring of which may lead to biased and inconsistent estimates of the hedonic model parameters [2].

To overcome these problems spatial error model (SEM) and the spatial lag model (SAR) are widely used (see, for ex., [2,9,12]). To account for the unobserved quality of the surrounding environment price model includes factors of spatial price lag or spatial correlation of errors, where the distance is measured as the distance in geographic space. In this research, we apply the SAR model because of the assumption that the marketed apartment price is set according to other apartments' prices. That is exactly what spatial lag model considers explicitly.

Learning how apartment sellers set prices on their apartments or considering papers in other fields that use spatial econometrics it is reasonable to assume that similarity of different objects can be calculated not only as proximity in geographical space. Researchers reckon that there is no reason why they cannot use any notion of nearness that makes theoretical sense. As a result, Beck *et al.* [5] use two definitions of connectivity in their research on international trade. The first connectivity criteria are based on the geographical distance between states, the second criteria is relative trade. In addition, the apartment sales practice shows that sellers usually adjust an apartment price for the prices of similar objects, for example, apartments with the same number of rooms and year of construction, located approximately at the same distance from a city center. Thus, the only work to date, in which it was shown that sellers set prices, having information about the assessment of their apartment, obtained on the basis of estimates of apartments-analogs. Lee and Sasaki [10] in this work show that in the presence of an online price calculator on the website of the multi-listing system, sellers, putting the price of their ads, use information about the price from the calculator, along with the characteristics of the exhibited apartment. Based on this fact and the theoretical basis we can use not only closeness in geographic space but also closeness in characteristics space.

The traditional spatial autoregressive model presumes that there is a geographical form of dependence, which can be calculated, for example, as Euclidean distance on geographic coordinates. To measure the proximity between objects ensemble clustering can be considered. Ensemble clustering helps to find robust distance measure which is a better fit in some sense that each individual cluster dividing. Ensemble clustering is one of the realization of the common methodology in machine learning. For instance, it is possible to use a cluster ensemble where the decision is found in three steps. First of all, a number of clusters are obtained. For each partition, the co-association boolean $n \times n$ matrix is calculated, where n is a number of objects in a sample. The matrix elements reveal if the pair belongs to the same cluster or not. Secondly, the weighted co-association matrix is found taking into account all the obtained partitions. The weights can be equal or reflect some quality and diversity estimates of the partitions. Finally, the matrix elements are used as new similarity measures between data points to construct the consensus clustering [6].

In this research, we follow an ensemble clustering approach to measure objects proximity and test whether the closeness in characteristics space is considered to be a significant factor in explaining an apartment price. This research methodology is based on the hedonic price model for a particular real estate object where price is determined by internal and external property characteristics. This model is extended to spatial autoregression model, where the distance weighted prices of other marketed objects are included to explain the price. Along with the well-established geographic distance matrix, we also include the prices weighted by the distance matrix in characteristics space.

We use data on apartments sales of the secondary property in Perm that is taken from the *Metrosphera.ru* website, which is considered as one of the most popular sources of information about the Perm real estate market. The data include 14200 ads for apartments between October 2014 and February 2015. The apartment listings represent the following information: district, address, number of rooms, price, floor level, total square footage, house type and house material. The unit of observation presents a declared price for a particular apartment.

We reveal that the proximity in characteristics space plays a significant role in the quality improvement of the spatial autoregressive model. In this project, we study the choice of measures of distance between objects. Geographical distance proximity is less important than the proximity of objects in characteristics space while using both geographic coordinates and characteristics of objects to obtain distance measure provides the best model fit. This kind of proximity explains the real pricing mechanism and, accordingly, allow to obtain more accurate prediction of the real estate price.

2 Data

In order to test the predictive ability of various proximity measures, we collected primary data on advertisements about the selling of secondary property in Perm, Russia. Perm is the 11th largest city in Russia and has approximately 1 million residents. We use *Metrosphera.ru* listing website as the most extensive source concerning housing market information in the city with, expertly, near 80% of all selling ads placed on it. The apartment listings represent the following information: district, address, number of rooms, price, floor level, total square footage, house type and house material. It should be noted that the initial declared price is used as a unit of observation. Price revisions are not taken into account. In order to extract the data, we develop a scraping tool for daily downloading of all advertisements available in the period from October 2014 to February 2015.

To add the geographical coordinates, we take the GIS data of all buildings in Perm with their addresses and geographic coordinates and match it with prices data by the address of a building. This allows constructing the geocoded data on prices and calculating the geographic distances between listed objects. The GIS dataset also contains the full list of firms and its industry belonging.

In order to train predictive models in an "honest" fashion we randomly split a dataset of 14200 observations into three sets: two auxiliary sets with 4200, 5000 and one main estimation set of 5000 observations respectively. The training procedure description is provided in the following section. Descriptive statistics of property characteristics taken into estimation set are presented in Table 1.

Table 1. Descriptive statistics for continuous variables

Statistic	Mean	SD	Min	Median	Max
Price, thous. rub.	2920	1635.84	295	0.189	24700
Price, per m^2, thous. rub.	56.52	11.45	7.88	55.37	110.87
Number of rooms	2.02	0.92	2	2	8
Floor level	4.57	3.81	1	4	25
Total area, m^2	45.45	24.20	12	51.95	346
Number of supermarkets in 200 m radius	1.26	1.62	0	1	21
Number of supermarkets in 500 m radius	4.35	3.53	0	4	28
Number of supermarkets in 1000 m radius	12.65	7.62	0	11	41
Number of observations	5000				

Table above represents that the differences in price and footage of apartments are essential. The maximum price per square meter is almost twice the average price, and the maximum area of the apartment is seven times higher than its average. In addition, one of the mainly sold apartments is two-bedroom. However, it can relate to the fact that two-bedroom apartments are the most widespread in the real estate market.

Regarding the type of house, more than a quarter of the analyzed apartments have Improved space planning (Table 2). A smaller share is occupied by apartments with Khrushchev-era, Brezhnev-era and Individual projects. Other house types make up less than 10% of the sample. Apartment districts are distributed almost evenly. The only exception is the Leninsky district where we have only 246 flats located there (about 5% of the sample), which may be explained by the fact that the Leninsky district is the smallest district of Perm with the lowest number of inhabitants.

Table 2. Descriptive statistics for categorical variables

Variable	Obs	Share
Material:		
Brick	2629	53
Panel	2371	47
House type:		
Lenin's project	125	2.50
Stalin-era	43	0.86
Khrushchev-era	1004	20.08
Brezhnev-era	874	17.48
Grey panel	305	6.10
Full-size	327	6.54
Small family	218	4.36
Small-size	5	0.10
Improved space planning	1304	26.08
Individual project	795	15.90
House type:		
Leninskiy	246	4.92
Kirovskiy	843	16.86
Motovilikhinskiy	822	16.44
Ordzhonikidzevskiy	601	12.02
Dzerzhinskiy	799	15.98
Industrial	772	15.44
Sverdlovskiy	917	18.34

3   Methodology

In order to reveal whether the closeness in characteristics space is considered to be a significant factor in explaining an apartment price, we take advantage of the traditional spatial autoregressive model.

A spatial autocorrelation model generally has the form:

$$\ln P_i = X_i \beta + \rho W_i \ln P + \epsilon_i \tag{1}$$

where P_i is an asking price for the property i; $\{X_i\}_{1 \times m}$ is a vector of property i's characteristics; $\beta_{m \times 1}$ is a vector of characteristics effects; $P_{\times 1}$ is a vector of objects' prices; $W_{n \times n}$ is a proximity (weighting) matrix in geographical space; ρ is a parameter of geographic spatial correlation of prices; ϵ_i is an error term.

Traditionally, the elements of the matrix W are calculated as the inverse distance between objects. However, in the data used for this study, there are apartments located at the same building. These observations will provide a zero distance when calculating the elements of the weighting matrix, and, as a consequence, an infinitely large value of price weight. To overcome this problem the elements of the weighting matrix are calculated using the formula proposed in (Getis and Aldstadt, 2010).

$$w_{ij} = 1 - [\frac{d_{ij}}{\max\limits_{i,j} d_{ij}}]^{\gamma} \qquad (2)$$

where $d_{i,j}$ is the Euclidean distance between apartments i and j; $\gamma > 0$ is the compression parameter of the distance space.

The maximum distance between two apartments in the data is approximately 50 km. The closer apartments in geographical space, the closer the $w_{i,j}$ value to 1. The closer the γ parameter to zero, the greater relative weight is given to the closer objects.

Anselin (1988) shows that in the estimation of models with a spatial lag by the maximum likelihood method the residuals can be on average not equal to zero, so the R^2 calculated by the explained sum of squares and the total sum of squares of residuals cannot be interpreted in the usual way. In maximum likelihood estimation context the common measure of fit is out-of-sample value of a likelihood or likelihood-based measures of fit with a penalty on the number of parameters estimated. The best known is the Akaike information criterion (AIC). Then the optimal value of the parameter γ is the value that maximize a likelihood. To find the optimal value of γ we search over grid and train model (1) on auxiliary sample of 5000 observations. Estimation results are provided in the next section.

Along with the specification including well-established geographic-based weighting matrix, we estimate a model specification including the prices weighted by the proximity matrix in characteristics space. Each element of proximity (weighting) matrix W, w_{ij}, in characteristics space is a function of characteristics of properties i and j.

We follow [6] ensemble clustering approach to measure objects proximity in characteristics space. An algorithm for calculation of the proximity matrix in characteristics space is as follows:

1. Reveal the optimal number of clusters;
2. Run k-means clustering 100 times taking random subset of standardized property characteristics as clustering variables;
3. Calculate matrix W where each element $w_{ij} = d_{i,j}^{\gamma}$ represents the shrinked by γ share of the joint enterings $d_{i,j}$ of apartments i and j in the same cluster.

We estimate γ in the same fashion searching over grid on a train sample.

The last model specification use the same algorithm for constructing a proximity space. This time we cluster objects by both standardized geographic coordinates and property characteristics, calculate a number of enterings of the objects in the same clusters and shrink this measure to obtain weights in a mixed space.

4   Results

The number of clusters are revealed using averaged over clustering replications Calinski-Harabasz index (CHI) for each proximity space. We train clustering models on a auxiliary set with 4200 observations and calculate CHI on a 5-fold cross-validation. We reveal an 8 optimal clusters number for characteristics space and 7 for a mixed space.

Then we estimate optimal values of γ using search over grid. Estimation is performed on a second auxiliary set of 5000 observations. We use 5-fold cross validation to obtain "honest" value of space shrinkage parameter γ each specification of SAR model. Results are provided in Table 3. The optimal value of γ for SAR models in geographic and characteristics space is $\frac{1}{50}$ while optimal value for a mixed space is $\frac{1}{25}$.

Table 3. CV LL values depending on γ

The value of γ	SAR in geo. space (1)	SAR in char. space (2)	SAR in mixed space (3)
$\gamma = \frac{1}{2}$	1848.8	2103.0	1967.5
$\gamma = \frac{1}{5}$	1774.6	1876.7	1837.2
$\gamma = \frac{1}{10}$	2123.2	2039.5	1920.0
$\gamma = \frac{1}{25}$	2037.8	2105.7	2224.3
$\gamma = \frac{1}{50}$	2131.5	2155.5	2150.8
$\gamma = \frac{1}{100}$	1858.4	1957.2	1972.1

The estimation results for spatial autoregressive models are presented in Table 4. Estimation was performed on the main estimation set of 5000 observations with ensemble clustering trained on a first auxiliary set and with optimal γ values obtained on a second auxiliary set. The specification (1) is simple linear regression of price on apartment characteristics. The specification (2) takes into account the spatial autocorrelation of prices in geographical space, the specification (3) takes into account the spatial autocorrelation of prices in the space of characteristics, while specification (4) accounts for a spatial correlation in mixed space of geographical coordinates and property characteristics. The estimation is carried out by the maximization of likelihood on a full set. Model fit is estimated as a mean value of log. likelihood on a 5-fold cross-validation. Controls include the number of rooms, floor level, total square, the dummy on the first floor, the material of a building, number of supermarkets in a specific radius as well as dummies for house type and the district where the house is located.

The first thing to note is that the results for the SAR models in Table 4 show clear evidence of spatial clustering. The spatial correlation parameter ρ in models (2)–(4) is statistically significant and has a positive sign. This result

Table 4. Comparison of models predictive power

	OLS (1)	SAR in geo. space (2)	SAR in char. space (3)	SAR in mixed space (4)
ρ	–	0.0001*** (0.0000)	0.822*** (0.002)	0.226*** (0.001)
No. of obs.	5000	5000	5000	5000
No. of params.	23	24	24	24
CV LL	2157.2	2230.1	2348.6	2366.5
CV AIC	−4268.4	−4412.2	−4649.2	−4685.0

Notes: Standard errors in parentheses, ***indicates significance on 1% level. All model specifications include controls on property and surroundings characteristics.

is consistent with the intuitive idea that the price of the apartment positively depends on the prices of the surrounding apartments. Consequently, the inclusion of a spatial lag in the model is justified. The ρ value in the models (3) and (4) is also statistically significant and has a positive sign, suggesting that a real estate price is positively correlated with prices of apartments having similar characteristics. This suggests that characteristics relations contain information independent of geographical distance.

The comparison of models (1)–(4) shows that the spatial autoregressive model in characteristics space has higher likelihood and lower Akaike information criterion values. Thus, the closeness in characteristics space is considered to be a significant factor in explaining an apartment price. Moreover, we can argue that apartments located in different districts, but having the same characteristics have some sort of price convergence. Thus, we determine that the distances between objects can be measured not only through the coordinates between objects. Inclusion of the proximity matrix in characteristics space improves price model fit.

A new model of a residential property price provides a new understanding of how the market works. For example, taking into account spatial interactions between objects we show that characteristics are more important for people and has a greater influence on a price than geographical location. This information may let house sellers to set price optimally, because price movement of similar objects should be taken into account when the property listed and during price revisions.

5   Conclusion

In this research, we analyze the spatial effect of marketed objects' prices on residential property price. We employ data on listed objects obtained from the *Metrosphera.ru* website. Data contain information about selling ads for 5000 apartments in Perm from October 2014 to February 2015. The special feature of the data is that all the real estate objects have the information about geographical location and a set of characteristics, therefore the closeness can be defined in different ways. Previous works usually concentrate on a studying of a geographical distance and the spatial effects related to it.

In this research, we estimate the traditional spatial autoregressive model using a geographical space dependence, which is calculated as Euclidean distance on coordinates. Then using an ensemble clustering approach we develop an algorithm to measure the distance between objects and to construct a characteristic space matrix and a mixed space using both geographical coordinates and property characteristics.

The construction of models with different measurements of the distance between objects (in geographical space, space of characteristics and mixed one) shows that apartment prices are linked together in various directions. Construction of a mixed space provides the best price model fit.

A new approach of proximity measurement reflects the unique features of the market, gives new insights about the behavior of sellers and in the future, it can be used to obtain the highest predictive power of the price models in the real estate market.

References

1. Ahlfeldt, G.M., Maennig, W.: Assessing External Effects of City Airports: Land Values in Berlin. Hamburg Contemporary Economic Discussion Paper, vol. 11 (2008)
2. Anselin, L.: Spatial Econometrics: Methods and Models. Kluwer Academic Publishers, Norwell (1988)
3. Bayer, P., Ferreira, F., McMillan, R.: A unified framework for measuring preferences for schools and neighborhoods. J. Polit. Econ. **115**(4), 588–638 (2007)
4. Gibbons, S.: The costs of urban property crime. Econ. J. **114**(499), 441–463 (1992)
5. Beck, N., Gleditsch, K.S., Beardsley, K.: Space is more than geography: using spatial econometrics in the study of political economy. Int. Stud. Quart. **50**(1), 27–44 (2006)
6. Berikov, V., Vinogradova, T.: Regression analysis with cluster ensemble and kernel function. In: van der Aalst, W.M.P., et al. (eds.) AIST 2018. LNCS, vol. 11179, pp. 211–220. Springer, Cham (2018). https://doi.org/10.1007/978-3-030-11027-7_21
7. Hess, D.B., Almeida, T.M.: Impact of proximity to light rail rapid transit on station-area property values in buffalo, New York. Urban Stud. **44**, 1041–1068 (2007)
8. Hoshino, T., Kuriyama, K.: Measuring the benefits of neighbourhood park amenities: application and comparison of spatial hedonic approaches. Environ. Resour. Econ. **45**(3), 429–444 (2010)
9. Montero, J.M., Mínguez, R., Fernández-Aviles, G.: Housing price prediction: parametric vs. semi-parametric spatial hedonic models. J. Geogr. Syst. **20**(1), 27–55 (2018)
10. Lee, Y.S., Sasaki, Y.: How sensitive are sales prices to online price estimates in the real estate market? In: AREUEA-ASSA 2015 Conference Papers (2014)
11. Linden, L., Rockoff, J.E.: Estimates of the impact of crime risk on property values from Megan's laws. Am. Econ. Rev. **98**(3), 1103–1127 (2008)
12. Osland, L.: The importance of unobserved attributes in hedonic house price models. Int. J. Hous. Markets Anal. **6**, 63–78 (2013)
13. Tse, R.Y.: Estimating neighbourhood effects in house prices: towards a new hedonic model approach. Urban Stud. **39**(7), 1165–1180 (2002)

14. Troy, A., Grove, J.M.: Property values, parks, and crime: a hedonic analysis in Baltimore, MD. Landscape Urban Plan. **87**(3), 233–245 (2008)
15. Wang, Y., Potoglou, D., Orford, S., Gong, Y.: Bus stop, property price and land value tax: a multilevel hedonic analysis with quantile calibration. Land Use Policy **42**, 381–391 (2015)
16. Zhang, L., Zhou, J., Hui, E.C., Wen, H.: The effects of a shopping mall on housing prices: a case study in Hangzhou. Int. J. Strateg. Property Manag. **23**(1), 65–80 (2019)

Fast Nearest-Neighbor Classifier Based on Sequential Analysis of Principal Components

Anastasiia D. Sokolova$^{(\boxtimes)}$ and Andrey V. Savchenko

Laboratory of Algorithms and Technologies for Network Analysis, National Research University Higher School of Economics, Nizhny Novgorod, Russian Federation
adsokolova96@mail.ru

Abstract. In this paper we improve the speed of the nearest neighbor classifiers of a set of points based on sequential analysis of high-dimensional feature vectors. Each input object is associated with a sequence of principal component scores of aggregated features extracted by deep neural network. The number of components in each element of this sequence is dynamically chosen based on explained proportion of total variance for the training set. We propose to process the next element with higher explained variance only if the decision for the current element is unreliable. This reliability is estimated by matching of the ratio of the minimum distance and all other distances with a certain threshold. Experimental study for face recognition with the Labeled Faces in the Wild and YouTube Faces datasets demonstrates the decrease of running time up to 10 times when compared to conventional instance-based learning.

Keywords: Sequential analysis · Principal component analysis · High-dimensional features · Deep neural network · Nearest neighbor classifier · Face recognition

1 Introduction

Nowadays, pattern recognition plays significant role in many automatic systems because of the deep learning achievements of a near human-level performance in various areas [1]. If the available dataset contains a small number of training instances for each category, the situation becomes much more complicated [2,3]. Here domain adaptation and transfer learning techniques are used for extraction of off-the-shelf feature vector with pre-trained deep neural networks [4,5]. These vectors are usually classified by the instance-based learning approach, e.g. the k-nearest neighbour (k-NN). However, as such feature vectors are high-dimensional, classification speed can be too low for many practical applications, especially if an observed object contains a *set or a sequence of points* [6]. Here more information can be exploited from multiple observed points [7], but the

W. M. P. van der Aalst et al. (Eds.) AIST 2019, LNCS 11832, pp. 73–80, 2019.
https://doi.org/10.1007/978-3-030-37334-4_7

requirement of real-time processing prevents an implementation of too sophisticated recognition algorithms, especially, when the number of categories is rather large [8]. One practically important example is unconstrained face recognition in video surveillance systems, where hundreds of frames should be processed in a few seconds. Therefore, a lot of researches try to accelerate the recognition speed.

In this paper we propose a modification of the k-NN with superior performance for instances represented by high-dimensional feature vectors. Each object in an observed set of points is represented as a sequence of principal component scores of deep embeddings. These features are aggregated into a single descriptor [9]. During the decision-making process, the next principal components are processed only when the decision for the current number of components is unreliable. We claim that such method allows to sufficiently increase the decision speed without significantly degradation in accuracy.

The rest of the paper is organized in the following way: in Sect. 2 we formulate proposed recognition method based on sequential analysis of principal components. In Sect. 3, experimental results are presented for the face recognition. In Sect. 4, we give concluding comments and future plans.

2 Materials and Methods

The task of set-of-points classification is formulated as follows. It is required to assign an input set of T feature vectors $\mathbf{x}(t) = [x_1(t), ..., x_D(t)]$ into one of C classes. They are specified by a training set of $R \geq C$ points $\mathbf{x}_r = [x_{r;1}, ..., x_{r;D}], r \in \{1, ..., R\}$, which class label $c(r) \in \{1, ..., C\}$ is known. We assume that dimensionality D is rather high, e.g., when deep convolutional neural network (CNN) is used for feature extraction.

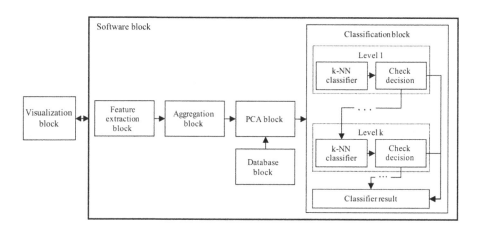

Fig. 1. The architecture of the proposed software

The architecture of the developed system is shown in Fig. 1. Here, firstly, the features are extracted from image using convolutional neural network. The features of all input points are aggregated into a single descriptor $\mathbf{x}$. If the number of sequential images in homogeneous segment is more than two, the straightforward aggregation techniques should be used, such as medoids, average features vector (AvePool) or median features [3]. Next, the NN method can be used to make decision in favor of the closest class. Unfortunately, the run-time complexity of the NN rule $O(RD)$ may be insufficient if the number R of instances and dimensionality D are high. Hence, we use sequential analysis [10,11] to efficiently match parts of features. In particular, we perform PCA (principal component analysis) for all training vectors $\{\mathbf{x}_r\}$ [3], and extract all $\tilde{D} = \min(D, R)$ principal component scores. In the result for each r-th reference image there is $[\tilde{x}_{r;1}, ..., \tilde{x}_{r;\tilde{D}}]$ component vector. The same transformation is applied for an input descriptor $\mathbf{x}$ in order to obtain the principal component scores $[\tilde{x}_1, ..., \tilde{x}_{\tilde{D}}]$.

In this paper we propose to sequentially process a hierarchy of features, so that the l-th level of a hierarchy ($l \in \{1, ..., L\}$) is represented by a vector $\tilde{\mathbf{x}}^{(l)}$ of the first d_l principal components. The number $d_l \in \{1, ..., \tilde{D}\}$ is chosen in order to explain variance rate $\sigma_l^2 \in (0; 1]$ for the training set so that $\sigma_1^2 < \sigma_2^2 < ... < \sigma_L^2 = 1$. At each hierarchy level l only first d_l scores are matched. If the dissimilarity measure is additive (such as Euclidean distance), then we can utilize the distances from the previous level to speed-up the matching:

$$\rho(\tilde{\mathbf{x}}^{(l)}, \tilde{\mathbf{x}}_r^{(l)}) = \rho(\tilde{\mathbf{x}}^{(l-1)}, \tilde{\mathbf{x}}_r^{(l-1)}) + \sum_{d=d_{(l-1)}+1}^{d_l} \rho(\tilde{x}_d, \tilde{x}_{r;d}), \tag{1}$$

where $d_0 = 0$ so that $\rho(\tilde{\mathbf{x}}^{(0)}, \tilde{\mathbf{x}}_r^{(0)}) = 0$. In such case, it is possible to compute only the distances between new components and use the computed distances on the previous levels [12]. Next, the distance $\rho_c(\tilde{\mathbf{x}}^{(l)})$ between input object and each class c at the l-the level is computed, and the NN class is obtained

$$c_l^* = \underset{c \in C_l}{\operatorname{argmin}} \rho_c(\tilde{\mathbf{x}}^{(l)}). \tag{2}$$

from a set of candidates C_l. This set contains initially all subjects: $C_1 = \{1, ..., C\}$. At each l-th step it is refined in order to further check only rather reliable categories. In order to obtain them, we treat the feature vectors as the estimates of probability distributions of (hypothetical) random variables with D possible values. The task is reduced to a problem of testing of hypothesis about class distributions. Its maximal likelihood decision is obtained using 1-NN rule with the Kullback-Leibler (KL) divergence [13]. It is known [13] that the scaled KL divergence between an input sample and class c has the asymptotically non-central chi-squared distribution with $(\tilde{D} - 1)$ degrees of freedom and the KL divergence between distributions of true class and c-th class. Hence, if the null hypothesis is true (label c_l^* is correct), the ratio $\frac{\rho_c(\tilde{\mathbf{x}}^{(l)})}{\rho_{c_l^*}(\tilde{\mathbf{x}}^{(l)})}$ has the non-central F-distribution $F(\tilde{D}-1, \tilde{D}-1; \rho(c; c_l^*))$, where $\rho(c; c_l^*)$ is the KL divergence between class c and the nearest neighbor. Thus, we propose to refine the set of candidates as follows:

$$C_{l+1} = \left\{ c \in C_l \left| \frac{\rho_c(\tilde{\mathbf{x}}^{(l)})}{\rho_{c_l^*}(\tilde{\mathbf{x}}^{(l)})} \leq \delta \right. \right\}, \tag{3}$$

where $\delta \geq 1$ is a fixed threshold, which can be chosen as the quantile of the non-central F-distribution for the KL divergence. According to this expression, at the next hierarchical level only those classes are checked, distances to which are not much higher than the minimal distance to the NN c_l^*. Otherwise, we increase the level and compute the distances between new features. It can continue as long as we process the full feature vector. It is important to emphasize that the rule (3) can be used with an arbitrary dissimilarity measure. Indeed, similar procedure was used in the SIFT method in order to obtain the reliable matches of the visual keypoints [14].

3   Experimental Results

The experimental study is devoted to unconstrained face recognition. We implemented a special C++ software prototype. The faces were detected using the pre-trained TensorFlow Model (MobileNet SSD) trained on the WiderFace dataset [15]. Four publicly available CNNs were used for feature extraction:

1. VGGFace [16]: extracts $D = 4096$ features vector in the output of "fc7" layer from 224×224 RGB images;
2. Lightened CNN version C (LCNN) [17]: extracts $D = 256$ features vector ("eltwise fc2" layer) is computed from 128×128 grayscale image;
3. VGGFace2 [18]: ResNet50 model, which extracts $D = 2048$ features vector from "pool5_7 $\times$ 7_s1";
4. FaceNet [19]: extracts $D = 512$ features vector from "embeddings" layer.

To make extracted features robust to observation conditions L_2-normalization [3] and a square of Euclidean distance to match principal components (1) are used. Sequential analysis was implemented in two variants, namely, with (1) fixed number of components (64) on each hierarchical level [12]; and (2) variable number of components depending on the explained variance. In the latter case, we use $L = 6$ levels so that $\sigma_1^2 = 0.5, \sigma_2^2 = 0.6, ..., \sigma_L^2 = 1$. The proposed approach (Fig. 1) was compared with conventional k-NN classifier for all D features and first 64 principal component scores.

The experiments were carried on a machine with 2 GHz Core i7 CPU, 6Gb RAM. The first experiment deals with the processing of single images ($T = 1$) from the Labeled Faces in the Wild (LFW) dataset [20] with more than 13000 facial photos. Photos of $= 1680$ different subjects with more than two images were split into train/test sets using 50% ratio. We should note that in such case the training sample contains rather small number of photos per subject. Hence, the accuracy of conventional classifiers (support vector machines, random forests, etc.) is 2–5% lower when compared to the instance-based learning (k-NN) [12]. The results are shown in Table 1.

Table 1. Face recognition results for the k-NN classifier, LFW dataset

Metric	Classifier	VGGFace	LCNN	VGGFace2	FaceNet
Accuracy (%)	k-NN, all components	96.31	97.48	98.66	98.15
	k-NN, 64 components	94.10	96.21	96.95	97.36
	Sequential k-NN, fixed no. of components	95.97	96.97	98.32	98.15
	Sequential k-NN, variable no. of components	95.80	96.81	97.98	97.94
Time (ms)	k-NN, all features	50.39	4.88	34.78	9.37
	k-NN, 64 components	2.53	2.50	2.53	2.52
	Sequential k-NN, fixed no. of components	2.92	2.54	2.84	2.69
	Sequential k-NN, variable no. of component	3.20	2.23	2.81	1.87

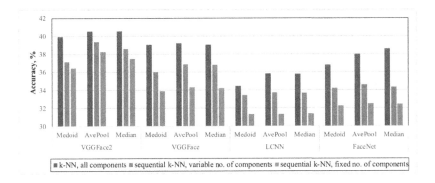

Fig. 2. k-NN classification, accuracy, YTF dataset

Here the proposed approach is only 0.3–1% less accurate when compared to the traditional k-NN for all features, though 2–7-times speed-up is obtained. Conventional matching of only fixed number of principal components is 0.6–1.7% less accurate than the proposed implementation (1)–(3) of sequential analysis.

The next experiment is focused on video-based face identification with YouTubeFaces (YTF) dataset [21]. It contains 3425 videos of $C = 1595$ different people. An average of 2.15 videos are available for each subject. The shortest track duration is $T = 48$ frames, the longest track contains $T = 6070$ frames, and the average length of a video clip is 181.3 frames. The dependencies of recognition accuracy and running time on various aggregation techniques [3] are shown in Figs. 2 and 3, respectively.

Here the accuracy is much lower when compared to the previous experiment (Table 1) because the LFW dataset contains more images per one subject, while YTF dataset includes a lot of similar video frames. Nevertheless, our sequential procedure again enables to significantly reduce the running time without noticeable accuracy degradation. The most effective results were demonstrated

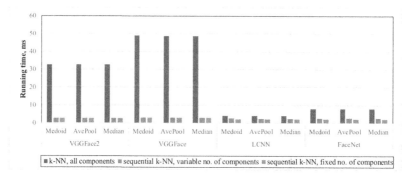

Fig. 3. k-NN classification, time (ms), YTF dataset

by PCA with variable number of components in each level obtained by increasing explained variance for features extracted by VGGFace2 CNN. The decision-making time in the proposed approach is 10-times lower than the conventional implementation of the k-NN.

4   Conclusion

In this paper we proposed the modification of the k-NN classification method based on sequential analysis of high-dimensional features (1)–(3). Experiments with face recognition and contemporary deep features demonstrated that our approach is up to 8–10-times faster when compared to the original k-NN method, while the accuracy decreases only by 0.1–0.9%.

The main direction for further research is to examine sophisticated classifiers in order to increase the recognition accuracy. Moreover, it is important to use more difficult dissimilarity measures, e.g., implement metric learning to point-to-set and set-to-set distance learning [6]. Finally, it is necessary to study contemporary aggregation techniques with learnable pooling [7] instead of simple averaging of individual features.

Acknowledgments. The article was prepared within the framework of the Basic Research Program at the National Research University Higher School of Economics (HSE).

References

1. Goodfellow, I., Bengio, Y., Courville, A.: Deep Learning. MIT Press, Cambridge (2016)
2. Savchenko, A.V.: Sequential three-way decisions in multi-category image recognition with deep features based on distance factor. Inf. Sci. **489**, 18–36 (2019)

3. Sokolova, A.D., Kharchevnikova, A.S., Savchenko, A.V.: Organizing multimedia data in video surveillance systems based on face verification with convolutional neural networks. In: van der Aalst, W.M.P., et al. (eds.) AIST 2017. LNCS, vol. 10716, pp. 223–230. Springer, Cham (2018). https://doi.org/10.1007/978-3-319-73013-4_20

4. Frome, A., Corrado, G.S., Shlens, J., Bengio, S., Dean, J., Mikolov, T.: Devise: a deep visual-semantic embedding model. In: Advances in Neural Information Processing Systems (NIPS), pp. 2121–2129 (2013)

5. Sharif Razavian, A., Azizpour, H., Sullivan, J., Carlsson, S.: CNN features off-the-shelf: an astounding baseline for recognition. In: Proceedings of the IEEE Conference on Computer Vision and Pattern Recognition Workshops (CVPRW), pp. 806–813 (2014)

6. Lu, J., Wang, G., Deng, W., Moulin, P., Zhou, J.: Multi-manifold deep metric learning for image set classification. In: Proceedings of the IEEE Conference on Computer Vision and Pattern Recognition (CVPR), pp. 1137–1145 (2015)

7. Yang, J., et al.: Neural aggregation network for video face recognition. In: Proceedings of the IEEE Conference on Computer Vision and Pattern Recognition (CVPR), pp. 4362–4371 (2017)

8. Savchenko, A.V., Milov, V.R., Belova, N.S.: Sequential hierarchical image recognition based on the pyramid histograms of oriented gradients with small samples. In: Khachay, M.Y., Konstantinova, N., Panchenko, A., Ignatov, D.I., Labunets, V.G. (eds.) AIST 2015. CCIS, vol. 542, pp. 14–23. Springer, Cham (2015). https://doi.org/10.1007/978-3-319-26123-2_2

9. Sokolova, A.D., Savchenko, A.V.: Cluster analysis of facial video data in video surveillance systems using deep learning. In: Kalyagin, V., Pardalos, P., Prokopyev, O., Utkina, I. (eds.) NET 2016. Springer Proceedings in Mathematics & Statistics, vol. 247. Springer, Cham (2018). https://doi.org/10.1007/978-3-319-96247-4_7

10. Yao, Y.: Granular computing and sequential three-way decisions. In: Lingras, P., Wolski, M., Cornelis, C., Mitra, S., Wasilewski, P. (eds.) RSKT 2013. LNCS (LNAI), vol. 8171, pp. 16–27. Springer, Heidelberg (2013). https://doi.org/10.1007/978-3-642-41299-8_3

11. Savchenko, A.V.: Fast multi-class recognition of piecewise regular objects based on sequential three-way decisions and granular computing. Knowl.-Based Syst. **91**, 252–262 (2016)

12. Savchenko, A.V.: Granular computing and sequential analysis of deep embeddings in fast still-to-video face recognition. In: IEEE 12th International Symposium on Applied Computational Intelligence and Informatics (SACI), pp. 515–520 (2018)

13. Kullback, S.: Information Theory and Statistics. Courier Corporation, New York (1997)

14. Lowe, D.G.: Distinctive image features from scale-invariant keypoints. Int. J. Comput. Vis. **60**(2), 91–110 (2004)

15. Yang, S., Luo, P., Loy, C.C., Tang, X.: Wider face: a face detection benchmark. In: IEEE Conference on Computer Vision and Pattern Recognition (CVPR), pp. 5525–5533 (2016)

16. Parkhi, O.M., Vedaldi, A., Zisserman, A.: Deep face recognition. In: Proceedings of the British Machine Vision, vol. 1, no. 3, pp. 6–17 (2015)

17. Wu, X., He, R., Sun, Z., Tan, T.: A light CNN for deep face representation with noisy labels. IEEE Trans. Inf. Forensics Secur. **13**(11), 2884–2896 (2018)

18. Cao, Q., Shen, L., Xie, W., Parkhi, O.M., Zisserman, A.: VGGface2: a dataset for recognising faces across pose and age. In: 13th IEEE International Conference on Automatic Face & Gesture Recognition (FG 2018), pp. 67–74 (2017)

19. Schroff, F., Kalenichenko, D., Philbin, J.: FaceNet: a unified embedding for face recognition and clustering. In: Proceedings of the IEEE Conference on Computer Vision and Pattern Recognition (CVPR), pp. 815–823 (2015)
20. Learned-Miller, E., Huang, G.B., RoyChowdhury, A., Li, H., Hua, G.: Labeled faces in the wild: a survey. In: Kawulok, M., Celebi, M.E., Smolka, B. (eds.) Advances in Face Detection and Facial Image Analysis, pp. 189–248. Springer, Cham (2016). https://doi.org/10.1007/978-3-319-25958-1_8
21. Wolf, L., Hassner, T., Maoz, I.: Face recognition in unconstrained videos with matched background similarity. In: IEEE International Conference on Computer Vision and Pattern Recognition (CVPR), pp. 529–534 (2011)

A Simple Method to Evaluate Support Size and Non-uniformity of a Decoder-Based Generative Model

Kirill Struminsky[1,2(✉)] and Dmitry Vetrov[3]

[1] National Research University Higher School of Economics, Moscow, Russia
k.struminsky@gmail.com
[2] Skolkovo Institute of Science and Technology, Moscow, Russia
[3] Samsung-HSE Laboratory, National Research University
Higher School of Economics, Moscow, Russia

Abstract. Theoretical analysis in [1] suggested that adversarially trained generative models are naturally inclined to learn distribution with low support. In particular, this effect is caused by the limited capacity of the discriminator network. To verify this claim, [2] proposed a statistical test based on the birthday paradox that partially confirmed the analysis. In this paper, we continue this line of work and develop a parameter-free and straightforward method to estimate the support size of an arbitrary decoder-based generative model. Our approach considers the decoder network from a geometric viewpoint and evaluates the support size as the volume of the manifold containing the generative model samples. Additionally, we propose a method to measure non-uniformity of a generative model that can provide additional insight into the model's behavior. We then apply these tools to perform a quantitative comparison of common generative models.

1 Introduction

The ongoing growth of the area of generative modeling highlights the lack of tools needed to evaluate models and compare different approaches. Using just one metric for model comparison, although seemingly appealing, can be misleading in practice. For instance, test-likelihood is often used to differentiate between generative models [5,8], but its values may contradict our intuitive notion of model fidelity [18].

Various other attempts were taken to define a quality measure for generative models of natural images. Another example is the Inception score [16], which reportedly correlates with the perceptual image quality, but can be insensitive to highly-degenerate distributions.

The Frechet Inception Distance (FID) is a metric proposed in [7] to resolve this issue. FID is a measure of the difference in alignment of embeddings of the training set and generative model samples, used as a more intuitive proxy to model fidelity.

W. M. P. van der Aalst et al. (Eds.) AIST 2019, LNCS 11832, pp. 81–93, 2019.
https://doi.org/10.1007/978-3-030-37334-4_8

Recently, detailed analysis of adversarial training [1] indicated that GANs are prone to underestimating support size of the modeled distribution. This issue may severely decrease the model's performance, yet a general goodness-of-fit test would not directly reveal the issue. Thus, a specific test to estimate this effect is needed. For instance, [2] proposed a test based on the birthday paradox to detect this issue. However, the test relies on an additional procedure that detects coincidental samples and cannot differentiate between distributions on low support and highly uneven distributions. Since the real data distribution can be highly uneven in practice, the birthday paradox test can falsely indicate low support. Additionally, the test is highly sensitive to the definition of coincidental samples, which can be problematic to introduce for data from an arbitrary domain.

One way to define support size of a generative model is to assume that the underlying distribution is a discrete distribution on finite support and then compute the number of elements in the support. For example, an image with D pixels can be represented as $\{1, \ldots, 256\}^{3D}$ and the support size of a generative model is the number of realizable images. However, even for images of moderate sizes, the potential number of realizable images is vast, and the enumeration of images can be difficult. A common remedy, in this case, is to resort to lower-dimensional sample embeddings [14] or to coarsen the notion of distinct samples ([2]), which, in turn, can introduce additional bias to the estimates.

In contrast, we view the decoder as a differentiable mapping and estimate the support size as the volume of the decoder output. When the decoder output lies on a two-dimensional plane, our notion of the support size coincides with the familiar notions of length and surface area for one- and two-dimensional decoder input respectively. As a by-product of our method, we get an estimate for the Kullback-Leibler divergence between the generative model distribution and the uniform distribution on the same support. Even though there is no reason to expect the outcome distribution to be uniform, an extreme non-uniformity of the model distribution can lead to low sample diversity just like the small support does. In practice, these tools can be used to study and compare different decoder-based generative models.

Our contributions: (1) we propose a simple parameter-free method to estimate support size of a generative model, (2) a metric for the degree of distribution unevenness, (3) with these tools we verify several previously made claims: (3.1) WGAN-GP is less prone to mode collapse than GAN, (3.2) the support size of GAN is related to the discriminator network capacity, (3.3) variational auto-encoders have bigger support than GANs.

2   Related Work

The generative model evaluation methods advance together with the models. Test likelihood was among the first approaches to the problem. Some of the models admit likelihood computation by design [5,13], others have intractable, yet well-defined likelihood [9] and some have ill-defined likelihood [6]. In [19] the authors proposed an approach to define and evaluate test likelihood of an arbitrary decoder-based generative model. The definition gave a computationally heavy method to estimate test likelihood that required additional hyperparameter tuning. Further consideration of likelihood [18] raise doubts that likelihood can be treated as a reliable measure of model fidelity in a broad sense.

As a result, various evaluation methods that avoid likelihood computation have been proposed. Typically, these methods rely on auxiliary knowledge about the modeled distribution, such as data labels or low-dimensional embeddings. For example, the Inception score [16], a popular metric for adversarial networks, is known to correlate highly with visual sample quality. It uses an additional classifier network to compute mutual information between generative model samples. The main drawback of inception score is that it can assign high scores to degenerate distributions. As an improvement of the Inception score, the Frechet Inception Distance [7] has been proposed. The method computes the difference between real data and synthetic data statistics that are also defined using auxiliary embeddings.

All of the evaluation approaches mentioned before are general-purpose and aim at evaluating overall model fidelity. Deviations in their values can indicate that a model does not approximate the data well, but the deviations cannot differentiate between different failure modes. In [1] it was shown that adversarially trained model tend to underestimate support of the true distribution. A test based on the birthday paradox has been proposed in [2] to measure this effect, referred to as *the mode collapse*. However, this method has several drawbacks. Firstly, it does not differentiate between highly uneven distributions and distributions with low support. These are the two types of mode collapse, collapse to a subset of modes and collapse to a sub-manifold within the data distribution, underlined in [11]. Secondly, the test relies on an additional function that detects coincidental samples that, in turn, can be hard to define for an arbitrary domain.

In recent work [14] proposed an evaluation procedure to disentangle sample fidelity and mode coverage of a generative model using a precision-recall curve defined for distributions. The procedure compares histograms obtained after clustering a set of real and fake (generated) samples and is reliable as long as the clusters are representative. On the other hand, similarly to Inception Score, the precision-recall curve can be misleading if generator covers all clusters but outputs only a few distinctive samples. As a complement to the precision-recall curve, we can identify non-uniform distributions with few distinctive modes before the clustering is done.

3   Estimates of Support Volume and Non-uniformity

3.1   Notation and Assumptions

In what follows we denote the decoder as $f : Z \to X$, where $Z \subseteq \mathbb{R}^d$ is an open subset of $\mathbb{R}^d$ (typically $(0,1)^d$ or $\mathbb{R}^d$) and $X = \mathrm{Im}\, f \subseteq \mathbb{R}^D$ is the decoder image, **the support** of the distribution defined by $f(\cdot)$. Our goal is to estimate the volume of X, which we denote as $\mu(X)$, the latent variable distribution as $p_z(\cdot)$.

Our analysis is carried out within the following assumptions:

1. Generative model samples x are obtained by a deterministic transformation $f : Z \subseteq \mathbb{R}^d \to X \subseteq \mathbb{R}^D, d < D$ of a continuous random variable $z \sim p_z$;
2. Mapping $f(\cdot)$, the decoder, is differentiable almost everywhere;
3. Mapping $f(\cdot)$ is a bijection from Z to X.

A majority of generative adversarial networks satisfy the first two assumptions. By design, variational auto-encoders have stochastic decoder: to generate a sample one has to generate a random vector $z \in Z$, apply deterministic decoder $f(\cdot)$ and then sample from a distribution defined by $f(z)$. Thus, strictly speaking, VAE does not satisfy the first requirement. Nevertheless, the second sampling step is often omitted because for the default fully-factorized distributions it effectively adds noise to the decoder output. Since the deterministic decoder $f(\cdot)$ of variational auto-encoder satisfies the first two assumptions and is often used in practice, we include it in the study. Generative models with discrete output, such as PixelRNN [13], or discrete latent variables, such as deep Boltzmann machines [15], do not satisfy the assumptions and thus lie outside the scope of applicability of the framework described below.

While the injectivity assumption needs additional validation, even for convolutional architectures with $ReLU$[1] non-linearity we observed that for all generated samples the Jacobi matrix of f had full rank. The full rank strongly suggests that the third assumption usually holds as well. This observation is possible because, typically, the generative network's narrowest part is its input. Specifically, the Jacobi matrix of a neural network is a product of weight matrices and the Jacobi matrices of the intermediate activations. The former matrices are likely to have full rank, while for $ReLU$ non-linearities the latter are diagonal matrices with ones and zeros on diagonal. If the Jacobi matrices of intermediate activations have sufficiently high rank, the Jacobi matrix of the decoder will have the rank of its narrowest part, the input.

3.2   Volume Form and Distribution Density on a Manifold

In this section we briefly cover the differential geometry concepts used in the paper. For more details on integration on surfaces see [3]. For more general overview refer to [17].

To define the notion of manifold volume $\mu(X)$, we have to start by defining the volume form $\mathrm{d}x$ of the manifold X. The volume form can be thought of as a function that for a point $x = f(z)$ on a manifold returns the volume of an infinitesimal neighborhood containing this point. In our case, the volume form is

$$\mathrm{d}x := \left| \frac{\partial f}{\partial z}^\mathsf{T} \frac{\partial f}{\partial z} \right|_{z=f^{-1}(x)}^{\frac{1}{2}} \mathrm{d}z, \tag{1}$$

---

[1] $ReLU(x) = \max(0, x)$.

where dz is the Lebesgue measure of the latent space $Z \subseteq \mathbb{R}^d$ and $\frac{\partial f}{\partial z}$ is the Jacobian matrix of the decoder f. This formula is a generalization of change of variables formula for integration ($dx = |\frac{\partial f}{\partial z}| dz$ for the dimensionality-preserving f) to the case when the substitution does not preserve the dimensionality.

The volume form allows us to introduce integration over the manifold

$$\int_X g(x)dx := \int_Z g(f(z)) \left| \frac{\partial f}{\partial z}^T \frac{\partial f}{\partial z} \right|^{\frac{1}{2}} dz, \tag{2}$$

so the overall support volume can be computed as an integral of volume form over the manifold:

$$\mu(X) = \int_X dx = \int_Z \left| \frac{\partial f}{\partial z}^T \frac{\partial f}{\partial z} \right|^{\frac{1}{2}} dz. \tag{3}$$

The notorious mode collapse effect can be caused either by small support size or by extreme irregularities of the model distribution: for example, an isotropic Gaussian distribution with a tiny scale parameter σ is highly irregular and has low sample diversity, although its support is the whole real line. To quantify **non-uniformity** of a distribution, we propose to compute Kullback-Leibler divergence between the generative model $p_x(\cdot)$ and uniform distribution $u_X(\cdot)$ on its support $X = \mathrm{Im}\, f$. The scale invariance of Kullback-Leibler divergence suggests that it can be used for comparison between manifolds of different volume. Formula for density of a generative model follows from Formula 2 for integration:

$$p_x(x) = p_z \left(f^{-1}(x) \right) \left| \frac{\partial f(x)}{\partial z}^T \frac{\partial f(x)}{\partial z} \right|^{-\frac{1}{2}}_{z=f^{-1}(x)} . \tag{4}$$

The uniform distribution on the manifold X has density $u_X(x) = \frac{1}{\mu(X)}$, therefore the Kullback-Leibler divergence between the generative model distribution $p_x(\cdot)$ and the uniform distribution $u_X(\cdot)$ on the same support can be computed as

$$D_{KL}(p_x \| u_X) = \int_Z p_z(z) \log \left(p_z(z) \left| \frac{\partial f}{\partial z}^T \frac{\partial f}{\partial z} \right|^{-\frac{1}{2}} \right) dz + \log \mu(X). \tag{5}$$

3.3 Stochastic Estimates for Model Evaluation

Both the support volume and the Kullback-Leibler divergence are integral metrics of the generative model. Their numerical computation with quadratures is impractical even for moderate manifold dimension d. To address the issue we introduce Monte-Carlo estimates for the manifold volume $\mu(X)$ and the divergence D_{KL}.

We use the latent variable distribution $p_z(\cdot)$ to compute estimates for the volume

$$\tilde{\mu}(X) = \frac{1}{N} \sum_{i=1}^{N} \frac{1}{p_z(z_i)} \left| \frac{\partial f}{\partial z}^T \frac{\partial f}{\partial z} \right|_{z=z_i}^{\frac{1}{2}} \tag{6}$$

and the Kullback-Leibler divergence

$$\tilde{D}_{KL}(p_x \| u_X) = \frac{1}{N} \sum_{i=1}^{N} \log \left(p_z(z_i) \left| \frac{\partial f}{\partial z}^T \frac{\partial f}{\partial z} \right|_{z=z_i}^{-\frac{1}{2}} \right) + \log(\tilde{\mu}(X)), \tag{7}$$

where $(z_1, \ldots, z_N) \overset{\text{i.i.d.}}{\sim} p_z$. In practice, we found that the variance of these estimates is relatively low and for reliable generative model evaluation a moderate number of samples is sufficient.

3.4 Jacobian Clamping Through Lens of Support Volume

The Jacobian clamping has been proposed in [12]. The authors discovered that Jacobians of generator generally become ill-conditioned and proposed a regularizer that promotes well-conditioned generators. The Jacobian clamping improves the mean FID score of GANs and reduces the inter-run variance of GANs. The method is a fast-to-compute heuristic that encourages all of the eigenvalues of the Jacobian $\frac{\partial f}{\partial z}$ to lie within segment $[\lambda_{\min}, \lambda_{\max}]$, where $\lambda_{\min}$ and $\lambda_{\max}$ are hyperparameters of the method.

At the same time, the determinant in the definition of the volume form in Eq. 1 can be rewritten as the product of Jacobian singular values $\lambda_1(z), \ldots, \lambda_d(z)$:

$$\left| \frac{\partial f}{\partial z}^T \frac{\partial f}{\partial z} \right|^{\frac{1}{2}} = \prod_{i=1}^{d} |\lambda_i(z)|. \tag{8}$$

Note that the singular values depend on z, the point at which the Jacobian is computed. As a result, the singular values regularization will affect the support volume of the generative model. For instance, if the generator input is uniformly sampled from the unit cube $Z = [0, 1]^d$, the Jacobian clamping would promote the support volume in the range $[\lambda_{min}^d, \lambda_{max}^d]$:

$$\lambda_{\min}^d \leq \mu(X) = \int_Z \prod_{i=1}^{d} |\lambda_i(z)| dz \leq \lambda_{max}^d. \tag{9}$$

In other words, the Jacobian clamping can be seen as the support volume regularization. In particular, as Fig. 4 in [12] indicates, the Jacobian clamping helps the decoder to avoid small support volume.

4   Experiments

We begin the experimental section with an illustration of the proposed tools by considering a toy setting where the support volume is known. Next, we proceed to evaluation and comparison of commonly used generative models in a regular setting. In the final experiment, we replicate an experiment from [2] to apply the evaluation method to models that exhibit mode collapse.

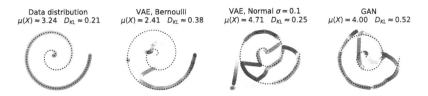

Fig. 1. Approximations of a low-dimensional distribution with various generative models. Color indicates distribution densities: high-density regions are red, low-density regions are blue. The black dashed line is the support of the data distribution. (Color figure online)

4.1   Toy Example: A Curve on a Plane

We start by considering a toy example, in which the data distribution is defined by a known decoder mapping $f_1 : [0,1] \to \mathbb{R}^2$ that maps a uniformly distributed random variable into a two-dimensional plane

$$f_1(z) = \left[\frac{1}{2}(1 + z\cos(4\pi z)); \frac{1}{2}(1 + z\sin(4\pi z))\right]. \tag{10}$$

The support of this distribution is a spiral, depicted in Fig. 1. For this example the support size is just the length of the curve defined by the decoder mapping.

Figure 1 depicts outputs of a fixed decoder for each generative model. VAE trained with Bernoulli likelihood $p(x|z,f) = \prod_{i=1}^{D} f_i(z)^{x_i}(1 - f_i(z))^{1-x_i}$ outputs samples on a manifold with small support (a short curve). On the contrary, VAE trained with Gaussian likelihood $p(x|z,f) = \prod_{i=1}^{D} \mathcal{N}(x_i|f_i(z), \sigma^2)$ overestimates the support size.

GAN recovers the manifold, but the distribution density has distinctive modes in the center of spiral. The high value of D_{KL} indicates this irregularity.

4.2   Comparison on Conventional Datasets

In this section, we applied the proposed tools to evaluate and compare a number of popular generative models. Our main motivation is to draw connections between the support volume of generative models and the diversity of its samples.

Support size as it is does not indicate neither sample quality, nor sample diversity. Therefore, along with an estimate of support size $\log \mu(X)$, we report Frechet Inception Distance [7] as a metric for visual quality of samples. We also report Kullback-Leibler divergence estimate (see Eq. 7) because highly non-uniform distributions can suffer from small sample diversity even when the support is large and the birthday paradox test B as an independent measure of sample diversity (for detailed description refer to [2]). To apply the birthday paradox test we say that to images coincide if the Euclidean distance between them does not exceed a dataset-specific threshold.

Table 1. Logarithm of support volume $\log \mu(X)$, Kullback-Leibler divergence between the generative model and a uniform distribution on its support, the birthday paradox test output B and Frechet Inception Distance (lower is better) computed for various generative models. Models with best FID scores are in bold. We make several conclusions. Firstly, for GANs the birthday paradox test is generally aligned with the support size estimate $\log \mu(x)$, however, results of the test exhibit high variance. Secondly, as the birthday paradox test and support volume indicate, WGAN-GP learns distributions with higher or equal to GANs sample diversity. Finally, the support volume of variational auto-encoders is sensitive to the choice of observation model.

	MNIST				Fashion MNIST			
Model	$\log \mu(X)$	D_{KL}	B	FID	$\log \mu(X)$	D_{KL}	B	FID
VAE Bern	-24.5 ± 1.8	46.0 ± 1.3	98.8 ± 20.0	31.6 ± 1.0	-28.9 ± 2.8	45.4 ± 2.7	17.2 ± 1.7	73.5 ± 1.2
VAE $\mathcal{N}$	86.6 ± 1.5	22.6 ± 1.0	>4096	114.4 ± 0.4	87.9 ± 1.4	24.1 ± 1.6	>4096	155.0 ± 1.0
GAN	$\mathbf{50.8 \pm 3.5}$	$\mathbf{78.5 \pm 2.3}$	$\mathbf{80.0 \pm 9.4}$	$\mathbf{14.7 \pm 1.2}$	$\mathbf{85.0 \pm 9.0}$	$\mathbf{74.4 \pm 5.7}$	$\mathbf{134.6 \pm 16.2}$	$\mathbf{47.4 \pm 7.3}$
WGAN-GP	45.0 ± 4.8	67.6 ± 5.3	186.2 ± 43.0	25.6 ± 1.8	$\mathbf{67.2 \pm 4.6}$	$\mathbf{62.2 \pm 5.8}$	$\mathbf{158.2 \pm 14.1}$	$\mathbf{51.5 \pm 1.4}$
	CIFAR10				CelebA			
Model	$\log \mu(X)$	D_{KL}	B	FID	$\log \mu(X)$	D_{KL}	B	FID
VAE Bern	-11.1 ± 3.3	39.6 ± 2.9	2.6 ± 0.5	186.7 ± 1.0	116.2 ± 7.1	29.2 ± 6.6	5.2 ± 0.4	82.4 ± 0.5
VAE $\mathcal{N}$	125.2 ± 2.6	26.0 ± 2.9	5.8 ± 0.7	112.2 ± 0.7	171.5 ± 2.9	18.6 ± 3.1	12.6 ± 1.0	92.5 ± 1.0
GAN	135.3 ± 4.6	35.5 ± 2.5	90.2 ± 13.9	98.1 ± 4.9	135.7 ± 11.3	34.6 ± 4.7	48.6 ± 15.0	74.4 ± 4.3
WGAN-GP	$\mathbf{154.2 \pm 5.9}$	$\mathbf{34.8 \pm 2.2}$	$\mathbf{135.4 \pm 21.2}$	$\mathbf{76.5 \pm 1.8}$	$\mathbf{183.5 \pm 6.9}$	$\mathbf{28.4 \pm 2.8}$	$\mathbf{313.2 \pm 26.2}$	$\mathbf{47.0 \pm 2.2}$

In the experiments, we use the same architecture as in [4] and the default hyperparameters from [10] open-source implementation for all generative networks. In Table 1 we report the results averaged over five independently trained models along with the standard deviations. Note that we chose an architecture of moderate complexity and do not expect the state of the art performance.

Support Size Comparison for Adversarial Networks. We start with comparing support sizes of GAN and WGAN-GP. We observe that, indeed, WGAN-GP outperforms GAN on bigger datasets (CIFAR10, CelebA). It has advantage both in terms of visual quality of samples measured by FID score and in terms of sample diversity measured by the birthday paradox test. Importantly, support sizes for all models are aligned with the results of the birthday paradox test: big support indicates higher sample diversity. We also note that $\log \mu(X)$ and

$D_{KL}(p_x\|u_X)$ estimates for a particular model are robust to the choice of random seed, while the birthday paradox test exhibits significantly higher variance.

On smaller datasets (MNIST, Fashion MNIST) GAN outperforms WGAN-GP in terms of FID score. Again, as the birthday paradox test indicates, WGAN-GP has higher sample diversity, but it has smaller support volume estimate $\log \mu(X)$. We found that difference in sample diversity in this case is mainly due to distribution inhomogenity: GAN has bigger support, but D_{KL} indicates that the corresponding distribution is more uneven than the distribution of WGAN-GP.

VAE Support Size. In the experiments we found that the resulting support volume for VAE decoder is sensitive to the choice of likelihood, which plays the role of reconstruction loss in VAEs.

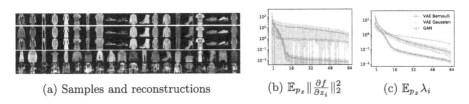

(a) Samples and reconstructions	(b) $\mathbb{E}_{p_z}\|\frac{\partial f}{\partial z_i}\|_2^2$	(c) $\mathbb{E}_{p_z}\lambda_i$

Fig. 2. *(Left, 2a)* From top to bottom: fashion MNIST test set samples, reconstructions with Bernoulli-VAE, reconstructions with Gaussian VAE, samples from Bernoulli VAE, samples from Gaussian VAE. *(Right, 2b, 2c)* The average Jacobian column norms $\mathbb{E}_{p_z}\|\frac{\partial f}{\partial z_i}\|_2^2$ show the decoder of variational auto-encoder with Bernoulli likelihood ignores some of the components of the input noise vector z.

Variational auto-encoder trained with Bernoulli likelihood, on average, had the smallest support volume $\mu(X)$ in all of the experiments.

The further analysis showed, that the Bernoulli variational auto-encoder did not use some components of the input noise vector z. As a result, the support almost collapsed into a lower-dimensional manifold and had diminishing volume. Figure 2b presents the ordered average norms $\mathbb{E}_{p_z}\|\frac{\partial f}{\partial z_i}\|_2^2$. Note that the majority of values for auto-encoder with Bernoulli likelihood is close to 10^{-2}, which indicates that output typically did not change significantly as the corresponding component z_i changed. The volume is an integral of product of Jacobian singular values (see Eqs. 3 and 8), which we show in Fig. 2c. The smaller singular values of Bernoulli variational auto-encoder resulted in smaller support volume when compared with GAN and variational auto-encoder with Gaussian likelihood. Note that a similar behavior has been already observed in Sect. 3 of [12], where for VAE the authors observed the small decoder eigenvalues with little variability.

On the contrary, the support volume of variational auto-encoder trained with Gaussian likelihood was the biggest in the majority of experiments, but the

Kullback-Leibler divergence $D_{KL}(p_x\|u_X)$ has the smallest average value compared to other generative models. The indicates that VAE with Gaussian likelihood achieves high sample diversity through the diversity in decoder outputs.

Nevertheless, the high support volume does not necessarily imply good FID scores. On all datasets except for CIFAR10, VAE with Gaussian likelihood had worse FID score than the VAE with Bernoulli likelihood. This happens because the decoder of Gaussian VAE defines a manifold that includes the target distribution support only as a small subset and, as Kullback-Leibler divergence $D_{KL}(p_x \mid u_X)$ indicates, spreads the probability mass almost evenly over the whole manifold. Figure 2a provides samples and reconstructions for both VAE models trained on Fashion MNIST. Although the samples from the Gaussian VAE do not look like pieces of clothing, the model has sharper reconstructions than the Bernoulli-VAE.

Importantly, the results for Gaussian VAE do not imply that the model has been trained poorly. The loss function of VAE is a proxy to the model likelihood. It has been previously observed in [18] that the likelihood of a mixture-model that outputs a significant portion of irrelevant samples along with samples from target distribution can still be close to the likelihood of a good target distribution model.

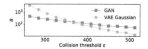

Fig. 3. The results of the birthday paradox test for variational auto-encoder with Gaussian likelihood and GAN for different collision tests on CelebA.

Birthday Paradox Test Results for VAE. Values of the birthday paradox test B for VAE models are abnormally low for some datasets. In [2] the authors conjectured that this happens because L_2 metric is prone to falsely detect collisions on blurry VAE samples, thus the birthday paradox test is not suitable for VAE.

In fact, we observed that the change of the hyperparameter ε, i.e. the maximum squared Euclidean distance between a pair of samples that we identify as coincidental, can lead to contradictive conclusions. Figure 3 presents the results of the birthday paradox test for vanilla GAN and variational auto-encoder with Gaussian likelihood on CelebA dataset. For moderate values of epsilon the birthday paradox test indicates that VAE has smaller support size. However, as ε gets smaller, the outcome of the birthday paradox test grows faster for VAE than for GAN. Finally, for the small values of ε the fine-grained outcome of the birthday paradox test is consistent with our estimates of the support sizes.

Table 2. GAN support size and distribution non-uniformity depending on the capacity of discriminator network. This experiment directly confirms the conjecture from [1], showing that the support size drops as the capacity of discriminator network is reduced.

dim	8	32	128	1024
FID	168.9 ± 83.7	71.3 ± 2.5	68.4 ± 3.8	74.4 ± 4.3
$\log \mu$	55.8 ± 14.1	106.5 ± 4.8	124.2 ± 13.3	135.7 ± 11.3
D_{KL}	34.4 ± 4.6	31.4 ± 3.0	32.0 ± 1.4	34.6 ± 4.7
B	14.6 ± 1.6	60.4 ± 6.7	55.4 ± 26.2	48.6 ± 15.0

Fig. 4. Random samples from models with the smallest (left) and the largest (right) support.

4.3   Mode Collapse Simulation

Theoretical analysis in [1] suggests the support size of GANs grows with the capacity of discriminator network. In [2] the birthday paradox test has been proposed to verify this conjecture. Although the test clearly indicates the low diversity of generative model samples, it does not distinguish between models with low support and models with highly uneven distribution.

In this subsection, we replicate an experiment from [2]. Similarly, we trained on CelebA dataset a generative network with a varying capacity of discriminator by changing the number of neurons dim in the last layer of the discriminator.

Table 2 contains statistics for GAN models, trained with $dim \in \{8, 32, 128, 1024\}$. The results of Birthday paradox test and the support volume estimates indicate that the diversity of samples increases with d, while the FID score increases as the sample diversity increases. At the same time, D_{KL} estimates suggest the distributions are equally uneven in all cases. From this we make a conclusion that the models with weaker discriminator learn distributions with smaller support.

The numerical findings are consistent with the visual characteristics presented in Fig. 4. Among the samples of the low support model a significant number of images share the same background, pose and set of facial features. On the other hand, collisions between samples of the big support model are not so easy to identify.

5   Conclusion

In this paper, we proposed a method to estimate the support size and the non-uniformity of a decoder-based generative model. To illustrate the use cases for these tools, we carried out a numerical comparison between generative networks and verified some of the common claims about deep generative models.

Acknowledgements. The authors would like to thank Ilya Tolstikhin for fruitful discussions. The work was partly supported by Ministry of Education and Science of the Russian Federation (grant 14.756.31.0001). Dmitry Vetrov was also partly supported by Samsung Research, Samsung Electronics.

References

1. Arora, S., Ge, R., Liang, Y., Ma, T., Zhang, Y.: Generalization and equilibrium in generative adversarial nets (GANs). arXiv preprint arXiv:1703.00573 (2017)
2. Arora, S., Zhang, Y.: Do GANs actually learn the distribution? An empirical study. arXiv preprint arXiv:1706.08224 (2017)
3. Ben-Israel, A.: The change-of-variables formula using matrix volume. SIAM J. Matrix Anal. Appl. **21**(1), 300–312 (1999)
4. Chen, X., Duan, Y., Houthooft, R., Schulman, J., Sutskever, I., Abbeel, P.: InfoGAN: interpretable representation learning by information maximizing generative adversarial nets. In: Advances in Neural Information Processing Systems, pp. 2172–2180 (2016)
5. Dinh, L., Sohl-Dickstein, J., Bengio, S.: Density estimation using real NVP. arXiv preprint arXiv:1605.08803 (2016)
6. Goodfellow, I., et al.: Generative adversarial nets. In: Advances in Neural Information Processing Systems, pp. 2672–2680 (2014)
7. Heusel, M., Ramsauer, H., Unterthiner, T., Nessler, B., Klambauer, G., Hochreiter, S.: GANs trained by a two time-scale update rule converge to a nash equilibrium. arXiv preprint arXiv:1706.08500 (2017)
8. Kingma, D.P., Salimans, T., Jozefowicz, R., Chen, X., Sutskever, I., Welling, M.: Improved variational inference with inverse autoregressive flow. In: Advances in Neural Information Processing Systems, pp. 4743–4751 (2016)
9. Kingma, D.P., Welling, M.: Auto-encoding variational Bayes. arXiv preprint arXiv:1312.6114 (2013)
10. Lucic, M., Kurach, K., Michalski, M., Gelly, S., Bousquet, O.: Are GANs created equal. A large-scale study. arXiv e-prints (2017)
11. Metz, L., Poole, B., Pfau, D., Sohl-Dickstein, J.: Unrolled generative adversarial networks. arXiv preprint arXiv:1611.02163 (2016)
12. Odena, A., et al.: Is generator conditioning causally related to GAN performance? arXiv preprint arXiv:1802.08768 (2018)
13. Oord, A.V.D., Kalchbrenner, N., Kavukcuoglu, K.: Pixel recurrent neural networks. arXiv preprint arXiv:1601.06759 (2016)
14. Sajjadi, M.S., Bachem, O., Lucic, M., Bousquet, O., Gelly, S.: Assessing generative models via precision and recall. arXiv preprint arXiv:1806.00035 (2018)
15. Salakhutdinov, R., Larochelle, H.: Efficient learning of deep Boltzmann machines. In: Proceedings of the Thirteenth International Conference on Artificial Intelligence and Statistics, pp. 693–700 (2010)

16. Salimans, T., Goodfellow, I., Zaremba, W., Cheung, V., Radford, A., Chen, X.: Improved techniques for training GANs. In: Advances in Neural Information Processing Systems, pp. 2234–2242 (2016)
17. Spivak, M.: Calculus on Manifolds. W.A. Benjamin Inc., Massachusetts (1965)
18. Theis, L., Oord, A.V.D., Bethge, M.: A note on the evaluation of generative models. arXiv preprint arXiv:1511.01844 (2015)
19. Wu, Y., Burda, Y., Salakhutdinov, R., Grosse, R.: On the quantitative analysis of decoder-based generative models. arXiv preprint arXiv:1611.04273 (2016)

Natural Language Processing

Biomedical Entities Impact on Rating Prediction for Psychiatric Drugs

Elena Tutubalina[1,2], Ilseyar Alimova[1(✉)], and Valery Solovyev[1]

[1] Kazan (Volga Region) Federal University, Kazan, Russia
alimovailseyar@gmail.com

[2] Samsung-PDMI Joint AI Center, PDMI RAS, St. Petersburg, Russia

Abstract. There is a growing body of research on biomedical information extraction on social media texts, patient narratives, and scientific abstracts. Relatively less research has focused on the relationship between medical concepts mentioned in a given user review and patient satisfaction. In this work, we investigate the effect of health-related entities such as adverse drug reactions and drug indications on rating prediction. We present a method based on a supervised regression approach leveraging medical concepts of different types and their mechanisms. The experiments on a collection of reviews demonstrate that features based on biomedical entities mentioned in reviews result in performance gains of up to 8% in mean squared error. Moreover, we compute feature importance in the regression models and find that the most important features for predicting patients' negative attitudes are adverse drug reactions associated with functional problems.

Keywords: Rating prediction · Opinion mining · Adverse drug reactions · Natural Language Processing · Health social media · Machine learning

1 Introduction

User-generated texts on social media present a wide variety of facts and opinions on numerous topics, and this treasure trove of information is currently severely underexplored. In 2015, the Pew Social Media Fact Sheet revealed that 37% of adults identify social media discussions about "health and medicine" as the most interesting topic [8]. Public health monitoring and research studies have utilized social media for various biomedical tasks including detection of side effects [3,12], search of active accounts for disseminating flu information via tweets [13], life extension [5], investigation of the use of social media by doctors [6], etc.

Extracting information in biomedical domain is non-trivial and requires elaborate Natural Language Processing (NLP) methods. These methods go beyond the simple matching of natural language texts and vocabulary entries: string matching approaches are not able to link social media language to medical concepts since the words may not overlap at all. To the best of our knowledge,

© Springer Nature Switzerland AG 2019
W. M. P. van der Aalst et al. (Eds.) AIST 2019, LNCS 11832, pp. 97–104, 2019.
https://doi.org/10.1007/978-3-030-37334-4_9

the research on the relationship between medical concepts mentioned on a given user review and its rating has not received much attention in the literature. Several studies utilized side effects, extracted manually [4] or automatically [11], for quantitative evaluation of social media posts given a target drug. However, the main limitation of these studies is the lack of publicly available data that makes a large-scale investigation for several drugs an issue of concern.

The aim of our study was to investigate the effect of health-related entities on review rating prediction. We seek to answer the following research questions: **RQ1**: *Does the review representation by entities of different types have better rating prediction abilities than the bag-of-words representation?* **RQ2**: *Which side effects of drugs influence patient satisfaction?*

We utilize a gradient boosting regression and SVM models based on a set of features including entity types, their medical concepts, and entities' mechanisms. We conduct a set of experiments using consumer reviews on four drugs: Zoloft, Lexapro, Cymbalta, and Effexor XR, collected from Askapatient.com. In particular, we utilize a psychiatric treatment adverse reactions (PsyTAR) corpus [15], where each review is manually annotated with four types of named entities. We show that two-fold review representation by a bag of words and a set of entities are effective for rating prediction on a collection of reviews about medications.

2 Data

Psychiatric Treatment Adverse Reactions (PsyTAR) corpus [15] is the first open-source corpus of user-generated posts about psychiatric drugs taken from Askapatient.com. This dataset includes 887 posts (6004 sentences) about four psychiatric medications from two classes: (i) Zoloft and Lexapro from the Selective Serotonin Reuptake Inhibitor (SSRI) class; (ii) Effexor and Cymbalta from the Serotonin-Norepinephrine Reuptake Inhibitor (SNRI) class. The average number of words in each post is 110.

All posts were annotated manually for 4 types of entities: (i) adverse drug reactions (ADR); (ii) withdrawal symptoms (WD); (iii) drug indications (DI); (iv) sign/symptoms/illness (SSI). The total number of entities is 7415. Besides, the entities were classified (based on mechanisms) as: (i) physiological problems; (ii) psychological problems; (iii) cognitive problems; (iv) functional (social impact) problems.

Four annotators participated in the annotation process. Two annotators were pharmacy students, and two annotators had a background in health sciences. The annotators performed terminology association using the Unified Medical Language System (UMLS). Entities mentioned in the reviews are mapped to 918 unique concepts. A review from this website consists of its author, date and time, rating (1–5), and text body. Statistics of ratings are summarized in Table 1. As shown in Table 1, 48% reviews are associated with positive ratings (4 or 5 stars). The average rating is 3.15. Please see the detailed statistics in [14,15].

3   Method

We view this task in two aspects, such as a regression an classification problems. We utilize two machine learning models: Light Gradient Boosting Machine (LightGBM) and support vector machine (SVM). LightGBM model was developed with LightGBM library [7]. This supervised model was trained with 100 iterations and Mean Squared Error (MSE) as the loss function; a number of leaves equal to 100 and learning rate equal to 0.05. For SVM classifier We utilized scikit learn library [10]. For both models we applied the following features:

Table 1. Summary statistics of the PsyTAR corpus.

	Rating 1	Rating 2	Rating 3	Rating 4	Rating 5	All
Reviews	201	104	157	210	219	891
ADR entities	1376	702	1032	1038	665	4813
WD entities	252	73	77	105	83	590
DI entities	46	55	132	272	288	793
SSI entities	133	125	202	355	404	1219

- *bow*: the bag-of-words representation (word unigrams) of a review;
- *bow without entities*: the bag-of-words representation of a review, where all entities are cut;
- *{entity type}_bow*: the bag-of-words representation of entities of a given type (ADR, WD, DI, SSI);
- *{entity type}_umls*: the bag-of-concepts representation of entities a given type based on UMLS;
- *{entity type}_class*: these class-based features represent the number of entity of a given type for each class (physiological, psychological, cognitive, functional).

4   Experiments

The overall performance is evaluated in terms of the standard Mean Squared Error (MSE), which measures deviation from the target (human-labeled) score and is more tolerant to small discrepancies (less than 1.0). We applied two *baseline approaches*: we compute the standard Mean Squared Error (MSE) between the target ratings and a predefined score: the maximum score (5.0) and the average score (3.15). We also presented rating prediction results of a state-of-the-art neural model AspeRa [9] trained with default parameters on Google News embeddings; we note that the dataset size is not large enough for the neural networks to shine.

We used 5-folds cross-validation for evaluation. Results of experiments are summarized in Table 2. The MSE of LightGBM is lower in comparison to SVM classifier for all feature sets. The simple LightGBM with the *bow* representations outperformed three baselines, while SVM outperformed the only baseline with the maximum score (5.0). The LightGBM model based on the *bow* review representation, {*entity type*}_*umls* and {*entity type*}_*class* for all four types (ADR, WD, DI, SSI) achieves the best performance (1.415 MSE). Despite the differences in achieved results between the models under consideration, the overall trends are traced in the results. We provide a more detailed analysis of results and give answers to research questions below.

Table 2. Comparison of LightGBM and SVM based on different sets of features in terms of the Mean Squared Error (MSE). Lower results are better.

Features	LightGBM	SVM
bow	1.534	2.355
bow without entities	1.609	2.220
bow, ADR_umls	1.504	2.334
bow, WD_umls	1.529	2.324
bow, DI_umls	1.515	2.262
bow, SSI_umls	1.522	2.319
bow, ADR_bow	1.481	2.249
bow, WD_bow	1.531	2.406
bow, DI_bow	1.478	2.304
bow, SSI_bow	1.521	2.284
bow, all {entity type}_umls, all {entity type}_class	1.415	2.095
Baselines		
baseline, max. score 5.0	5.600	
baseline, avg. score 3.15	2.212	
AspeRa, 15 aspects	2.152	

Method	LightGBM	SVM
ADR_umls	2.072	3.742
WD_umls	2.144	5.277
DI_umls	2.193	4.631
SSI_umls	2.142	5.259
ADR_bow	2.096	3.484
WD_bow	2.187	5.164
DI_bow	2.098	4.751
SSI_bow	2.188	4.918
ADR_class	1.867	3.470
WD_class	2.203	5.094
DI_class	2.014	4.226
SSI_class	2.128	4.929

To answer the RQ1, we compare the results of models with *entities-based features* and model with the *bow* representation. Among {*entity type*}_*bow* models, the best results for both LightGBM and SVM models were achieved with ADR type (2.096 and 3.484 MSE respectively). These MSE scores are worse than the scores of the *bow, ADR_bow* model's on 0.562 and 1.272, respectively. This lead to the conclusion that entities itself are not enough to review representation. The review text could contain sentiment words which play a central role in opinion mining tasks. Still, the comparison of LightGBM *bag-of-words* and *bag-of-words without entities* results prove that the presence of entities contributes to improving rating prediction. Thus, the MSE of the model with bag-of-words without entities is higher on 0.75 than the model with the bag-of-words feature. For SVM model such tendency isn't repeated and *bag-of-words without entities* model outperformed the *bow* model on 0,135 of MSE. We assume that these results were influenced by a decrease in the number of features, which allowed the SVM to train better.

As shown in the second table, the results of {*entity type*}_*bow* and {*entity type*}_*umls* methods are similar. The LightGBM with *ADR_umls* feature achieved the best performance (2.072 MSE) against LightGBM based on a single set of features without *bow*. The SVM with *ADR_bow* feature obtained the highest results (3.484 MSE) against SVM based on a single set of features without *bow*. Similar, the LightGBM and SVM with ADR_class feature obtained the best results of 1.867 and 3.470 MSE respectively against models based on a set of entity types features without *bow*. This leads to the conclusion that the *ADR* entity type based features influence on review rating more than *WD, DI, SSI* entity type based features. Similar to {*entity type*}_*bow* features the combination of {*entity type*}_*umls* features with the *bow* representations lead to better prediction. The best improvement in comparison with *bow* model were obtained with ADR entity type for both LightGBM (−0.03) and SVM (−0,135) models (see the "bow, ADR_umls" row in Table 2).

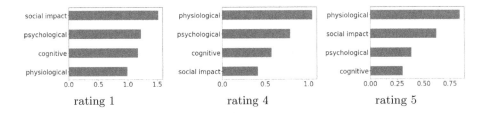

rating 1   rating 4   rating 5

Fig. 1. The feature importance of ADR_class.

In order to answer the RQ2, we compare the feature importance of classification model with ADR_class feature. The results are presented on Fig. 1. The graphics show that social impact side effects such as 'not have been able to work' (restricted work performance), 'unable to function' (difficulty in daily functioning), 'could not get out of bed' (bed-ridden) associated with low rating reviews, while high rating reviews are characterized by physiological side effects such as 'fatigue' (fatigue), 'dry mouth' (xerostomia), 'disturbed sleep' (sleep disturbances). This observation reflects quite natural attributes that match well with medical and commonsense intuition. For the rest of the ratings, the importance of side effects classes is approximately on the same level.

5   Related Work

The PsyTAR corpus was developed in 2019, therefore, little work has been done to explore this source of social media posts. Zolnoori et al. applied a dictionary look-up algorithm within from cTAKES for automatic extraction of annotated entities from the PsyTAR corpus [15]. cTAKES achieves 0.24 F1 score for the strict match extraction, while cTAKES based on additional PsyTAR's dictionaries achieves 0.49 F1 score. Clearly, the dictionaries gathered from laypersons language can efficiently improve automatic performance.

The works that are close to ours in terms of investigation of ADRs mentioned in social media are [11] and [4]. Antipov and Pokryshevskaya studied the association between side effects and patient satisfaction via review rating [4]. The focus of this study was a popular antiviral drug Tamiflu. The dataset consisted of 1325 user reviews about Tamiflu. 25 side effects were extracted using word frequency analysis. There ADRs were classified into 5 groups. The authors applied logistic regression and classification trees in order to recognize the association between side effects and rating. The authors concluded that neuropsychiatric reactions almost ensure a very negative rating. The key limitation of this study is a small number of extracted ADRs without corresponding medical concepts and manual effort on ADR classification. Smith et al. [11] explored NLP methods to elucidate the similarities and differences in ADRs mentioned in Twitter with those in traditional sources. The focus of this study was Adalimumab which is a top-selling drug in the USA. The authors utilized supervised methods for extraction of ADR mentions, followed by manual mapping of these mentions to standard nomenclature. The dataset consisted of 10,188 tweets with 2617 potential ADRs extracted from tweets. 801 ADRs were mapped to 232 unique UMLS concepts. UMLS concepts were aggregated into 16 categories of biologic systems. Disproportionality analysis of drug–ADR pairs showed that it is reasonable to compare ADRs in different sources. However, large-scale analysis of multiple medications is unrealizable due to the high level of effort on manual annotation. We note that both studies discussed above did not publish annotated datasets for researchers.

Akay et al. applied a neural network to investigate cancer patient satisfaction [2]. The authors conducted experiments on the texts from cancer-related forums. According to presented in this article case studies of user posts on Erlotinib drug, the most pressing concern from both satisfied and dissatisfied with treatment patients was the side effects. There is also a study which aims to gauge the experience of the drug Sitagliptin (trade name Januvia) by patients with diabetes mellitus type 2 [1]. The cluster-based analysis of user reviews in the forum DiabetesDaily showed that one source of negative opinion stems from the side effects of the drug.

6   Conclusion

In this paper, we have experimentally analyzed the impact of biomedical entities such as adverse drug reactions on review rating prediction. We empirically evaluated a regression model based on a set of features such as entity types, their medical concepts, and mechanisms against baselines based on the average rating and bag-of-words representation. Experiments on a dataset of user reviews about psychiatric drugs showed that the regression model that takes into accounts extracted entities performs better at rating prediction. Moreover, the qualitative analysis showed that social impact ADRs are truly substantial for prediction of negative satisfaction, while physiological ADRs tend to have a mild influence on the overall review rating. One potential future direction is to annotate a larger collection of user reviews about psychiatric drugs using modern neural architectures trained on the PsyTAR corpus. Another direction is to compare rating prediction results on two datasets which annotated manually and automatically. Additionally, comparison of extracted ADRs and withdrawal symptoms between drugs with different therapeutic effects and their impact on patient satisfaction can be investigated.

Acknowledgments. This research was supported by the Russian Science Foundation grant no. 18-11-00284.

References

1. Akay, A., Dragomir, A., Erlandsson, B.E.: A novel data-mining approach leveraging social media to monitor consumer opinion of sitagliptin. IEEE J. Biomed. Health Inform. **19**(1), 389–396 (2013)
2. Akay, A., Dragomir, A., Erlandsson, B.E.: Network-based modeling and intelligent data mining of social media for improving care. IEEE J. Biomed. Health Inform. **19**(1), 210–218 (2014)
3. Alimova, I., Solovyev, V.: Interactive attention network for adverse drug reaction classification. In: Ustalov, D., Filchenkov, A., Pivovarova, L., Žižka, J. (eds.) AINL 2018. CCIS, vol. 930, pp. 185–196. Springer, Cham (2018). https://doi.org/10.1007/978-3-030-01204-5_18
4. Antipov, E.A., Pokryshevskaya, E.B.: The effects of adverse drug reactions on patients' satisfaction: evidence from publicly available data on tamiflu (oseltamivir). Int. J. Med. Inform. **125**, 30–36 (2019)
5. Batin, M., Turchin, A., Sergey, M., Zhila, A., Denkenberger, D.: Artificial intelligence in life extension: from deep learning to superintelligence. Informatica **41**(4) (2017)
6. Brown, J., Ryan, C., Harris, A.: How doctors view and use social media: a national survey. J. Med. Internet Res. **16**(12), e267 (2014)
7. Ke, G., et al.: LightGBM: a highly efficient gradient boosting decision tree. In: Advances in Neural Information Processing Systems 30: Annual Conference on Neural Information Processing Systems 2017, Long Beach, CA, USA, 4–9 December 2017, pp. 3149–3157 (2017)
8. Kennedy B, F.C.: Public interest in science and health linked to gender, age and personality. Pew research center. http://www.pewinternet.org/2015/12/11/public-interest-in-science-and-health-linked-to-gender-age-and-personality/
9. Nikolenko, S.I., Tutubalina, E., Malykh, V., Shenbin, I., Alekseev, A.: AspeRa: aspect-based rating prediction model. In: Azzopardi, L., Stein, B., Fuhr, N., Mayr, P., Hauff, C., Hiemstra, D. (eds.) ECIR 2019. LNCS, vol. 11438, pp. 163–171. Springer, Cham (2019). https://doi.org/10.1007/978-3-030-15719-7_21
10. Pedregosa, F., et al.: Scikit-learn: machine learning in python. J. Mach. Learn. Res. **12**, 2825–2830 (2011)
11. Smith, K., Golder, S., Sarker, A., Loke, Y., O'Connor, K., Gonzalez-Hernandez, G.: Methods to compare adverse events in twitter to faers, drug information databases, and systematic reviews: proof of concept with adalimumab. Drug Saf. **41**(12), 1397–1410 (2018)
12. Yates, A., Goharian, N., Frieder, O.: Extracting adverse drug reactions from social media. In: Twenty-Ninth AAAI Conference on Artificial Intelligence (2015)
13. Yun, G.W., et al.: Social media and flu: media twitter accounts as agenda setters. Int. J. Med. Inform. **91**, 67–73 (2016)

14. Zolnoori, M., et al.: The psytar dataset: From patients generated narratives to a corpus of adverse drug events and effectiveness of psychiatric medications. Data Brief **24**, 103838 (2019)
15. Zolnoori, M., et al.: A systematic approach for developing a corpus of patient reported adverse drug events: a case study for SSRI and SNRI medications. J. Biomed. Inform. **90**, 103091 (2019)

Combining Neural Language Models for Word Sense Induction

Nikolay Arefyev[1,2(✉)], Boris Sheludko[1,2], and Tatiana Aleksashina[3]

[1] Samsung R&D Institute Russia, Moscow, Russia
[2] Lomonosov Moscow State University, Moscow, Russia
nick.arefyev@gmail.com
[3] SlickJump, Moscow, Russia

Abstract. Word sense induction (WSI) is the problem of grouping occurrences of an ambiguous word according to the expressed sense of this word. Recently a new approach to this task was proposed, which generates possible substitutes for the ambiguous word in a particular context using neural language models, and then clusters sparse bag-of-words vectors built from these substitutes. In this work, we apply this approach to the Russian language and improve it in two ways. First, we propose methods of combining left and right contexts, resulting in better substitutes generated. Second, instead of fixed number of clusters for all ambiguous words we propose a technique for selecting individual number of clusters for each word. Our approach established new state-of-the-art level, improving current best results of WSI for the Russian language on two RUSSE 2018 datasets by a large margin.

Keywords: Word sense induction · Contextual substitutes · Neural language models

1 Introduction

Ambiguity, including lexical ambiguity, when single word has several meanings, is one of the fundamental properties of natural languages, and is among the most challenging problems for NLP. For instance, modern neural machine translation systems are still surprisingly bad at translating ambiguous words [5], although there is some progress in the latest Transformer-based systems compared to previously popular RNN-based ones [19]. Word Sense Induction (WSI) task can be seen as clustering of occurrences of an ambiguous word according to their meaning. A dataset for WSI consists of text fragments containing ambiguous words. Each occurrence of these words is hand-labeled with the expressed sense according to some sense inventory (a dictionary or a lexical ontology). A WSI system gets a list of ambiguous words and text fragments as an input. For each ambiguous word the system should cluster its occurrences into unknown number of clusters corresponding to this word's senses. This is in contrast to Word Sense Disambiguation (WSD) task where systems are also given the sense

© Springer Nature Switzerland AG 2019
W. M. P. van der Aalst et al. (Eds.) AIST 2019, LNCS 11832, pp. 105–121, 2019.
https://doi.org/10.1007/978-3-030-37334-4_10

inventory used by annotators, so both the number of senses and some contextual information about these senses (in the form of their definitions or related words) are available.

Recently [3] has proposed a new approach to WSI that generates contextual substitutes (i.e. words, which can appear instead of the ambiguous word in a particular context) using bidirectional neural language models (LMs), and clusters TFIDF-scaled bag-of-words vectors of the substitutes. This approach showed SOTA results in SemEval 2013 WSI shared task for English. For instance, for the word *build* substitutes like *manufacture, make, assemble, ship, export* are generated when it is used in *Manufacturing some goods* sense and *erect, rebuild, open* are generated for the *Constructing a building* sense which allows distinguishing these senses. We improved this approach in two ways. First, the base approach simply unites substitutes retrieved from probability distributions estimated by forward and backward LMs each given only one-sided context. This results in noisy substitutes when either left or right context is short or non-informative. We explored several methods of combining forward and backward distributions and show that substitutes retrieved from a combined distribution perform much better. Second, the base method used the same number of clusters for each word (average number of senses per word was found to be optimal). We show that using a fixed number of clusters for all words has a huge negative effect on the WSI results and propose a method for selecting individual number of clusters for each ambiguous word. Our approach has achieved new SOTA results on the RUSSE 2018 WSI shared task for the Russian language [16] with a large improvement over previous best results according to the official leaderboard[1]. Also we compare performance of several pretrained Russian neural LMs in WSI.

2   Related Work

Existing WSI methods can be roughly categorized by their relatedness to one of the following lines of research. **Latent Variable Methods** define a probabilistic model of a text generation process that includes latent variables corresponding to word senses. Posterior probability given unlabeled corpus is estimated to solve WSI task. For instance, [13] relies on the Hierarchical Dirichlet Process and [6] employs the Stick-breaking Process. In [2] a rather complicated custom graphical model is proposed which aims at solving the sense granularity problem. **Graph Clustering** methods like [9,21] first build a graph with nodes corresponding to words, and weighted edges representing semantic similarity strength or co-occurrence frequency. Then graph clustering algorithms are applied to split neighbours of an ambiguous word into clusters interpreted as this word's senses. **Context clustering** methods represent each occurrence of an ambiguous word as a vector that encodes its context. For example, in [5,12] a weighted average of

---

[1] https://competitions.codalab.org/competitions/public_submissions/17806, https://competitions.codalab.org/competitions/public_submissions/17809,   see   post-competition tabs.

the context word embeddings is calculated, then a general clustering algorithm is applied.

Our work is mostly related to the line of research, which exploits **contextual substitutes** for the ambiguous word to differentiate between its senses. One of the best performing systems at SemEval 2013 generated substitutes using n-gram language models [7]. Later [3] proposed using neural language models and a few other tweaks, establishing a new SOTA on this dataset. In Sect. 3 we describe their method with slight modifications (adapting it to the RUSSE 2018 task) as our base approach and then propose several improvements. As an alternative to language models, [1] propose employing context2vec model [15] to generate substitutes and building from them the average word2vec representation instead of the bag-of-words representation. Context2vec model merges information from left and right context internally, which may result in better substitutes. However, context2vec requires lots of resources to train and is not readily available for many languages, including Russian. In contrast, neural LMs have become a standard resource available for many languages. Thus, in the current work we focus on using pretrained LMs and improving results for WSI by externally combining information from left and right context. In the preliminary experiments we tried using multilingual BERT model [8], which also combines left and right context internally and is pretrained on texts in different languages, including Russian. However, this model's vocabulary consists mainly of subwords (frequently occurring pieces of words similar to morphemes). Using BERT in a naive way (similarly to LMs) to generate substitutes results in small subwords generated that perform poorly for WSI. More sophisticated algorithms like [22] are required to generate whole words with BERT and we leave it for the future work.

Several competitions were organized to compare approaches to WSI. SemEval 2010 task 14 [14] and SemEval 2013 task 13 [11] are the most popular ones for English. For the Russian language RUSSE 2018 competition [16] has been held recently. Three datasets varying in context length and sense granularity have been built for this competition. In this paper we evaluate our methods on these datasets and compare our results with the best results of this competition. RUSSE 2018 requires hard clustering of text fragments containing an ambiguous word, i.e. each example shall appear in one and only one cluster. In this aspect it is similar to SemEval 2010 and, unlike SemEval 2013, which requires soft clustering (i.e. a probability distribution over clusters for each example). Adjusted Rand Index (ARI) was used as a quality measure of a clustering for each word. The weighted average of ARIs across all ambiguous words with weights proportional to the number of examples per word was used as the final metric. The winner of the competition didn't submit a paper, so little is known regarding the best approach, except that it used algebraic operations on word2vec embeddings to identify word senses [16]. However, the 2nd and the 3rd best results on all datasets were achieved by calculating weighted average of word2vec embeddings for context words and clustering them with either agglomerative clustering or affinity propagation [5, 12].

In the post-competition period [18] proposed pretraining the Transformer sequence transduction model [20] to recover ambiguous words hidden from it's input (an approach similar to the BERT model pretraining proposed later in [8]) and using outputs as well as hidden states from this model to represent the ambiguous word in a context. They achieved new SOTA on one of the datasets. According to the official post-evaluation leaderboard no other improvements has been achieved on RUSSE 2018 datasets yet (by the time of May 05 2019 and excluding this paper's results).

3   Approach

3.1   Baselines

We use the method from [3] as our baseline, with slight modifications to account for the differences in the datasets. Suppose our examples look like $l\,c\,r$, where c is the target ambiguous word and l, r are its left and right contexts. The original method does the following:

1. Use pretrained forward and backward LMs to estimate probabilities for each word w to be a substitute for c given only the left context $P_{fwd}(w|l)$ or only the right context $P_{bwd}(w|r)$. To provide more information to the LMs and bias it towards generating co-hyponyms of the target word, the target word can be included in the context using dynamic symmetric pattern "T and _"/"_ and T". For instance, for the sentence *These apples are sold everywhere* instead of *"These _"* we pass *"These apples and _"* to the forward LM, and instead of *"_ are sold everywhere"* we pass *"_ and apples are sold everywhere"* to the backward LM. The underscore denotes the position for which the model predicts possible words.

2. Take top K predictions from the forward and the backward LM independently, renormalize their probabilities so that they sum to one, and sample L substitutes from each distribution, resulting in 2L substitutes. Do it S times. Each of S sets of substitutes is called a representative of the original example. Build TFIDF BOW vectors for all representatives of all examples for a particular ambiguous word. Additionally, all substitutes are lemmatized to get rid of the grammatical bias (LMs can generate only plural or only singular substitutes depending on the grammatical form of the ambiguous word, so these substitutes will never intersect even for the same sense of the word unless they are lemmatized).

3. Finally, TFIDF BOW vectors are clustered using agglomerative clustering with cosine distance, average linkage and the same number of clusters for all words (we select it on the train set of each RUSSE 2018 dataset separately while [3] used the average number of senses in the test set of SemEval 2013). The required probability distribution over clusters for each example is estimated from the number of representatives of this example found in each cluster.

To obtain hard clustering required by RUSSE 2018 we can simply select the most probable cluster for each example (this method is denoted as **sampling** in our experiments). However, we have found that skipping sampling and using S = 1 representative consisting of top K predictions from each LM, while being conceptually simpler and deterministic, also delivers better results (**base** method). Also we have found that dynamic symmetric patterns (we have simply translated "T and _" to Russian) sometimes help a little but generally hurt a lot for our best models, so we don't use them by default. This is in line with the ablation study from [3] showing that the patterns are useful for verbs and adjectives but almost useless for nouns, which RUSSE 2018 datasets consist of. We leave integration of the patterns into our best models and experimenting on other datasets containing other parts of speech for the future work.

3.2   Combining LMs

During preliminary experiments we have found that using substitutes retrieved from forward and backward LMs independently results in lots of substitutes not related to the target word. For instance, consider the case when the ambiguous word is the first word of a sentence. The forward LM will simply predict all words which can appear as the first word, and these words will be unrelated to the meaning of the target word. Using patterns like "T and _"/"_ and T" improves this situation to some extent, at least the context will always contain the target word. However, we noticed other problems related to these patterns. Often after *and* the model generates not nouns which are meant to be co-hyponyms of the target word but verbs instead. Probably this is related to the agreement in number between the noun and its syntactically related words. For instance, the LM cannot generate coordinated subjects for a singular predicate, so it tries to generate a coordinate clause instead.

To solve these problems we propose taking top K words from a combination of distributions predicted by forward and backward LMs. We experiment with the following combinations.

Average (avg): $(P_{fwd} + P_{bwd})/2$

Positionally weighted average (pos-weight-avg): $\alpha P_{fwd} + (1 - \alpha)P_{bwd}$, where α is a function of normalized (divided by the example length) position of the ambiguous word in the example:

$$\alpha(pos) = max(min(0.5, 0.5\beta^{-1}pos), 0.5\beta^{-1}(pos - 1 + 2\beta))$$

This allows to weigh forward and backward LMs equally when both left and right context is larger than β times example length words and underweigh one of them when corresponding context becomes short. β is a hyperparameter to be selected.

Bayesian combination (bayes-comb): using Bayes' rule and supposing left and right context are independent given the target word we estimate the probability we are interested in as

$$P(w|l, r) = \frac{P(l, r|w)P(w)}{P(l, r)} = \frac{P(l|w)P(r|w)P(w)}{P(l, r)} \propto \frac{P(w|l)P(w|r)}{P(w)}$$

The numerator is estimated using P_{fwd} and P_{bwd}, but pretrained LMs that we use don't contain word frequencies in their vocabularies, so we cannot directly estimate the denominator. However, their vocabularies are sorted by frequency, so we can estimate word frequency ranks and approximate the denominator with Zipf distribution: $P(w) \propto 1/(rank(w))^z$. So finally we approximate the conditional probability of a substitute given context as:

$$\hat{P}(w|l,r) \propto P_{fwd}P_{bwd}(rank(w))^z$$

Interestingly, pointwise mutual information (PMI) which is another popular measure of relatedness between a word and a context can be approximated by exactly the same formula, but with different value of z:

$$PMI(w,(l,r)) = \frac{P(w|l,r)}{P(w)} \propto \frac{P(w|l)P(w|r)}{(P(w))^2}$$

$$\widehat{PMI}(w,(l,r)) \propto P_{fwd}P_{bwd}(rank(w))^{2z}$$

In contrast to conditional probability, PMI discounts frequent words and promotes rare ones. When we select z as a hyperparameter on the train set, effectively we select from the family of relatedness metrics of the form $P(w|l,r)/(P(w))^k$ one, which is optimal regarding the final evaluation metric (ARI).

3.3   Clustering

Following the original method, we exploit agglomerative clustering, but for each word select individual number of clusters. This approach is not only linguistically more plausible than using the same number of clusters for all words, but also resulted in significant improvement of the final results.

Agglomerative clustering has three hyperparameters: the function defining the distance between examples (affinity), the function defining the distance between clusters (linkage), and the number of clusters. Initially each example is put into a separate cluster. At each iteration, two nearest clusters are merged, until the specified number of clusters is reached. We use cosine affinity and average linkage (the distance between clusters is the average cosine distance between their members). For each word we select the number of clusters, which maximizes the silhouette score:

$$\frac{1}{n}\sum_{i=1}^{n}\frac{b_i - a_i}{max(a_i, b_i)},$$

where a_i is the mean distance from the i-th example to all examples from the same cluster, and b_i is the mean distance from the i-th example to all examples from the nearest different cluster. We compare this approach to another one, which sets for all words a single number of clusters selected on the train set.

4   Experiments and Results

4.1   Datasets and Evaluation

We evaluate our methods on the three datasets from RUSSE 2018 WSI shared task for the Russian language [16] with different sense granularity, context length and number of examples. Both sense inventory and examples for the active-dict dataset were extracted from the Active Dictionary of Russian [4], which is an explanatory dictionary of the contemporary Russian language. It contains 253 ambiguous words with 3.5 senses per word on average. The bts-rnc dataset contains examples for 81 ambiguous words extracted from the Russian National Corpus[2] and labeled with their senses according to the Large Explanatory Dictionary[3], has 3.1 senses per word on average. There are more examples per word compared to the active-dict (124 vs. 23 on average) and the examples are twice as long. The wiki-wiki dataset is the smallest one, it contains only 9 ambiguous words with 2.2 senses and 109 examples per word on average. This dataset contains only homonyms (words having several unrelated senses), so the senses are coarse-grained. Each dataset of RUSSE 2018 is split into train, public test and private test parts. We held out the private test parts to compare our best methods with previous SOTA results, and used the train and the public test parts for models and hyperparameters selection. We didn't use wiki-wiki for development and comparison of LM combination methods due to its small size. However, we do compare our best models to previously best results on all three datasets.

The metric used in RUSSE 2018 competition for the final ranking of participants is Adjusted Rand Index (ARI). However, different clustering evaluation metrics may exhibit different biases. For better comparison of our results to the previous best results we also adopt two complementary metrics used in the SemEval 2010 competition which also requires hard clustering, namely V-measure and paired F-score [14]. The former metric is biased towards a large number of clusters, while the latter metric prefers small number of clusters, so the geometric mean of these two metrics (AVG) proposed by [23] is usually reported. However, all aforementioned metrics are affected both by vectorization and clustering methods along with their hyperparameters which makes it difficult to compare vectorization methods alone. To estimate vectorization quality while abstracting from clustering hyperparameters selection, we propose and exploit maxARI metric, which is maximum possible ARI achieved by agglomerative clustering with all possible hyperparameter values (varying distance metric, linkage and, most importantly, the number of clusters). Clearly, this is an overestimation of possible results for a particular vectorization when using agglomerative clustering because it selects hyperparameters using gold labels. However, we have found this metric very useful for intermediate comparison and selecting hyperparameters of vectorization methods.

---

[2] http://ruscorpora.ru.
[3] http://gramota.ru/slovari/info/bts.

4.2   Comparison of LMs for the Russian Language

We compare ELMo LMs [17] trained on three different corpora (Wikipedia, WMT News and Twitter) for the Russian language.[4] Also we try ULMFiT [10] LM trained on a subset of the Russian Taiga corpus[5] which is available for the Russian language only as a forward LM. To make the comparison fair we used forward and backward LMs separately. Table 1 contains main characteristics of these models taken from corresponding web pages.

Table 1. LMs for the Russian language

Model	Corpora size, tokens	Vocabulary size, words	Perplexity
elmo-news	946M	1M	49.9
elmo-twitter	810M	1M	94.1
elmo-wiki	386M	1M	43.7
ulmfit	208M	60K	21.98

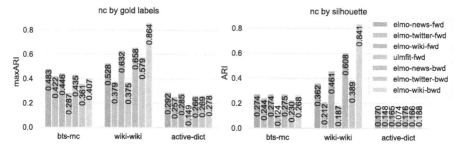

Fig. 1. Comparison of unidirectional LMs available for Russian. Number of clusters (nc) is selected for each word individually by maximizing either ARI using gold labels (for maxARI estimation) or silhouette score.

Figure 1 shows that ELMo LMs trained on the Russian Wikipedia outperform all other LMs by a large margin on the wiki-wiki dataset, which was also built from the Russian Wikipedia. All backward LMs are significantly better than their forward counterparts on this dataset due to longer right contexts. This is simply because contexts in wiki-wiki are rather large and frequently contain several occurrences of the ambiguous word, while we generate substitutes for the first occurrence only. For the other two datasets ELMo LMs trained on WMT News and Wikipedia give comparable results, the former having slightly

---

[4] http://docs.deeppavlov.ai/en/master/intro/pretrained_vectors.html.
[5] https://github.com/mamamot/Russian-ULMFit.

better maxARI on bts-rnc. ELMo LMs trained on Twitter perform slightly but consistently worse, and ULMFiT gives much worse results than other models. We suppose that bad performance of the ULMFiT model for WSI may be related to relatively small vocabulary (60K words compared to 1M in ELMo LMs), which may prevent generating substitutes that can discriminate different senses. Based on these results we have selected ELMo LMs trained on WMT News for all further experiments on bts-rnc and active-dict, and the ones trained on Wikipedia for wiki-wiki. It is also worth exploring combinations of different models trained on different corpora in the future work.

4.3 Comparison of LM Combination Methods

Figure 2 shows WSI results on the train sets depending on forward and backward LM combination method compared to the baselines and unidirectional LMs alone. Bayesian combination is the best combination method on all datasets according to both maxARI and ARI (excluding maxARI on bts-rnc for the fixed number of clusters which is worse than individual number of clusters anyway). Surprisingly, the original method (**sampling**) often performs worse than unidirectional LMs. However, our deterministic modification (**base**) performs better or comparable to them and all proposed combination methods improve its results.

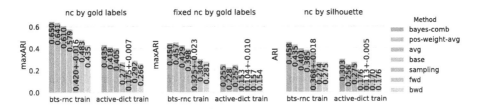

Fig. 2. Comparison of LM combination methods. *Fixed nc* denotes using the same number of clusters for all words.

It is trivial that performance upper bound when selecting individual number of clusters per word (nc by gold labels) is always better than using the same number of clusters for all words (fixed nc by gold labels). However, we would like to emphasize the large difference. Surprisingly, selecting individual number of clusters based on silhouette score (nc by silhouette), which doesn't exploit gold labels, often gives better or similar results to the upper bound for the fixed number of clusters. Unfortunately, the margin between ARI and maxARI is still large so it worth experimenting with other methods of selecting individual number of clusters.

4.4 Comparison to the Previous Best Results

Table 2 compares our baselines and best methods to the previous best methods according to the official leaderboard. The best previous results are improved on bts-rnc and active-dict by a large margin (which is especially large on bts-rnc presumably due to longer contexts which make neural LMs generating better substitutes). The results of our baseline method is also better than previous results on bts-rnc, but little worse on active-dict, which underlines the benefits of combining forward and backward LMs properly, especially when contexts are short. For both the baseline and the best vectorization method selecting the number of clusters based on silhouette score works significantly better than using fixed number of clusters selected on the train sets.

Table 2. Comparison with previous best results. Selecting the number of clusters individually using silhouette score (silnc) or as a hyperparameter on train (fixnc).

Model	bts-rnc		wiki-wiki		active-dict	
	Test ARI		Test ARI		Test ARI	
	Public	Private	Public	Private	Public	Private
bayes-comb-silnc	**0.502**	**0.451**	0.651	0.616	**0.331**	**0.298**
bayes-comb-fixnc	0.464	0.438	0.651	**0.682**	0.304	0.260
base-silnc	0.365	0.362	0.651	0.646	0.202	0.162
base-fixnc	0.328	0.298	0.651	0.394	0.143	0.141
post-competition improvement	–	–	–	–	**0.307**	0.234
competition best result	**0.351**	**0.338**	**1.0**	**0.962**	0.264	**0.248**
competition 2nd best result	0.281	0.281	1.0	0.659	0.236	0.227

In the public wiki-wiki test set there are only 2 words, one of them was clustered perfectly by all methods and another has only one sense while our methods split it into two clusters. The public test set contains 4 words for which our results are comparable to the competition 2nd best results but are much worse than the best results. The analysis of these results has revealed that such a large gap is mostly due to suboptimal number of clusters selected by our methods on wiki-wiki. Using our vectorization and clustering but with the same number of clusters as in the best submission improves our ARI on the private test set to 0.89 while maxARI is 0.95. Also it is possible that substitutes-based methods are suboptimal for homonyms because unrelated senses are likely to be accompanied by unrelated context words, so context clustering approaches may perform better in this scenario. However, experiments on larger datasets consisting of homonyms are needed to check this hypothesis.

To provide better comparison, in Tables 3 and 4 we report the metrics from SemEval 2010 Task 14 which is a similar competition for English. We have downloaded the previous best submissions from the RUSSE 2018 leaderboard and used the official SemEval 2010 evaluation script to calculate paired F-score (F-Sc) and V-measure (V-M). The results of two primitive baselines, one placing all examples into a single cluster and another allocating a separate cluster for each example, are calculated to show that both F-score and V-measure are

Table 3. Semeval 2010 metrics on bts-rnc.

Model	Public test				Private test			
	F-Sc	V-M	AVG	#Cl	F-Sc	V-M	AVG	#Cl
bayes-comb-silnc	0.795	**0.454**	**0.601**	4.0	**0.776**	**0.432**	**0.579**	3.8
bayes-comb-fixnc	**0.798**	0.404	0.568	4.0	0.772	0.421	0.570	4.0
base-silnc	0.721	0.351	0.503	2.9	0.702	0.363	0.505	2.9
base-fixnc	0.774	0.271	0.458	4.0	0.756	0.294	0.471	4.0
post-competition improvement	–	–	–	–	–	–	–	–
competition best result	0.731	0.271	0.445	2.4	0.710	0.3	0.462	2.3
competition 2nd best result	0.692	0.288	0.446	10.0	0.683	0.298	0.451	10.0
1 cluster per word	0.764	0.0	0.0	1.0	0.726	0.0	0.0	1.0
1 cluster per instance	0.0	0.221	0.004	130.6	0.0	0.244	0.005	127.5

highly biased towards small or large number of clusters and rather useless in isolation. Thus, we additionally report their geometric mean (AVG) and the average number of clusters per word (#Cl). On bts-rnc our methods outperform all previously best methods according to all metrics. On active-dict F-score of our methods is little worse, but V-measure and, most importantly, AVG are better. Worse F-Score can be explained by the larger number of clusters our methods produce. Section 4.5 provides more detailed analysis of pros and cons of our approach compared to the previous best submissions.

Table 4. Semeval 2010 metrics on active-dict.

Model	Public test				Private test			
	F-Sc	V-M	AVG	#Cl	F-Sc	V-M	AVG	#Cl
bayes-comb-silnc	0.484	0.538	**0.511**	5.2	0.459	0.505	**0.482**	5.4
bayes-comb-fixnc	0.453	0.527	0.489	6.0	0.421	0.479	0.449	6.0
base-silnc	0.401	0.395	0.398	4.6	0.362	0.365	0.363	5.1
base-fixnc	0.349	0.367	0.358	5.0	0.351	0.352	0.351	5.0
post-competition improvement	**0.513**	0.451	0.481	3.0	0.464	0.389	0.425	3.0
competition best result	0.489	0.411	0.445	3.2	0.467	0.392	0.428	3.4
competition 2nd best result	0.468	0.377	0.420	3.0	**0.468**	0.371	0.417	3.0
1 cluster per word	0.433	0.0	0.0	1.0	0.437	0.0	0.0	1.0
1 cluster per instance	0.0	**0.55**	0.014	21.9	0.0	**0.543**	0.010	22.4

4.5 Analysis of the Results

In this section we answer two questions: do we select the number of clusters better and can we cluster a target word occurrences into given number of clusters better than the previous best method? We compare our best submissions to the previous best submissions according to the private test ARI on bts-rnc and active-dict.

To answer the first question, we calculated the mean squared error (MSE) between the number of clusters in each submission and the true number of senses. Our method estimates the number of senses much worse (MSE is 3.41 versus 1.65 on bts-rnc and 8.48 versus 1.20 on active-dict), usually returning larger number of clusters. But is the optimal number of clusters equal to the true number of senses? In our case the data contains outliers (i.e. points which are far from all other points due to non-informative context or other vectorization problems). Thus, the number of clusters should be larger than the true number of senses. Otherwise, some senses will be merged with other senses while outliers will occupy their clusters. The results below and Appendix B provide some evidences that our method better estimates the optimal number of clusters.

It is hard to make comparison of vectorization and clustering algorithms separately, since we don't have vectors and cannot vary the number of clusters for the previous methods, only clustering results are available. One thing we can do is clustering examples of i-th word into the number of clusters P_i taken from the previous best submission, but using our vectors and clustering algorithm. Then the difference in performance cannot be due to the difference in the number of clusters selected. This is achieved by two techniques. The first one (bayes-comb-prevnc) simply sets the number of clusters equal to P_i in our agglomerative clustering, which is suboptimal due to outliers. Figure 3 shows that this works much worse than the number of clusters selected using silhouette score according to all metrics (except F-score on active-dict). Compared to the previous best submissions the results are mixed. Another technique (bayes-comb-prevnc2) first clusters the occurrences of each word into the number of clusters S_i selected by silhouette score. If $S_i > P_i$, then we leave P_i largest clusters intact and distribute all other examples among them by moving each example to the nearest cluster. Otherwise, we simply cluster examples into P_i clusters like before. This results in much better clustering compared to the previous best submissions given the same number of clusters according to all metrics. To conclude, compared to the previous best submissions our approach has selected the number of clusters that is better according to several evaluation metrics for our vectorization, but further from the true number of senses. It can also cluster the datasets better when the number of clusters is taken from another submissions and is not optimal given our vectorization.

5   Conclusion

Bidirectional neural LMs are a powerful instrument for different tasks including substitutes generation for WSI, but some tricks should be used to apply them to this task properly. We have proposed and compared several methods of combining forward and backward LMs for better substitutes generation. Also we have proposed a technique for selecting individual number of clusters per word and this improved results even further. We improved previous best results on

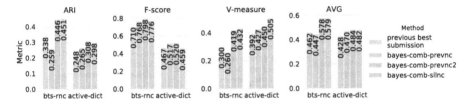

Fig. 3. Comparison of our and previous best submissions with different and equal number of clusters.

two datasets from RUSSE 2018 WSI shared task for the Russian language by a large margin. Finally, we have compared several Russian LMs regarding their performance for WSI.

Acknowledgements. We are grateful to Dima Lipin, Artem Grachev and Alex Nevidomsky for their valuable help.

A   Examples of Substitutes Generated

Table 5 provides examples of discriminative substitutes with their relative frequencies for each of two most frequent senses of several words. A substitute is called discriminative if it is frequently generated for one sense of an ambigusous word, but rarely for another. Formally, we take substitutes with the largest $\frac{P(w|sense_1)}{P(w|sense_2)}$, where $P(w|sense_i)$ is estimated using add-one smoothing:

$$P(w|sense_i) = \frac{cnt(w|sense_i) + 1}{cnt(sense_i) + |vocab|}$$

Additionally, we leave only substitutes which were generated at lest 10 times for one of the senses.

Table 6 lists ten most probable substitutes according to the combined distribution and according to the forward and the backward LM distributions separately for several examples. Substitutes from unidirectional distributions are very sensitive to the position of the target word. When either left or right context doesn't contain enough information at least halve of the substitutes will be not related to the target word. Combined distribution provides more relevant substitutes.

B   The Number of Clusters Selected

Figure 4 plots distributions of the differences between the true number of senses, the number of clusters in submissions and the optimal number of clusters. Silhouette score gives the number of clusters, which is usually larger than the number of senses, but is near the optimum with respect to ARI and given our vectors. The previous best submissions better estimate the true number of senses.

Table 5. Discriminative substitutes for several words from bts-rnc train

Балка	Штамп	Лавка
Number of examples: 81 Sense1: Горизонтальный опорный брус	Number of examples: 45 Sense1: Рельефное устройство для получения одинаковых графических оттисков	Number of examples: 67 Sense1: Скамья для сидения или лежания
перегородка 0.56/0.00	справка 0.58/0.00	корточки 0.49/0.00
люстра 0.52/0.00	печать 0.56/0.00	подушка 0.46/0.00
карниз 0.48/0.00	подпись 0.53/0.00	коврик 0.46/0.00
крышка 0.43/0.00	пометка 0.51/0.00	трибуна 0.39/0.00
панель 0.41/0.00	бирка 0.44/0.00	простыня 0.37/0.00
козырёк 0.35/0.00	талон 0.42/0.00	одеяло 0.36/0.00
каркас 0.33/0.00	ксерокопия 0.42/0.00	четвереньки 0.36/0.00
потолок 0.33/0.00	выписка 0.42/0.00	палуба 0.33/0.00
перекрытие 0.33/0.00	прописка 0.40/0.00	каталка 0.28/0.00
пластина 0.32/0.00	бланк 0.40/0.00	носилки 0.25/0.00
Number of examples: 38 Sense2: Длинный и широкий овраг	Number of examples: 47 Sense2: Принятый образец, которому следуют без размышлений	Number of examples: 82 Sense2: Небольшой магазин
деревня 0.01/0.47	гений 0.02/0.26	гостиница 0.00/0.28
степь 0.01/0.53	стереотип 0.04/0.40	закусочный 0.00/0.30
речушка 0.00/0.29	сюжет 0.02/0.28	типография 0.00/0.33
тайга 0.00/0.29	миф 0.02/0.28	касса 0.00/0.34
перевал 0.00/0.29	пафос 0.02/0.28	фабрика 0.00/0.34
бухта 0.00/0.29	ритм 0.00/0.23	супермаркет 0.00/0.35
озеро 0.00/0.29	скучный 0.00/0.23	бар 0.00/0.37
долина 0.00/0.34	канон 0.00/0.26	кофейня 0.00/0.37
речка 0.01/0.74	стиль 0.00/0.28	магазин 0.00/0.44
роща 0.00/0.42	жанр 0.00/0.30	аптека 0.00/0.48

Table 6. Substitutes generated for randomly selected examples.

Bayes-comb	Base forward	Base backward
Нет , я по-прежнему проживаю в своей квартире , и в паспорте есть нужный **штамп**. Просто сотни жителей в моем и соседних домах уже несколько месяцев живут		
штамп, этаж, номер, дом, ключ, абзац, прочерк, подпункт, пункт, пробел	штамп, номер, пункт, документ, знак	ул, м., обл, см, пр
Он был очень высок , наклонил голову , словно подпирая плечом потолочную **балку**, посмотрел на Сьянову серьезными черными глазами .		
занавеску, перегородку, ручку, подушку, стенку, табуретку, проволоку, стену, раму, плиту	стенку, подушку, перегородку, дверь, стену	увидел, помню, оглянулся, посмотрел, встал
Иногда туманным , осенним вечером он проходил вдоль **опушки** леса , шурша омертвевшими листьями , подобрав длинную , черную рясу ;		
опушки, кромки, заснеженного, живописного, посреди, соснового, цветущего, чащи, тропического, стога	забора, берега, стен, реки, кромки	посреди, глубь, опушке, вдоль, вглубь
Однажды , когда страховой агент заполнял мой **полис** , он допустил ошибку и написал меньшее количество лошадиных сил .		
бланк, талон, формуляр, полис, анкету, талончик, бак, протокол, водительский, профиль	бланк, номер, полис, паспорт, анкету	видимо, вероятно, возможно, значит, кажется
заранее старается выговорить для себя немало льгот . Так , например , у него остается **пост** почетного президента ФХР , солидная пенсия (около 100 тысяч рублей в месяц) , оплачиваемая		
стипендия, привилегия, оклад, виза, зарплата, диплом, значок, вакансия, лицензия, выслуга	право, возможность, шанс, квартира, масса	звание, приз, должность, стипендия, пост

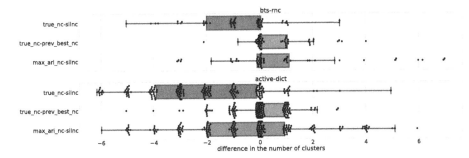

Fig. 4. Comparison of the number of clusters in our (silnc) and previous best submissions (prev_best_nc) with the true number of senses (true_nc) and the optimal number of clusters (max_ari_nc).

C   Hyperparameters

Table 7 shows the selected hyperparameters for the methods described in Sect. 3. For bts-rnc and active-dict datasets hyperparameters were selected using grid search on corresponding train sets. For wiki-wiki we used the hyperparameters from bts-rnc due to very small size of wiki-wiki train set. We selected the following hyperparameters.

1. **Add bias** (True/False). Ignoring bias in the softmax layer of the LM was proposed by [3] to improve substitutes, because adding bias results in prediction of frequency words instead of rare but relevant substitutes.
2. **Normalize output embeddings** (True/False). Similarly to ignoring bias, this may result in prediction of more relevant substitutes.

Table 7. Selected hyperparameters

Method	Add bias	Normalize output embeddings	K	Exclude Target	TF-IDF	S	L	z	β
bts-rnc									
bayes comb	False	False	200	True	False	–	–	2.0	–
pos weight avg	False	False	200	True	False	–	–	–	0.1
avg	False	False	150	True	False	–	–	–	–
base	False	False	200	True	False	–	–	–	–
sampling	False	False	200	True	True	20	15	–	–
forward	False	True	150	True	False	–	–	–	–
backward	False	False	300	True	False	–	–	–	–
active-dict									
bayes comb	False	False	200	True	True	–	–	2.0	–
pos weight avg	False	False	200	True	True	–	–	–	0.1
avg	False	False	150	True	True	–	–	–	–
base	False	False	200	True	True	–	–	–	–
sampling	False	False	200	True	True	20	10	–	–
forward	False	True	100	True	True	–	–	–	–
backward	False	True	300	True	True	–	–	–	–

3. **K** (10–400). The number of substitutes from each distribution.
4. **Exclude Target** (True/False). We want the substitutes for different senses of the target word to be non-overlapping. Thus, it may be beneficial to exclude the target word from the substitutes.
5. **TFIDF** (True/False). Applying TFIDF transformation to bag-of-words vectors of substitutes sometimes improve performance.
6. **S** (=20). The number of representatives for each example. It didn't affect the performance so we use the value from [3].
7. **L** (4–30). The number of substitutes to sample from top K predictions.
8. **z** (1.0–3.0). The parameter of Zipf distribution.
9. β (0.1–0.5). Relative length of the left or the right context after which the discounting of the corresponding LM begins.

References

1. Alagić, D., Šnajder, J., Padó, S.: Leveraging lexical substitutes for unsupervised word sense induction. In: Thirty-Second AAAI Conference on Artificial Intelligence (2018)
2. Amplayo, R.K., won Hwang, S., Song, M.: AutoSense model for word sense induction. In: AAAI (2019)
3. Amrami, A., Goldberg, Y.: Word sense induction with neural biLM and symmetric patterns. In: Proceedings of the 2018 Conference on Empirical Methods in Natural Language Processing, Brussels, Belgium, pp. 4860–4867. Association for Computational Linguistics (2018). https://www.aclweb.org/anthology/D18-1523
4. Apresjan, V.: Active dictionary of the Russian language: theory and practice. In: Meaning-Text Theory 2011, pp. 13–24 (2011)
5. Arefyev, N., Ermolaev, P., Panchenko, A.: How much does a word weigh? Weighting word embeddings for word sense induction. In: Computational Linguistics and Intellectual Technologies: Papers from the Annual International Conference "Dialogue", Moscow, Russia, pp. 68–84. RSUH (2018)
6. Bartunov, S., Kondrashkin, D., Osokin, A., Vetrov, D.: Breaking sticks and ambiguities with adaptive skip-gram. In: Proceedings of the International Conference on Artificial Intelligence and Statistics (AISTATS) (2016)
7. Baskaya, O., Sert, E., Cirik, V., Yuret, D.: AI-KU: using substitute vectors and co-occurrence modeling for word sense induction and disambiguation. In: Second Joint Conference on Lexical and Computational Semantics (* SEM), Volume 2: Proceedings of the Seventh International Workshop on Semantic Evaluation (SemEval 2013), vol. 2, pp. 300–306 (2013)
8. Devlin, J., Chang, M.W., Lee, K., Toutanova, K.: Bert: pre-training of deep bidirectional transformers for language understanding. arXiv preprint arXiv:1810.04805 (2018)
9. Hope, D., Keller, B.: UoS: a graph-based system for graded word sense induction. In: Second Joint Conference on Lexical and Computational Semantics (*SEM), Volume 2: Proceedings of the Seventh International Workshop on Semantic Evaluation (SemEval 2013), no. 1, Atlanta, Georgia, USA, pp. 689–694 (2013). http://www.aclweb.org/anthology/S13-2113

10. Howard, J., Ruder, S.: Universal language model fine-tuning for text classification. In: Proceedings of the 56th Annual Meeting of the Association for Computational Linguistics (Volume 1: Long Papers), Melbourne, Australia, pp. 328–339. Association for Computational Linguistics (2018). https://www.aclweb.org/anthology/P18-1031

11. Jurgens, D., Klapaftis, I.: Semeval-2013 task 13: word sense induction for graded and non-graded senses. In: Second Joint Conference on Lexical and Computational Semantics (* SEM), Volume 2: Proceedings of the Seventh International Workshop on Semantic Evaluation (SemEval 2013), vol. 2, pp. 290–299 (2013)

12. Kutuzov, A.: Russian word sense induction by clustering averaged word embeddings. CoRR abs/1805.02258 (2018). http://arxiv.org/abs/1805.02258

13. Lau, J.H., Cook, P., Baldwin, T.: unimelb: topic modelling-based word sense induction. In: Second Joint Conference on Lexical and Computational Semantics (*SEM): SemEval 2013), vol. 2, Atlanta, Georgia, USA, pp. 307–311 (2013). http://www.aclweb.org/anthology/S13-2051

14. Manandhar, S., Klapaftis, I.P., Dligach, D., Pradhan, S.S.: SemEval-2010 task 14: word sense induction & disambiguation. In: Proceedings of the 5th International Workshop on Semantic Evaluation, pp. 63–68. Association for Computational Linguistics (2010)

15. Melamud, O., Goldberger, J., Dagan, I.: context2vec: learning generic context embedding with bidirectional LSTM. In: Proceedings of The 20th SIGNLL Conference on Computational Natural Language Learning, pp. 51–61 (2016)

16. Panchenko, A., et al.: RUSSE'2018: a shared task on word sense induction for the Russian language. In: Computational Linguistics and Intellectual Technologies: Papers from the Annual International Conference "Dialogue", Moscow, Russia, pp. 547–564. RSUH (2018). http://www.dialog-21.ru/media/4324/panchenkoa.pdf

17. Peters, M.E., et al.: Deep contextualized word representations. In: Proceedings of the NAACL (2018)

18. Struyanskiy, O., Arefyev, N.: Neural networks with attention for word sense induction. In: Supplementary Proceedings of the Seventh International Conference on Analysis of Images, Social Networks and Texts (AIST 2018), Moscow, Russia, 5–7 July 2018, pp. 208–213 (2018). http://ceur-ws.org/Vol-2268/paper23.pdf

19. Tang, G., Müller, M., Rios, A., Sennrich, R.: Why self-attention? A targeted evaluation of neural machine translation architectures. arXiv preprint arXiv:1808.08946 (2018)

20. Vaswani, A., et al.: Attention is all you need. In: Advances in Neural Information Processing Systems, pp. 5998–6008 (2017)

21. Véronis, J.: HyperLex: lexical cartography for information retrieval. Comput. Speech Lang. **18**(3), 223–252 (2004)

22. Wang, A., Cho, K.: BERT has a mouth, and it must speak: BERT as a Markov random field language model. CoRR abs/1902.04094 (2019). http://arxiv.org/abs/1902.04094

23. Wang, J., Bansal, M., Gimpel, K., Ziebart, B.D., Yu, C.T.: A sense-topic model for word sense induction with unsupervised data enrichment. TACL **3**, 59–71 (2015)

Log-Based Reading Speed Prediction: A Case Study on *War and Peace*

Igor Tukh[1], Pavel Braslavski[1,2,3], and Kseniya Buraya[4(✉)]

[1] Higher School of Economics, Saint Petersburg, Russia
{igor-tukh,pbras}@yandex.ru
[2] Ural Federal University, Yekaterinburg, Russia
[3] JetBrains Research, Saint Petersburg, Russia
[4] ITMO University, Saint Petersburg, Russia
ks.buraya@gmail.com

Abstract. In this exploratory study, we analyze reading behavior using logs from an ebook reading app. The logs contain users' page turns along with time stamps and page sizes in characters. We focus on 17 readers of *War and Peace* by Leo Tolstoy, who read at least 80% of the novel. We aim at learning a regression model for reading speed based on shallow textual (e.g. word and sentence lengths) and contextual (e.g. time of the day and position in the book) features. Contextual features outperform textual ones and allow to predict reading speed with moderate quality. We share insights about the results and outline directions for future research. The analysis of reading behavior can be beneficial for school education, reading promotion, book recommendation, as well as for traditional creative writing and interactive fiction design.

Keywords: Reading speed · Text difficulty · Reading behavior · User modeling

1 Introduction

Reading speed is an indicator that is often used to assess reading skills in one's mother tongue or in a foreign language. Reading speed is commonly associated with cognitive abilities and personal effectiveness. There are numerous speed reading techniques aimed at increasing the speed of reading by several times. Reading speed is an indirect indicator of readers' interest in what they read. Most of the experiments and practical tests to measure reading speed are conducted in controlled settings, usually using short standardized educational or technical texts. One can assume that fiction reading is a more complex phenomenon, and reading speed of fiction works is affected by a variety of factors – besides superficial text complexity, plot, suspense, style, and reader's engagement may come into play. For example, an Argentine Canadian essayist and librarian Alberto Manguel wrote about different ways he used to read in his adolescence [11]: "First, by following breathlessly, the events and the characters

© Springer Nature Switzerland AG 2019
W. M. P. van der Aalst et al. (Eds.) AIST 2019, LNCS 11832, pp. 122–133, 2019.
https://doi.org/10.1007/978-3-030-37334-4_11

without stopping to notice the details, the quickening pace of reading sometimes hurtling the story beyond the last page... Secondly, by careful exploration, scrutinizing the text to understand its ravelled meaning, finding pleasure merely in the sound of the words or in the clues which the words did not wish to reveal, or in what I suspected was hidden deep in the story itself, something too terrible or too marvellous to be looked at". The former way of reading Manguel attributed to adventure fiction including *Odyssey*, while the latter – to works by Lewis Carrol, Dante, Kipling, and Borges. Despite many evidence like this, experimental investigation of fiction reading and analysis of reading 'in the wild' are few. This is partly because fiction reading remains mostly an intimate process avoiding external intrusion. Nowadays, ebooks reading logs can provide rich and fine-grained information about reading behavior.

In this study, we examine the reading speed of a novel by several readers over a considerable period of time. To this end, we use a log of a mobile reading application. In this exploratory study, we focus on 17 users, who read a significant portion (80% or above) of *War and Peace* by Leo Tolstoy. The log contains timestamps of users' turning pages on the mobile device and the sizes of these pages in characters. We align these log records with the text of the novel and split it into corresponding pages. For each reader we build a model that predicts their reading speed of a page based on its content or/and context and test the model on held-out data. As content features we employ various shallow features traditionally used in readability formulas such as word/sentences length and rare word ratio, whereas context features include time of the day, day of the week, and position in the book. The difficulty of reading speed prediction based on log data lies in the fact that, in contrast to laboratory settings, we can't control external conditions and know practically nothing about the readers whose behavior we study. Further obstacle is a non-perfect alignment of text and log entries. In addition, the range of speeds that we observe in the data is very wide, and corresponds not only to actual reading, but also to idle periods, skimming, and flipping-through the book.

The results show that we are able to obtain predictions of moderate quality if we narrow the range of considered speed values. Contextual features provide better predictions than textual ones. In future studies, we plan to expand the set of book titles and the population of readers, enrich the feature set, as well as address the problem of distinguishing actual reading from idling, skimming, and fast-forward browsing.

Analysis of real-life reading behavior can be of interest for school education, reading promotion, book recommendation, as well as for creative writing. In case we are able to connect reading speed to reader engagement, it can be used in interactive fiction, where a book's plot can change dynamically depending on the reader's explicit or implicit feedback.

2   Related Work

Standardized reading speed tests play an important role in educational practice and cognitive ability testing. An example of such a test is International Reading

Speed Texts (IReST) [17]. The authors develop a set of comparable texts in 17 languages. The authors use educational text intended primarily for school students. Based on the experiments involving 25 adult participants, they report average reading aloud speed and standard deviation for Russian – 986 (175) characters/min (white spaces and punctuation are not counted in). Another experiment on silent reading speed in Russian [5] involves 533 adult participants and uses a text on a philosophical topic. The authors report much higher average speed and surprisingly lower standard deviation – 1,596 (49) characters/min (it's not clear from the paper how white spaces and punctuation are treated). Eye tracking has been used for studying cognitive and psychological aspects of reading for over a century.

Early work focused mainly on the interpretation of eye movements and their relationship to cognitive processes of language processing and text comprehension. Different characteristics of the text (e.g., complexity and typographic values), specific language phenomena (different kinds of linguistic ambiguity, anaphora, semantic relationships between words, etc.), as well as individual characteristics of the reader (background knowledge, reading proficiency, etc.) were studied using eye tracking, see a comprehensive survey [15].

Kunze et al. [8] proposed using eye tracking for logging individual reading behavior analogously to fitness tracking or food logging mobile applications. The approach uses built-in cameras of the mobile devices for eye-tracking and delivers a summary of volume, speed and schedule of reading. Unfortunately, no results of such tracking studies have been reported yet.

A relevant task in the context of our study is eye tracking-based classifier into reading/skimming behavior [1]. In the skimming mode readers digest the text at speed up to two times higher than normal, at cost of missing details. Masson experimented with reading speeds beyond normal reading (200–300 words/min for English) [12]. He found out that at the speed of 375 words/min the reader can still comprehend main ideas expressed in the text, but at the speed of 600 words/min (about 3,000 chars/min) the reader is unable to follow even the main topics in the text. Such 'speed reading' can be still helpful in spotting concrete information in the text.

Nell [13] investigated reading speed in pleasure, or *ludic* reading. Within 30-min. reading sessions in the lab, he found a high degree of reading rate variability: average high/low per page speed ratio was 2.63 among 33 participants. The study also found out that readers slowed down on pages they liked most. However, since only page transitions were recorded within the experiment, it is not clear, whether this variability in due to slower 'linear' reading, or because most-liked passages were reread more than once. László and Cupchik [10] introduced two types of literary narratives – *action*- and *experience*-oriented – and measured reading speed, as well as subjective time perception during the reading of both. They showed that *action* fragments were read faster than *experience* fragments. The readers in the experiment associated *action* stories with lower reading difficulty and lower predictability. Brouwer et al. [4] used such physiological signals as EEG, ECG, skin conductance, heart rate, and respiration of readers to be able to distinguish emotionally intense vs. neutral sections of a novel.

Readability scores has been widely used to evaluate educational and instructional materials since 1920s. Traditional readability formulas are results of regression analysis that sees experimentally obtained text complexity as dependent variable and several computable text features as independent variables. Text complexity is measured for instance by reading time (normalized by the individual reading proficiency), post-reading questionnaires assessing text comprehension, or cloze tests. Text features used in readability formulas include various word lists (such as "easy", "hard", "abstract", "most frequent", etc. words), word and sentence length, number of prepositional phrases, etc.; see [7] for a comprehensive survey. In our study, we rely on these simple text difficulty signals. A more recent study [14] incorporates lexical, syntactic, and discourse features to predict text readability. We will consider more complex features for reading speed prediction in our future research.

There are some studies on news reading behavior on the Web. Constantinides et al. [6] use reading speed as a feature for news personalization. They distinguish three styles of news reading: detailed reading (up to 230 words/min), normal reading and skimming (230–700 words/min), and scanning (above 700 words/min). Lagun and Lalmas [9] analyze online news reading on sub-document level, which makes the study close to ours. The authors utilize viewport data and model user engagement with the content at sub-document level based on text features, as well as mine different news article reading patterns.

3   Data

In this study we use reading logs from Bookmate[1], a Russian mobile reading app. Upon installing the application, users get instant access to a free collection of several thousands of titles (mostly Russian classical novels); further they can choose from two subscription levels. Standard subscription grants a user access to the entire Russian book collection, excluding new arrivals, bestsellers, and business books. Premium subscription provides unlimited access to the entire collection. App logs used in the study correspond to almost 10 months – from January to October 2015. The data includes information about the users, books, and readings sessions. Detailed description of the dataset and reading behavior characteristics derived from the log can be found in [3].

Book data contains book ID, author, title and length in character.

User information includes their subscription type (paying/non-paying); some users indicated their gender and year of birth.

The main contents of the dataset are page turns.

Every record in the log contains the user and book IDs, the time stamp of a page turn, and the character range that corresponds to the turned page size, as well as additional auxiliary information. The size of the page in characters depends mainly on the screen size of the user's device.

---

[1] https://www.bookmate.com.

In this exploratory study we opted for focusing on a single book – *War and Peace* by Leo Tolstoy (*W&P*). The novel is considered as one of the Tolstoy's masterpieces and depicts life of several Russian noble families in the time of Napoleonic wars. The four-volume 1,300-page novel was first published in 1869. *War and Peace* is included in the Russian high-school curriculum; there are several film and television adaptations of the novel. It's interesting to note in the context of our study that there is an anecdotal evidence that female and male high-school students read the novel in different ways: girls skip battle scenes, while boys – balls and long dialogues.[2] *W&P* is also actively researched within the digital humanities, see for example [2].

For this study, we considered only readers, who have read at least 80% of the book, with earliest reading sessions within first 10% of the book.

We also removed users with the coverage above 120% and ended up with 17 readers.[3]

Then, we cleaned the data as follows.

We removed duplicate sessions, i.e. nearly simultaneous (up to five seconds apart from each other) sessions spanning the same text interval.

Readers don't read the novel linearly – we can observe backward and forward leaps in the data. Figure 1 illustrates reading sequence by one of the readers in the study: spikes below/above diagonal reflect backward/forward movements, respectively. Further, we removed distant 'backward jumps' within the book. We assume these records reflect navigational browsing, device orientation changing, waking up device, and similar behavior, thus contain no useful information in the context of the current study. We also removed sessions corresponding to the fragments shorter than 20 words (either headings or last lines of a chapter) that don't allow to calculate reliable content statistics.

On the next step, we calculate start/end positions for each page based on log data and mapped the intervals to the ebook file. This alignment can't produce perfect mapping with what the users saw on their screens and is a source of additional noise in the data. Based on two subsequent timestamps we calculate time the reader spent reading this page, except for the first page in the sequence. We set the reading time of the first page in the sequence to the session's average. Essentially, our final data consists of character ranges from *W&P* and such time intervals.

Table 1 describes main characteristics of the *W&P* readership in the study. It can be seen that the users vary greatly in terms of books accessed and time spent at reading in the app. Inspection of titles read by users with few books (e.g. #3, 7, 13, 15–17) suggests that these are high-school students reading free books from the list of assigned readings. This fact may influence results, since 'obligatory' reading behavior may differ from spontaneous leisure reading. Figure 2 shows overall distribution of reading speed for *W&P*, while Table 1 cites individual

---

[2] We were unable to test this hypothesis due to incomplete data.

[3] To calculate coverage we summed up all character ranges in the log entries for a particular reader. Coverage above 100% occurs, when the same text spans are read or just flipped through several times.

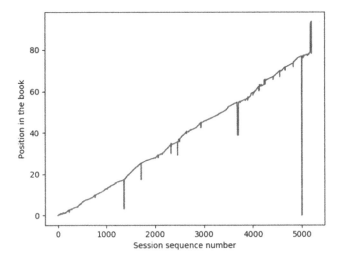

Fig. 1. Sequence of reading positions for reader #1

averaged speed values. It can be seen from the table that *unfiltered* average reading speed calculated upon log data is often significantly higher than normal speed (1,000–1,500 characters/min, see above). It suggests that user often skim or just flip through the book.

To visualize reading behavior of the 17 readers in the study, we produced a heatmap of reading speeds, see Fig. 3. To this end we divided the entire text of the *W&P* novel into 400 fragments of equal size (each fragment corresponds to about three pages of a printed book; we didn't account for volume and chapter division). Further, we normalized individual reading speeds by mapping them to [0..1] interval based on individual readers' min/max values and calculating new values for the fragments. Finally, we tried to rank the readers in such a way that readers with similar behavior are placed close to each other. As the figure shows, there is no common pattern in reading speeds. However, we can see that there are different types of readers: some of them read the novel with fewer speed variations, the other switch between slower and more accelerated reading.

Currently we have no reliable method to distinguish between idle periods (periods of user's inactivity followed by a page turn that appear as slow reading), normal reading, skimming, and fast-forward flipping. We set low/high thresholds for reading speed values 800/3,000 characters/min based on data reported in the literature and discard records outside this range. We assume that this span includes normal reading and skimming behavior. We investigate also the impact of threshold values on overall prediction quality (see below).

Table 1. 17 readers in the study: $ – subscription type; *gender* as indicated in the user's profile; *#books* – total different books accessed and *total time* in hours spent on the service during the log's period; unfiltered *average reading speed* for *War and Peace* and estimated *time to complete* the novel in hours when reading at this pace; total *#sessions* in the log and *#filtered* sessions; *?* indicates missing data.

UserID	$	gender	#books	total time	avg. speed	time to complete	#sessions	#filtered
1	premium	m	48	124.8	3,040.92	19.85	5,202	3,387
2	premium	m	70	171.4	2,781.33	21.70	2,303	1,425
3	?	?	5	65.1	2,234.29	27.01	5,158	2,525
4	free	m	36	190.6	2,415.27	24.99	3,261	1,901
5	free	m	28	89.2	3,824.17	15.78	1,616	937
6	standard	?	26	213.9	1,272.85	47.42	6,171	4,490
7	free	?	4	68.3	3,619.21	16.68	2,936	1,757
8	free	?	19	100.4	1,224.99	49.27	7,623	5,601
9	standard	m	17	78.6	1,282.31	47.07	6,862	4,484
10	free	?	17	70.7	4,091.45	14.75	2,324	1,247
11	free	f	10	64.4	7,948.70	7.59	1,784	497
12	premium	?	17	67.3	2,300.29	26.24	4,736	3,620
13	free	?	3	63.4	3,965.01	15.22	2,576	1,223
14	free	?	10	44.0	1,895.32	31.84	5,004	3,891
15	free	m	4	38.2	3,126.40	19.30	2,024	1,681
16	free	f	6	52.8	2,222.29	27.16	3,837	3,054
17	?	?	4	30.0	2,071.25	29.14	4,105	3,219

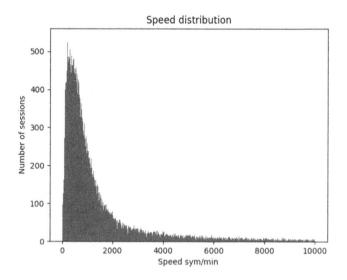

Fig. 2. Distribution of *W&P* reading speed

4   Reading Speed Prediction

So far, we have text fragments (pages), their lengths in characters, and time the app user spent reading this fragment (two latter parameters deliver reading speed). We represent each text fragment as a vector of features, split the data corresponding to each reader into train and test subsets, learn a regression model, and evaluate it.

We implement two groups of features:

1. Text features:
   - *#Words* – number of words on the page;
   - *#Sentences* – number of sentences on the page, incomplete sentences are still counted in;
   - *Average word length in characters*;
   - *Average sentence length in characters*;
   - *#Rare words* based on unigram frequencies in the Russian National Corpus[4]
   - *#Finite verbs* and *#Nouns* based on *mystem* POS tagger.[5]
2. Context features:
   - *Hour of the day*: 0..23;
   - *Is_weekend*: 0 if weekday; 1 otherwise;
   - *Position* of reading in percentage of the whole book.

We perform reading speed prediction based on text and context features separately, as well as on their combination.

We make predictions for each user independently. The data is split into train/test sets in the ratio 80/20 using two approaches:

- *Ordered*: sessions for a user are split in chronological order;
- *Random*: sessions for a user are split randomly.

We experiment with several regression methods implemented in *scikit-learn* library[6] and opted for Lasso [16] that delivered best results. We use Mean Absolute Error (MAE) as loss function. We also investigate the impact of different upper and lower speed thresholds on overall performance. We ran experiments for the lower threshold in the range from 100 to 800 characters/min and for the upper threshold in the range from 2,500 to 6,000 characters/min.

5   Results and Discussion

Table 2 summarizes the results of reading speed predictions for individual readers using different sets of features and data splits. As can be seen from the table, the quality of prediction differs significantly from reader to reader. One has to

---

[4] http://ruscorpora.ru/corpora-freq.html.
[5] https://tech.yandex.ru/mystem/.
[6] https://scikit-learn.org/.

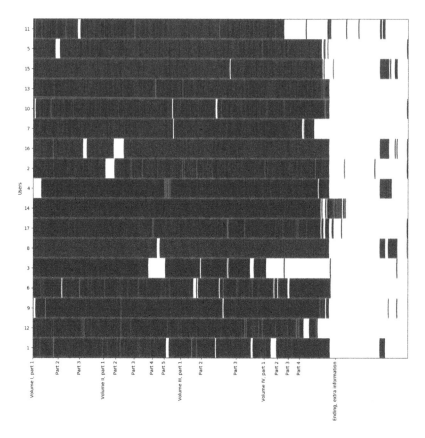

Fig. 3. Normalized reading speeds by 17 readers across *W&P*. The book is divided into 400 fragments of equal size; individual speed values are mapped to [0..1] interval; lower/higher speeds correspond to cold/hot parts of the spectrum, respectively; blanks correspond to missing data. Vertical ranking is aimed at placing readers with similar behavior close to each other.

keep in mind that speed distributions for individual readers also vary greatly (see Table 1, Fig. 3). In most cases, speed predictions based on simple contextual features are more accurate. In some cases, combination of textual and contextual features reduces the error (cf. readers #3, 15–17). However, these results are only marginally better than predictions based solely on contextual features. In the case when textual features provide a better prediction compared to contextual ones (#10), the gain is neglectable. The impact of different data splits is mixed: for some readers, the error is lower on the ordered split, for the others – on the random split. This probably reflects different reading styles: for some readers, the behavior is consistent throughout the book, for the others it changes from the beginning of the novel to the end. It should be noted that in the case of contextual/combined features and random split the position within the book is still taken into account through the corresponding feature.

Table 2. Mean absolute error (MAE) for reading speed predictions in characters/min for different feature sets (*text* only, *context* only, and *combined* text & context) and train/test splits (*ordered/random*). Low/high speed thresholds are fixed at 800/3000 char/min. Best values for each reader are in **bold**; best results based on *context* features are underlined.

UserID	Text		Context		Combined	
	Ordered	*Random*	*Ordered*	*Random*	*Ordered*	*Random*
1	487.08	480.36	**442.92**	451.41	491.76	443.42
2	245.73	213.71	**159.68**	173.45	286.81	245.47
3	464.85	479.98	452.41	519.26	**445.17**	497.89
4	449.22	742.82	**272.28**	273.31	441.30	546.80
5	183.55	190.31	**170.41**	238.93	230.28	195.72
6	213.09	247.29	**160.41**	169.20	242.75	301.02
7	523.77	467.71	556.72	**438.90**	477.06	557.49
8	319.76	337.85	242.19	**223.53**	304.79	281.13
9	553.60	511.86	**253.20**	288.58	550.86	572.51
10	**365.00**	741.51	394.38	365.47	460.32	513.33
11	1262.14	1347.58	**499.30**	524.66	1472.18	1894.76
12	581.26	809.61	374.96	**330.70**	474.70	756.10
13	589.69	412.75	**351.20**	352.27	613.75	859.96
14	664.30	307.85	297.49	**293.65**	837.69	295.41
15	245.60	220.81	240.76	223.66	230.00	**218.22**
16	395.09	291.42	266.35	261.30	265.07	**255.40**
17	312.45	318.45	424.45	303.71	438.03	**303.06**

We investigated the effect of cut-off thresholds on speed prediction; the results for the lower and upper thresholds are presented in Fig. 4. For the majority of readers, varying lower threshold from 100 to 800 characters/min has almost no effect. We can assume that their reading behavior doesn't change within this range. For four readers (#4, 12, 13, 17) the error drops with increasing the lower threshold. Supposedly, by increasing cut-off value we eliminate idle sessions that are different from actual reading. Results for the reader #11 demonstrate opposite behavior. These results suggest these differences must be taken into account when modeling reading behavior. We plan to tackle this problem in our future research.

6   Conclusions and Future Work

We conducted an exploratory study aimed at predicting reading speed of *War and Peace* fragments based on the log of an ebook application. We used two

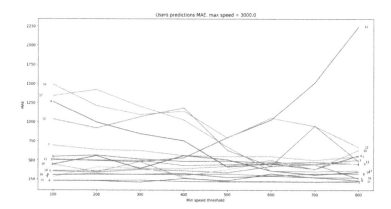

Fig. 4. Dependency of MAE on the lower speed thresholds. UserIDs are at the beginning/end of the curves. All values are for *combined* feature set and *ordered* train/test split.

sets of features – textual and contextual. Somewhat surprisingly, simple contextual features significantly outperformed predictions based on textual features traditionally used in readability formulas.

In the future, we plan to use a larger collection of books and a larger population of readers. We will perform a more thorough data cleansing and segmentation of the user base by their behavior. We will address a more accurate interpretation of the readers' behavior/actions based on the log data. We also plan to significantly expand the set of textual and contextual features.

Despite modest results we have obtained so far, we believe that the study of reading behavior and users' interactions with text documents enriches representations based solely on content analysis and can be beneficial for various domains. The results of the research may be of interest for school education, reading promotion, book recommendations, and creative writing.

Acknowledgements. We thank *Bookmate* for granting access to the dataset.

References

1. Biedert, R., Hees, J., Dengel, A., Buscher, G.: A robust realtime reading-skimming classifier. In: Proceedings of the Symposium on Eye Tracking Research and Applications, pp. 123–130 (2012)
2. Bonch-Osmolovskaya, A., Skorinkin, D.: Text mining War and Peace: automatic extraction of character traits from literary pieces. Digit. Sch. Hum. **32**(suppl_1), i17–i24 (2016)
3. Braslavski, P., Likhosherstov, V., Petras, V., Gäde, M.: Large-scale log analysis of digital reading. Proc. Assoc. Inf. Sci. Technol. **53**, 1–10 (2016)
4. Brouwer, A.-M., Hogervorst, M., Reuderink, B., van der Werf, Y., van Erp, J.: Physiological signals distinguish between reading emotional and non-emotional sections in a novel. Brain Comput. Interfaces **2**(2–3), 76–89 (2015)

5. Chmykhova, E., Davydov, D., Lavrova, T.: Experimental study of factors of speed reading (Eksperimental'noe issledovanie faktorov skorosti chteniya). Russ. Psyhologiya Obucheniya **9**, 26–36 (2014)
6. Constantinides, M., Dowell, J., Johnson, D., Malacria, S.: Exploring mobile news reading interactions for news app personalisation. In: Proceedings of the 17th International Conference on Human-Computer Interaction with Mobile Devices and Services, pp. 457–462 (2015)
7. DuBay, W.H.: The principles of readability (2004). http://www.impact-information.com/impactinfo/readability02.pdf
8. Kunze, K., et al.: Quantifying reading habits: counting how many words you read. In: Proceedings of the 2015 ACM International Joint Conference on Pervasive and Ubiquitous Computing, pp. 87–96 (2015)
9. Lagun, D., Lalmas, M.: Understanding user attention and engagement in online news reading. In: Proceedings of the Ninth ACM International Conference on Web Search and Data Mining, pp. 113–122 (2016)
10. László, J., Cupchik, G.C.: The role of affective processes in reading time and time experience during literary reception. Empir. Stud. Arts **13**(1), 25–37 (1995)
11. Manguel, A.: A History of Reading. Knopf Canada, New York (1996)
12. Masson, M.E.J.: Cognitive processes in skimming stories. J. Exp. Psychol. Learn. Mem. Cogn. **8**(5), 400–417 (1982)
13. Nell, V.: The psychology of reading for pleasure: needs and gratifications. Read. Res. Q. **23**(1), 6–50 (1988)
14. Pitler, E., Nenkova, A.: Revisiting readability: a unified framework for predicting text quality. In: Proceedings of the Conference on Empirical Methods in Natural Language Processing, pp. 186–195 (2008)
15. Rayner, K.: Eye movements in reading and information processing: 20 years of research. Psychol. Bull. **124**(3), 372 (1998)
16. Tibshirani, R.: Regression shrinkage and selection via the lasso. J. R. Stat. Soc. Ser. B (Methodol.) **58**(1), 267–288 (1996)
17. Trauzettel-Klosinski, S., Dietz, K.: Standardized assessment of reading performance: the new international reading speed texts IReST. Investig. Ophthalmol. Vis. Sci. **53**(9), 5452–5461 (2012)

Cross-Lingual Argumentation Mining
for Russian Texts

Irina Fishcheva (ID) and Evgeny Kotelnikov$^{(\boxtimes)}$ (ID)

Vyatka State University, Kirov, Russia
`fishchevain@gmail.com`, `kotelnikov.ev@gmail.com`

Abstract. Argumentation mining refers to automatic extraction of arguments and their relations from texts. This field has been evolving rapidly in recent years, but there is almost no research for the Russian language. The present study is an attempt to overcome this gap. Firstly, we create the first argument-annotated corpus of Russian based on Argumentative Microtext Corpus and make it publicly available. Secondly, we study the importance of various feature types. Contextual and lexical features turn out to be the most significant. Thirdly, we evaluate the performance of various classifiers for argumentation mining. Bagging and XGBoost classifiers give the best results. Fourthly, we assess the possibility of using several machine translation systems (Google Translate, Yandex.Translate and Promt) for automatic creating of argument-annotated corpora. Google Translate appears to be the best system to reach this goal.

Keywords: Argumentation mining · Machine translation · Parallel corpora · Feature selection

1 Introduction

A branch of computational linguistics called *argumentation mining* has evolved immensely over the past few years. Argumentation mining aims at automatically extracting structured arguments from unstructured textual documents [14]. There has been a special Workshop on Argument Mining (ArgMining)[1] since 2014. Reviews of recent research in this field can be found in [9, 14, 16, 27]. Besides computational linguistics, argumentation has been the subject of artificial intelligence analysis for a long time in which modelling arguments and the inference based on them are emphasized [5].

Van Eemeren and Grootendorst give the following definition: "Argument is a verbal, social, and rational activity aimed at convincing a reasonable critic of the acceptability of a standard by putting forward a constellation of propositions justifying or refuting the proposition expressed in the standard." [29, p. 1]. With this definition in mind, the following tasks are considered in argumentation mining [27, p. 6]:

---

[1] https://argmining19.webis.de.

W. M. P. van der Aalst et al. (Eds.) AIST 2019, LNCS 11832, pp. 134–144, 2019.
https://doi.org/10.1007/978-3-030-37334-4_12

1. detection of argumentative text;
2. selection of the base components of argumentative structure – argumentative discourse units (ADU);
3. determining the main claim reflecting the standpoint;
4. assignment of ADUs' roles – supporting ("pro") or rejecting ("contra");
5. identifying relations between ADUs (whether ADU are connected);
6. construction of full argumentative structure;
7. defining type and quality of argumentation.

In argumentation mining, the primary focus is on tasks 2–6.

Works on argumentation mining (and in most other areas of computational linguistics as well) are based on annotated corpora (legal documents [22], student essays [17, 26], Wikipedia articles [2], political discourse and debate [13], news articles [24], discussion fora [9], scientific writing [12], and others). So far, there have been quite a few, usually small corpora annotated using some annotation scheme [27, p. 51]. For example, AIFdb[2] has collected more than 170 corpora of various quality and size, about 14,000 texts in total.

However, high-quality annotated corpora mostly include only English texts. For other languages, such resources are considerably poor, to the best of our knowledge, there have been no annotated corpora for argumentation mining for the Russian language in particular. There appear to be two main ways to obtain annotated corpora of Russian. The first one is to create a corpus with plenty of arguments and subsequently annotate it[3]. The second way is to translate the existing annotated corpus (usually English) into Russian with the given annotation. However, both methods are considered to be rather time-consuming and expensive. For instance, Eger et al. [8] indicated that the translation of persuasive essays corpus (7, 141 sentences) [26] from English into German with the given argumentation annotation took 270 h and cost $3,000. In the present study, along with traditional human translation, machine translation is used.

The contribution of the present study is as follows: (1) we create the first argument-annotated corpus of the Russian language and make it publicly available; (2) we study the importance of various feature types for argumentation mining; (3) we study the performance of various classifiers for argumentation mining; (4) we assess the possibility of using different machine translation systems for automatic creating of argument-annotated corpora.

2 Previous Work

There is a non-negligible body of work on cross-lingual argumentation mining.

Aker and Zhang [4] studied English argument-annotated Wikipedia articles [2] using the corresponding Mandarin Chinese Wikipedia articles. The arguments from the English texts were manually mapped to the Chinese texts. In addition, machine

---

[2] http://corpora.aifdb.org.

[3] There exist special tools for manual text annotation, such as GraPAT [25], WebAnno [32] and Brat [28].

translation of the English texts into Chinese based on MOSES [10] and Google Translate was used, and the quality of translation was evaluated (24% for MOSES-based translation and 49% for Google-based translation). In the present study, in addition to translation (both machine and human), the automatic "pro" and "contra" classification of the sentences was carried out.

Eger et al. [8] proposed to use two cross-lingual techniques – direct transfer [15] and projection [31] – for cross-lingual token-level argumentation mining. They were used for Argumentative Microtext Corpus [20] and Persuasive Essays Corpus [26]. Annotation projection gave the best results (F1 = 0.6757 for Persuasive Essays Corpus). The present study is focused on sentence-level argumentation mining and machine translation. In addition, the importance of various feature types is studied.

Sliwa et al. [24] used parallel corpus of news articles in English, Turkish, Greek, Albanian, Croatian, Serbian, Macedonian, Bulgarian, Romanian and Arabic. First, the English version was annotated as argumentative or non-argumentative by voting of eight classifiers [3]. Afterwards the annotation was automatically transferred to corpora of other languages. In this study, the gold annotation of "pro" and "contra" sentences is automatically transferred to the translated texts. What is more, we study various feature types and classifiers.

3   Original and Translated Versions of Argumentative Corpus

In the present study, Argumentative Microtext Corpus[4] (ArgMicro) is used (Table 1). The corpus consists of two parts. Part 1 (2015) [20] contains 112 short texts in English and in German (from 3 to 10 sentences) on various topics (raising retirement age, health insurance, school uniform, etc.). Part 2 (2018) [23] contains 171 texts in English only (from 2 to 10 sentences) on various topics as well (poaching, cell phones and social media addiction, recycling, etc.). The texts are annotated according to the annotation scheme proposed by Peldszus and Stede [21]: each sentence is annotated as supporting ("pro" – 1,256) or rebutting ("contra" – 252) the main claim expressed in the given text. The relations between the sentences are established: "support" (730), "rebut" (245), "undercut" (140), "additional" (78) and "example" (32).

Table 1. Characteristics of ArgMicro corpus.

	Texts	ADU_pro	ADU_contra	ADU_total	ADU/text
Part 1	112	451	125	576	5.14
Part 2	171	805	127	932	5.45
Total	**283**	**1,256**	**252**	**1,508**	**5.33**

---

[4] http://angcl.ling.uni-potsdam.de/resources/argmicro.html.

At the initial stage of the present study, ArgMicro corpus was translated into Russian by a professional translator and using three machine translation systems: Google Translate[5], Yandex.Translate[6] and Promt[7]. We assessed the quality of machine translation in comparison with human translation using BLEU [18] and NIST [7] metrics from NLTK [6] (Table 2).

Table 2. BLEU and NIST metrics for machine translations of ArgMicro.

Machine translation	BLEU	NIST (n = 2)
Google Translate	0.2902	6.5004
Yandex.Translate	0.2177	5.6184
Promt	0.1803	5.2442

Table 2 shows that Google Translate is more closely to human translation, Promt is considered to be the most remote from it.

Therefore, we received four Russian versions of ArgMicro corpus – one using human translation and three using machine translation. These versions are publicly available[8].

4   Features and Classifiers

The morphological analysis was previously carried out with the help of Mystem[9] for all Russian versions of the corpus.

We studied the importance of various feature types:

- positional features – position of the sentence in the text: whether it is the first or the last one, and in which third of the text it is located;
- lexical features – discourse markers ("consequently", "I think", "eventually", etc.) and modal words ("need", "maybe", "necessarily", etc.), including negations, 255 features in total;
- punctuation features – comma, colon, semicolon, question and exclamation marks;
- morphosyntactic features – N-grams based on parts of speech (nouns, pronouns, verbs, adjectives and adverbs), N = {2, 3, 4}, and grammatical features of verbs: tense, mood, person;
- contextual features – lexical, punctuation and morphosyntactic features in the previous (contextual_prev) and the following (contextual_next) sentences, and in both of them (contextual);

---

[5] https://translate.google.com.

[6] https://translate.yandex.ru.

[7] https://www.translate.ru.

[8] https://github.com/kotelnikov-ev/ArgMicro_Russian.

[9] https://tech.yandex.ru/mystem.

- TF.IDF features – the model based on 4,687 unique words of the corpus was constructed;
- word2vec features – a 300-dimensional model trained on Russian Wikipedia and Russian National Corpus containing 384,764 words[10] was used [11]. The sentence vector was determined by averaging of word vectors.

TF.IDF and word2vec features turned out to be ineffective during the experiments (this finding correlates with Aker et al. [3] and Afantenos et al. [1]) and hereinafter only binary models based on positional, lexical, punctuation, morphosyntactic and contextual features are used.

We solved the task of sentence classification into two classes – "pro" and "contra." The task is highly imbalanced (see Table 1): "pro" – 1,256 sentences (83.3%), "contra" – 252 sentences (16.7%), thus, macro F1-measure was used to evaluate the performance.

The following models were used as classifiers (with the range of selected parameters):

- Bagging classifier: number of trees = [50, 75, 100, 150, 200, 300, 500];
- Random Forest classifier (RFC): number of trees = [50, 75, 100, 150, 300, 500], maximum depth of the tree = [2, 8, 12, 16, 20, 30];
- Complement Naive Bayes (CNB): smoothing alpha = [0, 0.2, 0.5, 1];
- Support Vector Machines (SVM): kernels = [linear, rbf, poly];
- AdaBoost: number of trees = [50, 75, 100, 150, 200, 300, 500];
- XGBoost: number of trees = [50, 75, 100, 150, 200]; maximum depth of the tree = [2, 6, 8, 10, 12, 16, 20, 30].

The scikit-learn implementation [19] was applied for all the classifiers except XGBoost. For XGBoost classifier the separated library was used[11].

To make an evaluation, 5-fold cross-validation was carried out on the whole ArgMicro corpus. 5-fold nested cross-validation was performed for selection of optimal classifier parameters for each fold.

5 Results and Discussion

5.1 Human Translation

Table 3 presents classification results for various feature groups and various classifiers.

Bagging classifier (F1 = 0.699) gives the best results, XGBoost (F1 = 0.692) ranks second. SVM (F1 = 0.685) takes third place, AdaBoost (F1 = 0.675) ranks fourth. On the whole, these four classifiers give fairly close results (the margin between the first and the fourth classifiers is 0.024). Random Forest Classifier and Complement Naive Bayes appear to have given lower results (F1 = 0.616 and F1 = 0.590 respectively).

---

[10] https://rusvectores.org/ru/models.

[11] https://xgboost.readthedocs.io.

We checked the statistical significance of differences of classifiers results based on one-tailed Wilcoxon signed-rank test at a significance level of 0.05. It turned that:

- Bagging = XGBoost, Bagging > (SVM, AdaBoost, Random Forest Classifier, Complement Naive Bayes);
- XGBoost = SVM, XGBoost > (AdaBoost, Random Forest Classifier, Complement Naive Bayes);
- SVM = AdaBoost, SVM > (Random Forest Classifier, Complement Naive Bayes);
- AdaBoost > (Random Forest Classifier, Complement Naive Bayes);
- Random Forest Classifier = Complement Naive Bayes.

Table 3. Results for "pro/con" sentence classification – human translation (macro F1-measure).

Feature group	Number of features	Bagging	RFC	CNB	SVM	AdaBoost	XG-Boost	Average
positional	5	0.454	0.454	0.144	0.454	0.454	0.454	0.402
lexical	255	0.618	0.610	0.491	0.606	0.605	0.606	0.589
punctuation	5	0.454	0.454	0.165	0.454	0.454	0.454	0.406
morphosyntactic	783	0.534	0.527	0.513	0.502	0.502	0.509	0.515
contextual_prev	1,043	0.481	0.479	0.458	0.507	0.511	0.475	0.485
contextual_next	1,043	0.655	**0.616**	0.499	0.658	0.638	0.631	0.616
contextual	2,086	0.657	0.556	0.556	0.613	0.639	0.631	0.609
all without positional	3,129	**0.699**	0.524	0.570	0.652	0.665	0.668	0.630
all without lexical	2,879	0.640	0.541	0.544	0.611	0.620	0.678	0.606
all without punctuation	3,129	0.675	0.548	0.580	0.652	0.668	0.690	0.636
all without morphosyntactic	2,351	0.657	0.563	**0.590**	0.635	0.648	0.673	0.628
all without contextual_prev	2,091	0.663	0.598	0.587	**0.685**	**0.675**	0.676	**0.647**
all without contextual_next	2,091	0.586	0.493	0.548	0.603	0.560	0.571	0.560
all without contextual	1,048	0.594	0.564	0.541	0.577	0.582	0.595	0.576
all	3,134	0.664	0.553	0.581	0.656	0.669	**0.692**	0.636

The best performance among separate feature groups is ensured by the contextual features, the contextual features in the next sentence in particular (F1 = 0.616 for all the classifiers on average). The lexical features generally give rather a good result (F1 = 0.589). The remaining features (positional, punctuation, morphosyntactic) separately do not allow us to successfully classify the sentences.

Moreover, we studied the influence of each feature group on the performance of the classification. For this purpose, we excluded each feature group from the overall feature set ("all without..."). The lowest performance (on average) is given and, therefore, the most importance is attached again if the contextual features in the next sentence are excluded (F1 is 0.076 lower in contrast to the overall feature set) and the contextual features on the whole (F1 is 0.06 lower). The lexical features constitute the second most important feature group, F1 is 0.03 lower if this feature group is excluded. The remaining feature groups have almost no influence on the classification performance.

The overall feature set ("all") ensures almost the maximum classification performance on average (F1 = 0.636), but if the contextual features in the previous sentence are excluded, F1 is slightly higher (0.012).

The analysis of the impact of feature selection on the classifiers reveals the following. Bagging classifier achieves the best result (F1 = 0.699) using the feature set without positional features. Compared to the overall feature set, there is an increase by 0.035. On the contrary, XGBoost gives the maximum performance using the overall feature set and it is 0.024 lower without positional features. SVM and AdaBoost reach the maximum F1-measure using the overall feature set without contextual features in the previous sentence. Random Forest classifier gives the maximum F1-measure value using the contextual features only in the next sentence, and Complement Naive Bayes gives a similar result using the overall feature set without morphosyntactic features.

We evaluated the importance of some features using Gini Index, which is well established in comparative studies of feature selection methods [30]. The ten most important features were as follows:

- "но" ("but") – contextual lexical feature for the next sentence;
- "хотя" ("though") – lexical feature;
- "однако" ("however") – contextual lexical feature for the next sentence;
- "некоторый" ("some") – lexical feature;
- "утверждать" ("assert") – lexical feature;
- "конечно" ("of course") – lexical feature;
- "поэтому" ("therefore") – contextual lexical feature for the previous sentence;
- "пора" ("it is time") – contextual lexical feature for the next sentence;
- "ADJ" – contextual morphosyntactic feature for the next sentence.

5.2   Machine Translation

After studying the classifier performance and the importance of features for human translation, we chose the best feature groups from Table 3 ("contextual_next", "all without positional", "all without morphosyntactic", "all without contextual_prev", "all") and trained these classifiers on ArgMicro corpus versions obtained by using machine translation (Table 4). The same splitting in the cross-validation procedure was used, thus, the results are comparable to Table 3. The classifiers were built on the training parts of machine translation corpora and evaluated on the test parts of human translation corpora.

The ranking of classifiers according to their performance gives the same results obtained in the human translation analysis. The best result is also obtained using bagging classifier with the same feature set ("all without positional"). It is important to note that this result exceeds the result obtained from human translation corpus: F1 = 0.714 vs. F1 = 0.699. XGBoost classifier ranks second again with F1 = 0.702, but another feature set is used – "all without positional" – instead of the overall feature set in human translation.

SVM and AdaBoost classifiers give almost similar results, F1 = 0.678 and F1 = 0.676 respectively. Random Forest classifier and Complement Naive Bayes rank last again (F1 = 0.624 and F1 = 0.606).

We also checked the statistical significance of results of classifiers based on one-tailed Wilcoxon signed-rank test at a significance level of 0.05: Bagging > XGBoost > (SVM, AdaBoost) > (Random Forest Classifier, Complement Naive Bayes).

Table 4. Results for "pro/con" sentence classification – machine translation.

Feature group	Number of features	Bagging	RFC	CNB	SVM	AdaBoost	XG-Boost	Average
Google Translate								0.643
contextual_next	1,043	0.662	0.618	0.493	**0.678**	0.655	0.700	0.634
all without positional	3,129	**0.714**	0.528	0.575	0.666	0.676	0.695	0.642
all without morphosyntactic	2,351	0.698	0.586	0.588	0.659	0.667	0.690	0.648
all without contextual_prev	2,091	0.696	0.597	0.598	0.667	0.672	0.681	**0.652**
all	3,134	0.694	0.549	0.575	0.666	0.665	0.680	0.638
Yandex.Translate								0.633
contextual_next	1,043	0.651	0.607	0.509	0.674	0.663	0.630	0.622
all without positional	3,129	0.709	0.553	0.584	0.621	0.650	0.666	0.630
all without morphosyntactic	2,351	0.680	0.572	0.588	0.660	0.676	0.663	0.640
all without contextual_prev	2,091	0.694	0.584	**0.606**	0.655	**0.676**	0.680	0.649
all	3,134	0.692	0.551	0.580	0.626	0.631	0.665	0.624
Promt								0.629
contextual_next	1,043	0.656	**0.624**	0.481	0.674	0.596	0.668	0.616
all without positional	3,129	0.709	0.537	0.562	0.632	0.645	**0.702**	0.631
all without morphosyntactic	2,351	0.698	0.569	0.570	0.642	0.645	0.686	0.635
all without contextual_prev	2,091	0.691	0.580	0.579	0.638	0.671	0.676	0.639
all	3,134	0.687	0.550	0.558	0.623	0.638	0.690	0.624

When comparing machine translation systems, Google Translate turns out to be the best both in terms of the maximum classification performance (bagging classifier, F1 = 0.714) and in terms of average classification performance for all classifiers and feature sets (F1 = 0.643).

Yandex.Translate and Promt give the same maximum classification performance (bagging classifier, F1 = 0.709), but in terms of average classification performance Yandex.Translate gives slightly better results (F1 = 0.633 vs. F1 = 0.629). Interestingly, the results of the ranking of machine translation systems in terms of average classification performance (Table 4) correlate with the results of the quality assessment of machine translation (Table 2).

6 Conclusion

Thus, we created the first argument-annotated corpus of the Russian language containing 283 texts and 1,508 sentences based on human and machine translations of the two parts of the well-known Argumentative Microtext Corpus (ArgMicro) and made it publicly available.

We studied the importance of various feature types for argumentation mining. The conclusion to be drawn is that contextual features in the next sentence and lexical features are of high importance. The features of TF.IDF and word2vec vector models are not likely to be used since they degrade the classification performance. These findings are consistent with those of other studies. As for positional, punctuation, morphosyntactic and contextual_prev feature groups, the results obtained are contradictory: in some cases, they improve the classification performance, in other cases, they decrease it.

The effectiveness of various classifiers for argumentation mining was analyzed. Ensembles and SVM allow us to achieve high classification performance.

We can compare our results to results from other papers. For example, Skeppstedt et al. [23] classified sentences from the ArgMicro corpus as "pro" and "contra" with the help of minimum spanning tree decoder. They got the F1-measure from 0.695 to 0.782 for different scenarios of using part 1 and part 2 of ArgMicro corpus as train and test sets. Our best value of F1-measure is 0.714, thus we can conclude that the results are comparable but English argumentation sentences are easier to classify.

The possibility of using various machine translation systems for automatic creating of argument-annotated corpora was estimated. The classifiers trained in machine translation are of the same quality as those trained in human translation, and sometimes even surpass it. Google Translate gave the best results of the three machine translation systems – Google Translate, Yandex.Translate and Promt. At the same time, the results of the classification performance correlate with the quality assessment of machine translation.

Thus, we discovered the way to create argument-annotated corpora of the Russian language for future research.

Acknowledgments. The reported study was jointly financed by the German Academic Exchange Service (DAAD) and the Ministry of Education and Science of the Russian Federation within the "Michail Lomonosov" programme (2018).

References

1. Afantenos, S., Peldszus, A., Stede, M.: Comparing decoding mechanisms for parsing argumentative structures. Argum. Comput. **9**(3), 177–192 (2018)
2. Aharoni, E., et al.: A benchmark dataset for automatic detection of claims and evidence in the context of controversial topics. In: Proceedings of the First Workshop on Argumentation Mining, Baltimore, Maryland, USA, pp. 64–68 (2014)
3. Aker, A., et al.: What works and what does not: classifier and feature analysis for argument mining. In: Proceedings of the 4th Workshop on Argument Mining, Copenhagen, Denmark, pp. 91–96 (2017)
4. Aker, A., Zhang, H.: Projection of argumentative corpora from source to target languages. In: Proceedings of the 4th Workshop on Argument Mining, Copenhagen, Denmark, pp. 67–72 (2017)
5. Baroni, P., Gabbay, D., Giacomin, M., van der Torre, L. (eds.): Handbook of Formal Argumentation. College Publications, London (2018)
6. Bird, S., Loper, E., Klein, E.: Natural Language Processing with Python. O'Reilly Media Inc., Sebastopol (2009)
7. Doddington, G.: Automatic evaluation of machine translation quality using N-gram co-occurrence statistics. In: Proceedings of the 2nd International Conference on Human Language Technology Research, San Diego, California, USA, pp. 138–145 (2002)
8. Eger, S., Daxenberger, J., Stab, C., Gurevych, I.: Cross-lingual argumentation mining: machine translation (and a bit of projection) is all you need! In: Proceedings of the 27th International Conference on Computational Linguistics, Santa Fe, New Mexico, USA, pp. 831–844 (2018)
9. Habernal, I., Gurevych, I.: Argumentation mining in user-generated web discourse. Comput. Linguist. **43**(1), 125–179 (2017)
10. Koehn, P., et al.: Moses: open source toolkit for statistical machine translation. In: Proceedings of the 45th Annual Meeting of the ACL on Interactive Poster and Demonstration Sessions, Prague, Czech Republic, pp. 177–180 (2007)
11. Kutuzov, A., Kuzmenko, E.: WebVectors: a toolkit for building web interfaces for vector semantic models. In: Ignatov, D.I., et al. (eds.) AIST 2016. CCIS, vol. 661, pp. 155–161. Springer, Cham (2017). https://doi.org/10.1007/978-3-319-52920-2_15
12. Lauscher, A., Glavaš, G., Ponzetto, S. P., Eckert, K.: Investigating the role of argumentation in the rhetorical analysis of scientific publications with neural multi-task learning models. In: Proceedings of the 2018 Conference on Empirical Methods in Natural Language Processing, Brussels, Belgium, pp. 3326–3338 (2018)
13. Lippi, M., Torroni, P.: Argument mining from speech: detecting claims in political debates. In: Proceedings of the Thirtieth AAAI Conference on Artificial Intelligence, Phoenix, Arizona, USA, pp. 2979–2985 (2016)
14. Lippi, M., Torroni, P.: Argumentation mining: state of the art and emerging trends. ACM Trans. Internet Technol. **16**(2), 1–25 (2016)
15. McDonald, R., Petrov, S., Hall, K.: Multi-source transfer of delexicalized dependency parsers. In: Proceedings of the Conference on Empirical Methods in Natural Language Processing (EMNLP 2011), Stroudsburg, PA, USA, pp. 62–72. (2011)
16. Moens, M.-F.: Argumentation mining: how can a machine acquire common sense and world knowledge? Argum. Comput. **9**, 1–14 (2018)
17. Nguyen, H.V., Litman, D.J.: Argument mining for improving the automated scoring of persuasive essays. In: Thirty-Second AAAI Conference on Artificial Intelligence, New Orleans, Louisiana, USA, pp. 5892–5899 (2018)

18. Papineni, K., Roukos, S., Ward, T., Zhu, W.J.: BLEU: a method for automatic evaluation of machine translation. In: Proceedings of the 40th Annual meeting of the Association for Computational Linguistics (ACL-2002), Philadelphia, Pennsylvania, USA, pp. 311–318 (2002)
19. Pedregosa, F., Varoquaux, G., Gramfort, A., Michel, V., Thirion, B., et al.: Scikit-learn: machine learning in python. J. Mach. Learn. Res. **12**, 2825–2830 (2011)
20. Peldszus, A., Stede, M.: An annotated corpus of argumentative microtexts. In: Argumentation and Reasoned Action: Proceedings of the 1st European Conference on Argumentation, Lisbon, Portugal, pp. 801–815 (2015)
21. Peldszus, A., Stede, M.: From argument diagrams to argumentation mining in texts: a survey. Int. J. Cogn. Inform. Nat. Intell. **7**(1), 1–31 (2013)
22. Reed, C., Palau, R.M., Rowe, G., Moens, M.-F.: Language resources for studying argument. In: Proceedings of the 6th Conference on Language Resources and Evaluation (LREC 2008), Marrakech, Morocco, pp. 91–100. ELRA (2008)
23. Skeppstedt, M., Peldszus, A., Stede, M.: More or less controlled elicitation of argumentative text: enlarging a microtext corpus via crowdsourcing. In: Proceedings of the 5th Workshop in Argumentation Mining, Brussels, Belgium, pp. 155–163 (2018)
24. Sliwa, A., et al.: Multi-lingual argumentative corpora in English, Turkish, Greek, Albanian, Croatian, Serbian, Macedonian, Bulgarian, Romanian and Arabic. In: Proceedings of the Eleventh International Conference on Language Resources and Evaluation (LREC 2018), Miyazaki, Japan, pp. 3908–3911 (2018)
25. Sonntag, J., Stede, M.: GraPAT: a tool for graph annotations. In: Proceedings of the 9th International Conference on Language Resources and Evaluation (LREC 2014), Reykjavik, Iceland, pp. 4147–4151 (2014)
26. Stab, C., Gurevych, I.: Parsing argumentation structure in persuasive essays. Comput. Linguist. **43**(3), 619–659 (2017)
27. Stede, M., Schneider, J.: Argumentation Mining. Synthesis Lectures on Human Language Technologies. Morgan & Claypool Publishers (2018)
28. Stenetorp, P., Pyysalo, S., Topić, G., Ohta, T., Ananiadou, S., Tsujii, J.: BRAT: a web-based tool for NLP-assisted text annotation. In: Proceedings of the Demonstrations at the 13th Conference of the European Chapter of the Association for Computational Linguistics, Avignon, France, pp. 102–107 (2012)
29. van Eemeren, F.H., Grootendorst, R.: A Systematic Theory of Argumentation: The Pragma-dialectical Approach. Cambridge University Press, Cambridge (2004)
30. Vora, S., Yang, H.: A comprehensive study of eleven feature selection algorithms and their impact on text classification. In: Proceedings of the Computing Conference, London, UK, pp. 440–449 (2017)
31. Yarowsky, D., Ngai, G., Wicentowski, R.: Inducing multilingual text analysis tools via robust projection across aligned corpora. In: Proceedings of the First International Conference on Human Language Technology Research (HLT 2001), Stroudsburg, PA, USA, pp. 1–8 (2001)
32. Yimam, S.M., Gurevych, I., Eckart de Castilho, R., Biemann, C.: WebAnno: a flexible, web-based and visually supported system for distributed annotations. In: Proceedings of the 51st Annual Meeting of the Association for Computational Linguistics: System Demonstrations, Sofia, Bulgaria, pp. 1–6 (2013)

Dynamic Topic Models for Retrospective Event Detection: A Study on Soviet Opposition-Leaning Media

Anna Glazkova[1]([🖂])[iD], Valery Kruzhinov[2][iD], and Zinaida Sokova[2][iD]

[1] Institute of Mathematics and Computer Sciences, University of Tyumen,
Perekopskaya Street 15a, 625003 Tyumen, Russian Federation
anna_glazkova@yahoo.com
[2] Institute of Social Sciences and Humanities, University of Tyumen,
Lenina Street 23, 625003 Tyumen, Russian Federation
{v.m.kruzhinov,z.n.sokova}@utmn.ru

Abstract. In recent years, there has been an increasing interest in digital humanities. This interest is justified by the development of natural language processing tools and the emergence of digitized text collections of documents in different fields of knowledge, for example, literature, art, philosophy, and history. In this paper, we applied unsupervised topic modeling to the Bulletin of Opposition, the journal of Soviet opposition published by Trotskyists in Paris from 1929 to 1941, to analyze the main trends in the Russian opposition-leaning media. We identified topic classes using models based on Latent Dirichlet Allocation and examined Dynamic Topic Models as a tool to single out the main issues of interest for historical research. Applying topic modeling and statistical methods, we proposed an approach to Retrospective Event Detection that was evaluated on a human-annotated set of historical news items. The present study may help to improve event detection on smaller text corpora.

Keywords: Event detection · Topic modeling · Latent Dirichlet Allocation · Historical research · Media · Digital humanities · Statistical methods

1 Introduction

Periodicals are one of the richest information resources for a retrospective study. Indeed, they set the stage for political, economic, and social news, manage their agenda, express and shape public opinion and attitude, and comprehensively cover the life of society. Hence, it is no surprise that media sources attract the researchers' attention as an exceptionally crucial and multi-topic historical resource. Although newspapers and journals contain large amounts of valuable information, they still remain one of the least studied types of historical documents. This is due to the fact that it is often difficult to look through a massive amount of information to get the answer to the stated research problem.

© Springer Nature Switzerland AG 2019
W. M. P. van der Aalst et al. (Eds.) AIST 2019, LNCS 11832, pp. 145–154, 2019.
https://doi.org/10.1007/978-3-030-37334-4_13

The ability to process large volumes of data can be provided by modern computer technologies, as some historical media sources are available in a digital form as online-based archives. Web archives make it possible to apply contemporary natural language methods to historical documents processing. Thus, researchers can get a wide array of valuable scientific resources.

In this paper, we constructed topic models based on the news items published in the Bulletin of the Opposition, a Russian-language journal running in Paris from 1929 to 1941. It was launched by Leon Trotsky who employed it after his expulsion from the USSR as a platform to struggle with Joseph Stalin. The Bulletin of the Opposition reflected and expressed the viewpoints of the left oppositional movement regarding noteworthy events of Soviet life and the basic concepts of socialism – Marxism, revolution, and the working class. The authors of the articles were well-known oppositionists-Marxist thinkers, as well as anonymous authors who had to hide their names for security reasons. Applying a statistical assessment of the frequency of topic mentions, we propose an approach to detecting significant events.

The paper is organized as follows. Section 2 gives a brief overview of the recent research on the application of topic models for event extraction. Section 3 is concerned with the dataset and methodology used for this study. Section 4 then deals with the experimental results for our text collection. Finally, Sect. 5 presents our conclusions.

2   Related Work

Topic modeling is a type of statistical modeling for recognizing main topics in a collection of documents. As a rule, topic modeling is based on Latent Dirichlet Allocation (LDA), a hierarchical network that relates words and documents through latent topics [1]. Topics are characterized by diverse frequency of words. The document is presented as a bag-of-words approach, and the topic looks like a set of words ranked in decreasing order of their probabilities.

The LDA model imposes constraints on temporal relationships between topics. The possible way to model temporal relationships is to utilize the Dynamic Topic Model (DTM) [2]. The algorithm organizes the corpus on time slices; therefore, a topic is a sequence of distributions over the words. DTM uses Gaussian (normal) distributions to detect the evolution sequence. The estimation of the topics on a time slice depends on the estimation from the previous time slice. DTM helps to observe how the keywords of a topic change over time [3].

Some researchers have attempted to exploit topic models to extract events from a text collection. Jelodar et al. [4] described the main types of topic models and their applications in natural language processing tasks. Chen et al. proposed the TH-LDA model (Hashtag, Time and LDA model) for the detection of hot news events from microblogs [5]. The scholars implemented the algorithm to detect subevents, and to track the evolution and development of events. Keane et al. worked out an algorithm for event extraction from large text collections based on the cosine similarity score for daily topic models [6]. The approach to event extraction from social media using LDA and a scoring function is depicted in [7]. Ge et al. initiated the method of opinion extraction based on TF-IDF and topic modeling [8].

3 Retrospective Event Detection

3.1 Dataset

We used the text data from the online collection of digitized historical newspapers, including the issues of the Bulletin of the Opposition, published from 1929 to 1941 [9]. The issues were segmented by articles. Each article was subjected to computer analysis. Tables 1 and 2 show the main properties of the text collection.

Table 1. Properties of the text collection.

Property	Value
Number of issues	87
Number of articles	794
Number of tokens	316462

The dataset raised three challenges:

1. Since the journal was rather sporadic in publishing issues, the number of texts was relatively small. For instance, in 1930 and 1941 the periodicity was as follows: 11 and 3 issues respectively. Besides, the issues included a different number of articles.
2. A small number of time slices made it more difficult to use statistical methods to interpret the DTM model.
3. There was no a marked up test sample to evaluate event detection quality for our obtained dataset.

Table 2. Number of articles per year.

Year	1929	1930	1931	1932	1933	1934	1935	1936	1937	1938	1939	1949	1941
Value	50	145	116	81	76	19	43	64	57	65	55	16	7

Taking into account that the number of articles in different issues was not the same, it made sense to combine documents in time slices by year of publishing but not by issues. Therefore, the texts in the collection were grouped into 13 time slices, representing the number of years of journal running. Next, we automatically marked up domestic and international political affairs, mentioned in the journal. As a result, we got 324 event mentions. All of them were roughly divided into topic groups at the data preprocessing stage.

The data preprocessing included the following steps:

- removing line breaks, numbers and other special characters;
- lemmatizing words using Pymorphy [10];
- representing the text as a vector of words;
- building the bigram and trigram topic models;
- removing stopwords.

3.2   DTM

The textual data loaded for DTM included preprocessed texts and the list of numbers of articles on each time slice. Our model consisted of 13 time slices.

We designed a topic model for 10 topics, which reflect the main content of the journal. Our approach to choosing the optimal number of topics was to build several topic models with different values of number of topics (K) and pick the number that gives the highest coherence value.

Table 3 shows the identified topics and their explanations from a historical viewpoint. We manually indicated a short title for each topic in the Heading column in order for they are easily identified. It should be noted that some articles commented on events (Political Repressions, World War II), while the others discussed ideological and abstract issues and contained reflections on the goals of socialism and criticism of Stalinism (Ideology, Revolution). The DTM model was implemented by using Gensim Python library [11].

3.3   Interest Growth Detection

We calculated the topic distribution for each document in the collection. It is a vector of length equal to the number of topics in the DTM model. The topic distribution vector Dis_i shows how a document D_i extracted from a set of documents D corresponds to each topic T_k from a set T. Each element w_k of the vector Dis_i is a number from 0 to 1, where 0 is a complete non-match and 1 is an absolute match.

$$Dis_i = \{w_1, ..., w_K\}, \sum_{k=1}^{K} w_k = 1, i \in [1, I], \tag{1}$$

K is a number of topics (in our experiments we used $K = 10$), I is a number of documents.

Then, we got the averaged topic distribution vector for each time slice to obtain the average values of topic distribution for a group of documents.

$$\overline{Dis_p} = \{\overline{w_1}, ..., \overline{w_K}\}, p \in [1, P], \tag{2}$$

P is a number of time slices (the value for our dataset was equal to 13).

Table 3. Topics and explanations.

№	Heading	Keywords	Explanation
1	Ideology	Trotskyism, Marx, Bolshevik, Lenin, the October Revolution (троцкизм, Маркс, большевик, Ленин, октябрьская революция)	Discussions on Marxism theoretical grounds and their interpretations in Trotskyism and Leninism.
2	Workers	worker, factory, work, ruble, household (рабочий, завод, работа, рубль, хозяйство)	Analysis of workers' status. In the Soviet Union the working class was a privileged social stratum and ruling class. In 1929, the Soviet government decreed the transition to a 7-hour working day.
3	Conflict with Stalin	Stalin, Trotsky, Lenin, Voroshilov, Moscow, revolution (Сталин, Троцкий, Ленин, Ворошилов, Москва, революция)	Trotsky's anti-Stalinist articles. Trotsky was called the voice of the Russian opposition to Stalin.
4	World War I	Soviet Union, World War, America, Europe, Capitalism (Советский Союз, мировая война, Америка, Европа, капитализм)	Description of World War I (1914-1918). Lenin considered imperialist wars as an absolute evil. However, he strongly believed that the World War would obviously weaken the Russian Empire, so the Bolsheviks would be able to establish the dictatorship of the proletariat and overthrow the monarchy.
5	Political Repressions	process, defendant, case, Moscow, Blumkin (процесс, подсудимый, дело, Москва, Блюмкин)	Politically driven show trials were a lamentable tradition in the Soviet Union. Very often Trotskyists were charged with espionage and attempts to restore capitalism. In 1929, Jacob Blumkin was sentenced to death for trying to establish contacts with Trotsky abroad. This murder was the precursor of the mass killings in 1936-1938.
6	Economy	industry, collective farm, collectivization, plan, five-year plan (промышленность, колхоз, коллективизация, план, пятилетка)	In the late 1920s, Stalin initiated the industrialization of the country. Five-year plans were nationwide economic plans aimed at the rapid development of the Soviet economy. The Soviet state planning committee (Gosplan) worked out the plans based on the theory of the productive forces, which was an essential part of the Communist Party's ideology. At the same time, privately-owned farms were grouped in collective farms. Collectivization allowed the state to recover agricultural outputs and establish food supply plans.
7	Trotsky about Stalin	Trotsky, Stalin, article, press, newspaper (Троцкий, Сталин, статья, пресса, газета)	Trotsky's comments on the phenomenon of Stalinism. Trotsky attacked Stalin's absolute power over the Communist Party and the USSR, his current policy and public reaction. Since the spring of 1938, Trotsky had almost entirely focused on compiling a book about Stalin and then he began to publish articles related to the subject of the future book.

Table 3. (*continued*)

№	Heading	Keywords	Explanation
8	Revolution	revolution, proletariat, worker, bourgeoisie, dictatorship (революция, пролетариат, рабочий, буржуазия, диктатура)	Reflections on the revolution and its nature, the interclass antagonism. Trotskyism was the theory of a "permanent" (continuous) revolution or the revolution denying the peasantry as a revolutionary force able to seize social, economic and political power. On the contrary, Leninism recognized the peasantry as an ally of the proletariat. This controversy was the basis of the struggle between Leninism and Trotskyism since 1905.
9	Radek's Case	oppositionist, party, fight, Radek, comrade (оппозиционер, партия, борьба, Радек, товарищ)	Karl Radek was an active supporter of Trotsky and the Left Opposition. Later, he was expelled from the Party in the USSR. However, in 1929 he publicly declared his solidarity with Stalin and the break with Trotskyism. In 1936, after the Second Moscow trial, Karl Radek was sentenced to jail and then killed in a labor camp at the order of Beriya.
10	World War II	fascism, Germany, Hitler, war, Europe (фашизм, Германия, Гитлер, война, Европа)	Description of World War II timeline, including the events leading up to the war: Hitler's appointment as the Chancellor of Germany (1933); the beginning of the Second World War (1939); the beginning of the Great Patriotic War (1941).

Hence, the corresponding values of the averaged vector $\overline{Dis_p}$ elements were established. The given vectors reflect the mentions dynamics of the topic in a text collection.

$$\overline{Dyn_k} = \{y_1, ..., y_P\}, p \in [1, P], \tag{3}$$

y_p is an averaged number of mentions for the topic T_k at the time slice p.

The values of the vector $\overline{Dyn_k}$ elements allowed us to conclude that the interest to the topic had changed over time. We assumed that if the interest to the topic had greatly increased for a certain period, then there had been crucial events related to the given topic during that period. Therefore, we calculated the vector of interest growth for the topic by the following formula:

$$G_k = \{g_1, ..., g_{P-1}\}, \tag{4}$$

$$g_j = \begin{cases} y_{j+1} - y_j, y_{j+1} - y_j > 0 \\ 0, y_{j+1} - y_j \leq 0 \end{cases},$$

$$j \in [1, P-1].$$

3.4 Event Detection

We are convinced that if the value g_j is recognized as an outlier, then a surge of interest to the given topic is detected at the corresponding time interval. Since in our case the length of the G_k vector is rather small, certain statistical methods cannot be applied (for example, the three-sigma rule). Therefore, we decided to use the Romanovsky criteria to determine outliers. The Romanovsky criteria can be exploited, if $4 \leq P \leq 20$:

$$\beta_j = \frac{g_j - \overline{g}}{S_g},\tag{5}$$

$\overline{g}$ and S_g are respectively the mean value and standard deviation for all elements of the vector G_k, excluding g_j.

If the value of β_j is greater than the threshold of β, then g_j is an outlier. β threshold is a standard value for accuracy α.

The event detection algorithm works as follows:

1. Check the maximum element g_m in the sequence $g_1, ..., g_{P-1}$.
2. If g_m is an outlier, we assume that an event was detected at time slice $m + 1$; otherwise, it means the end of the search.
3. Exclude g_m from consideration. If the number of elements in the sequence is still ≥ 4, go to step 1. If not, the search ends.

The source code and the preprocessed text corpus are available at [12].

4 Results

Figures 1 and 2 demonstrate diagrams constructed to illustrate the event detection results. World War II topic (Fig. 1a) contains three points identified as outliers (1933, 1939 and 1941). Memorable events related to the topic were detected at these time slice. The greatest outlier (1933) is obviously connected with Hitler's appointment as German Chancellor, while the others are coupled with the beginning of World War II (1939) and the Great Patriotic War (1941).

Two outliers were discovered for Political Repressions topic (Fig. 1b): the Second Moscow trial, which extremely affected the Trotskyists (1936) and the large-scale Stalinist political repressions (1937). Table 4 presents the keywords with the highest frequency increase identified by the DTM model for outlier time slices in World War II and Political Repressions topics.

Figure 2 illustrates the diagram for the Revolution topic as an example of a non-event topic. No outliers were detected in this topic. The Revolution topic mainly contains the texts of essays and analytical articles. These texts usually focus on reflections of the opposition on the necessity for the revolution and working-class dictatorship, on proclamation October revolution as a great success for Russia. They did not describe the events that took place at that time. Thereby, the interest to the topic did not change abruptly, but always remains at about the same level.

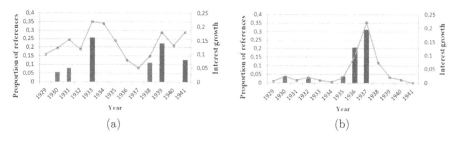

Fig. 1. Examples of topics containing significant events: (a) World War II; (b) Political Repressions. Results obtained with $\alpha = 0.1$.

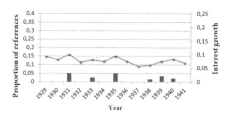

Fig. 2. Example of non-event topic. $\alpha = 0.1$.

Table 4. Keywords for outlier time slices.

Topic	Year	Keywords
Political Repressions	1936	Process, testimony, defendant, Moscow, Pyatakov (процесс, показания, подсудимый, Москва, Пятаков)
Political Repressions	1937	Case, testimony, Stalin, communication, interrogation (дело, показания, Сталин, связь, допрос)
World War II	1933	Party, fascism, politics, leader, Germany (партия, фашизм, политика, вождь, Германия)
World War II	1939	Fascism, war, party, bourgeoisie, Hitler (фашизм, война, партия, буржуазия, Гитлер)
World War II	1941	Struggle, proletariat, fascism, Hitler, victory (борьба, пролетариат, фашизм, Гитлер, победа)

We manually marked up the journal texts to validate the findings. The markup was carried out as follows: we indicated the text affiliation to one of the topics obtained by the DTM model. Thus, 324 texts were allocated. We constructed vectors depicting the interest growth and frequency of the events related to this or that topic for each one. Further, we applied our approach to

Table 5. Values of correlation coefficient.

Event topic	Correlation
Workers	.751
Political repressions	.893
Economy	.914
Trotsky about Stalin	.681
Radek's case	.84
World War II	.928
Average	**.835**

event detection to the received vectors. Finally, we calculated the correlation coefficient between the interest growth vectors and G_k vectors obtained earlier (Table 5). On average, the correlation coefficient was equal to 0.835.

5   Conclusion

The proposed approach allows detecting events in datasets divided into a small number of time slices (no more than 20). The utility of this approach is achieved by the absence of necessity for manual setting a threshold value that determines the significance of the event. This value is automatically calculated on the basis of statistical values. The described approach detects retrospective events and cannot be exploited as such to process online stream of texts. Further research should focus on evaluating the form of the mentions dynamics and analyzing the interest evolution.

References

1. Blei, D.M., Ng, A.Y., Jordan, M.I.: Latent Dirichlet allocation. J. Mach. Learn. Res. **3**, 993–1022 (2003)
2. Blei, D.M., Lafferty, J.D.: Dynamic topic models. In: Proceedings of the 23rd International Conference on Machine Learning, pp. 113–120. ACM (2006)
3. Loukachevitch, N., Mischenko, N.: Evaluation of approaches for most frequent sense identification in Russian. In: van der Aalst, W.M.P., et al. (eds.) AIST 2018. LNCS, vol. 11179, pp. 99–110. Springer, Cham (2018). https://doi.org/10.1007/978-3-030-11027-7_10
4. Jelodar, H., et al.: Latent Dirichlet Allocation (LDA) and topic modeling: models, applications, a survey. Multimedia Tools Appl. **78**, 1–43 (2018)
5. Chen, J., Shang, Q., Xiong, H.: Hot events detection for chinese microblogs based on the TH-LDA model. In: Proceedings of the 2018 International Conference on Transportation & Logistics, Information & Communication, Smart City, TLICSC 2018. Atlantis Press (2018)
6. Keane, N., Yee, C., Zhou, L.: Using topic modeling and similarity thresholds to detect events. In: Proceedings of the 3rd Workshop on EVENTS at the NAACL-HLT 2015, pp. 34–42 (2015)

7. Gupta, M., Gupta, P.: Research and implementation of event extraction from Twitter using LDA and scoring function. Int. J. Inf. Technol. **11**(2), 365–371 (2019)

8. Ge, B., et al.: Microblog topic mining based on a combined TF-IDF and LDA topic model. In: Automatic Control, Mechatronics and Industrial Engineering: Proceedings of the International Conference on Automatic Control, Mechatronics and Industrial Engineering, ACMIE 2018, Suzhou, China, 29–31 October 2018, p. 291. CRC Press (2019)

9. ≪The Bulletin of Opposition≫. https://www.1917.com/Marxism/Trotsky/BO/Main.html. Accessed 13 Feb 2019

10. Morphological analyzer pymorphy2. https://radimrehurek.com/gensim/index.html. Accessed 13 Feb 2019

11. Gensim. Topic modeling for humans. https://radimrehurek.com/gensim/index.html. Accessed 13 Feb 2019

12. https://github.com/oldaandozerskaya/Bulletin_of_opposition. Accessed 03 June 2019

Deep Embeddings for Brand Detection in Product Titles

Andrey Kulagin[✉][ID], Yuriy Gavrilin[ID], and Yaroslav Kholodov[ID]

Innopolis University, Innopolis, Russia
{a.kulagin,y.gavrilin,ya.kholodov}@innopolis.ru

Abstract. In this paper, we compare various techniques to learn expressive product title embeddings starting from TF-IDF and ending with deep neural architectures. The problem is to recognize brands from noisy retail product names coming from different sources such as receipts and supply documents. In this work we consider product titles written in English and Russian. To determine the state-of-the-art on openly accessed "Universe-HTT barcode reference" dataset, traditional machine learning models, such as SVMs, were compared to Neural Networks with classical softmax activation and cross entropy loss. Furthermore, the scalable variant of the problem was studied, where new brands are recognized without retraining the model. The approach is based on k-Nearest Neighbors, where the search space could be represented by either TF-IDF vectors or deep embeddings. For the latter we have considered two solutions: (1) pretrained FastText embeddings followed by LSTM with Attention and (2) character-level Convolutional Neural Network. Our research shows that deep embeddings significantly outperform TF-IDF vectors. Classification error was reduced from 13.2% for TF-IDF approach to 8.9% and to 7.3% for LSTM embeddings and character-level CNN embeddings correspondingly.

Keywords: Text classification · FastText · LSTM with Attention · Embeddings · Triplet loss · Similarity learning · Brand classification

1 Introduction

Extracting brand information from product titles can be a challenging but rewarding task. This particular knowledge can be helpful in several business scenarios. For example, B2B companies that work with retail clients, can extract brand names from the text of receipts or from supply documents to get useful insights. These insights can answer questions like whether "Brand A" is more popular than "Brand B" in a particular region or a particular point of sales. Such insights, if provided, can influence the overall strategy of the company and bring out hidden profits. Another example can be found in e-commerce. As Majumder et al. [2] state, the process of extracting brand information from a noisy text is vital for an online marketplace as it could greatly enhance the user experience by enabling facets for product search.

W. M. P. van der Aalst et al. (Eds.) AIST 2019, LNCS 11832, pp. 155–165, 2019.
https://doi.org/10.1007/978-3-030-37334-4_14

Product records can contain Global Trade Item Number (GTIN) which is encoded in a barcode. If such number is available, the process of determining the brand is simple and straightforward - a single request to the knowledge base in which all GTINs have a corresponding brand. Unfortunately, product records rarely come alongside with GTINs, which raises the question of how brand recognition should be performed.

Brand information usually can be extracted from another field of product records - title. Product titles can be helpful in many ways. They usually contain either part of a brand name or the brand name can be easily recognized from the title. For example "HS shampoo 500" probably means Head&Shoulders. However, extracting desired information from product names is not a simple process. Product titles are subject to noise - they are full of typos, abbreviations, acronyms and contractions. These are the main challenges in determining the brand. Unfortunately, under these circumstances, simple heuristics like regular expressions are unable to solve this problem. In this paper, we are studying how we can utilize modern machine learning and deep learning methods to effectively solve this task.

2 Problem Formulation

We can define the problem as the following text classification task: Given a product title x as a string of arbitrary length in Russian or English, assign to it the brand label $b \in B$, where B is a set of all known brands presented in some knowledge base. The base itself consists of many labeled product titles. You can see some examples on the Fig. 1.

Moreover, we would like to extend our problem formulation to capture real-life conditions. Novel brands emerge every day and it would be useful if models could be *extensible* and classify them without being retrained. We propose determining whether a novel brand name is present in a product name and if so, labeling a few instances of this new brand. Thus, the original problem is reformulated with an addition of active-learning and few-shot learning constraints. Ultimately, we call a model *extensible* if it is able to recognize novel brands by just adding a few labeled examples and without changing the model's parameters.

It is worth to mention that initially we started with Named Entity Linking formulation of the problem, as it was proposed by [1]. But they worked with a closed dataset of Walmart products. Unfortunately, we are not aware of openly accessible Entity Linking datasets with product brands as entities. There is "The AIDA CoNLL-YAGO" [5] dataset which contains assignments of entities for the original CoNLL (2003) NER task. Among many other entity types, it contains a few organizations like Ford Motor Company or Apple Inc., but this is slightly different from the product brands we would like to work with (significant for business use cases mentioned in Sect. 1) like food, household chemicals, etc. For Russian language we could not find any applicable Entity Linking dataset. So, we decided to reformulate our task as a text classification problem, having found an appropriate, openly accessible dataset, which is described in Sect. 3.

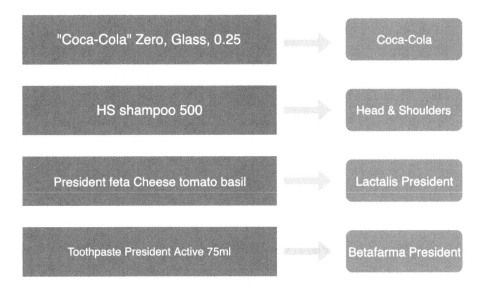

Fig. 1. Examples of brand classification

3   Dataset

To compare different algorithms we are going to use Universe-HTT barcode reference dataset (UHTT) [3]. We are going to work with the second version of this dataset, published in November 2018.

Initially UHTT dataset contained 3.5 million rows with the following attributes: **{GTIN, Product Title, Category, Brand}**. Product titles are written in both Russian and English. We removed 2% of rows which have invalid GTINs according to GS1 General Specification [4], and 50% of rows without brand information. Also, we have left only brands with at least 5 GTINs present in the dataset. As a result, we received 1666290 rows for 20545 brands.

We are not aware of any work, which uses the above mentioned dataset, so it seems like we could establish the state-of-the-art brand recognition method on it.

3.1   Cross-Validation

As we wish to test extensible models not only on brands, seen during training, we should properly split our dataset for cross-validation. We applied the following process (illustrated on Fig. 2). First of all, for every brand, 75% of all samples were marked as train (and 25% as test). This division is fixed. Then, during each cross-validation step (out of 4), 75% of all brands were marked as *known* brands and the rest 25% as *unknown*. *Unknown* brands are brands not seeing during training. Only *known train* part of samples will be available for a model during training time.

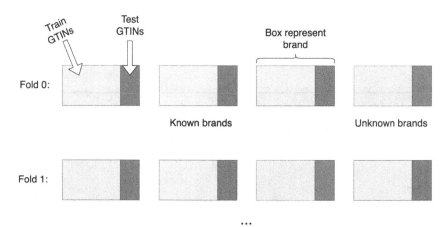

Fig. 2. Cross-validation process

4 Methodology

4.1 Preprocessing

Before each machine learning algorithm, every product title goes through the following steps:

1. Lowercasing
2. Replacing all the characters not present in the alphabet with spaces. Alphabet:
   ```
   0123456789
   abcdefghijklmnopqrstuvwxyz
   абвгдежзийклмнопрстуфхцчшщъыьэюяё
   ".,/-()%*℗<>_':+\&;
   ```

Then, we continue with some additional steps that are specific to the algorithms and will be described further.

4.2 TF-IDF + kNN

The first method we consider is TF-IDF on character n-gram features followed by k-Nearest Neighbors classifier. We have chosen it as a rather simple baseline since it has shown a relatively good performance on several text classification tasks in other studies [7].

Before passing product titles to TF-IDF, every product name goes through the process of transliteration, since we have 2 languages, and, for example, "Кока Кола", written in Russian, means essentially the same as "Coca-Cola", written in English. Despite having the same meaning without transliteration, different sets of n-grams would be extracted.

We observed that character 3-grams worked better than other combinations (we tried also 4-grams, and 3-grams together with 4-grams) and with k = 1 appeared to be the best choice (we also considered k = 3, k = 5), which is consistent with the findings in [1].

As a result of TF-IDF feature extraction, we obtained sparse vectors with ≈15000 dimensions. Cosine similarity is used as a distance metric for kNN classifier.

Since we are dealing with a dataset with hundreds of thousands of samples, brute force calculation to find nearest neighbors would be unfeasible in terms of computation time. In this work, for all algorithms that rely on Nearest Neighbors search, we used Hierarchical Navigable Small World graphs (HNSW) [6] as an Approximate Nearest Neighbors algorithm.

TF-IDF + kNN is a simple but strong baseline because it complies with extensibility property which we have defined earlier. kNN is able to detect novel brands and needs only to label a few titles with that brand to detect it later.

4.3   TF + SVM

As stated by Tang et al. [8] and Lai et al. [9] Linear Support Vector Machines is also considered as a strong baseline for text classification tasks and often compared to Deep Neural models. In our study we also use term frequencies of character n-grams as input features to SVM, since subword TF showed better performance if compared to word TF, because it helps the model overcome issues like typos and abbreviations.

Alternatively to TF-IDF + kNN, SVM does not need transliteration during preprocessing. This is because kNN relies heavily on distances in high-dimensional space of features, and the same is not true for SVM.

SVM is the model with the highest number of trainable parameters among all models we have tried. It has 924 millions of parameters covering 15400 known brands × 60000 n-gram features. Unfortunately, this approach is not extensible, since we need to retrain SVM every time we wish to be able to classify new brand.

4.4   LSTM + Attention

Majumder et al. in their work "Deep Recurrent Neural Networks for Product Attribute Extraction in eCommerce" [2] showed that Named Entity Recognition (NER) task on product titles can be successfully solved using one of the combinations: either bidirectional-LSTM with Attention or bidirectional-LSTM with Conditional Random Field. The authors have particularly focused on Brand attribute, so the domain of their work is the same as ours, although problem formulation is different as they are trying to solve a NER task.

Nevertheless, we decided to try bi-LSTM with Attention too, not only in pursuit of accuracy, but also for two useful properties: it can be modified to satisfy extensibility property which will be described in Sect. 4.6, and since neural network has Attention in it, it also has a built-in method of interpretability. We can interpret the model results by looking for words in a product title whose attention scores are the highest.

The model architecture we used in our study resembles the one described by Majumder et al. [2], except for two additional Dense layers in the end.

Here is the conceptual model architecture:

1. Word-level tokenization by space characters
2. FastText [10] embedding layer to utilize subword-level information and be able to work with unknown words. We used 300-dimensional vectors pretrained on Common Crawl from [11]
3. Bidirectional LSTM layer with 128 neurons in each direction
4. Attention mechanism suggested by Bahdanau et al. [13]
5. 2 Dense layers each with 512 neurons and ReLU activation followed by Dropout [12] with probability 0.5
6. Dense layer with softmax activation for brand classification with dimensionality ≈15400, depending on cross-validation fold.

To train this network we used Adam [14] optimizer and categorical cross-entropy loss.

4.5 Char-Level CNN

The second neural network architecture we have considered in this work is a character-level Convolutional Neural Network. It is a VDCNN-like [15] architecture where each conv-block consists of {1D convolutional layer, Batch normalization [16] and ReLU activation} combination.

Here is the char-level architecture we have considered:

1. Truncate/pad product titles to 120 characters (99.8% samples have length less than 120 characters)
2. Character-level embedding layer (dimensionality 16)
3. 3 conv-blocks with kernel size 64
4. Max Pooling with pooling size 2 and stride 2
5. 2 conv-blocks with kernel size 128
6. Max Pooling with pooling size 2 and stride 2
7. 1 conv-blocks with kernel size 256
8. Max Pooling with pooling size 2 and stride 2
9. 1 conv-blocks with kernel size 1024
10. Global Max Pooling
11. Dropout with probability 0.5
12. Dense layer with dimensionality 1024 and ReLU activation
13. Dropout with probability 0.5
14. Dense layer with Softmax activation for brand classification (dimensionality ≈15400, depending on cross-validation fold).

4.6   Distance Metric Learning for Neural Models

One of the biggest advantages of neural networks to classical machine learning algorithms for our problem is that we can easily make neural networks extensible. The key intuition behind that transition is the problem reformulation. Instead of predicting concrete brands, neural network output embeddings in some n-dimensional space, in our work $n = 32$, in which product titles with the same brand would be placed closer, and embeddings of different brands would be distant. In this case, after obtaining an embedding for a concrete product title, we could apply kNN and find the nearest neighbor in a train set with an associated brand.

To make both neural networks, those we have considered previously, become *extensible*, we need to change their last layers. Instead of Dense layer with softmax activation with the number of neurons equal to the number of known brands, we should place Dense layer with $n = 32$ neurons. Thus, neural network will output embedding in $n = 32$ dimensional space. Moreover, we need to change the loss function to facilitate the desired changes.

The loss function, designed to group together embeddings of one identity and at the same time put apart different ones, is called triplet loss. It was first suggested in the paper "FaceNet: A Unified Embedding for Face Recognition and Clustering" [17]. The authors used triplet loss for face recognition problem, where they have millions of identities, and proposed a system that was able to recognize new faces, not seen during training, just by adding a photo of a new face into the knowledge base, thus supporting one-shot learning property.

Triple loss can be described by the following equation:

$$\sum_{i}^{N} \left[||f(x_i^a) - f(x_i^p)||_2^2 - ||f(x_i^a) - f(x_i^n)||_2^2 + \alpha \right]_+$$

where f is a neural network that output embeddings, x_i^a - *anchor* sample, x_i^p - *positive* sample, x_i^n - *negative* sample. *anchor* and *positive* titles have the same brand, *anchor* and *negative* have different brands. α is a margin that is enforced between positive and negative pairs (0.2 in our work).

5   Results and Discussion

The results of our experiments can be found in Table 1. In this table, "Known brands" column describes the accuracy of classification titles from the test set of known brands. Further columns indicate the performance of classifiers in situations when we have added a few labeled instances of each novel brand to the search space. The word *global* means that the search space for nearest neighbors consists of both known and unknown brands. There are gaps in the rows corresponding to not extensible models. "Known in global" column indicates the performance of searching titles of known brands in global space, and "Unknown in global" displays the accuracy of the methods on novel, few-shot labeled brands.

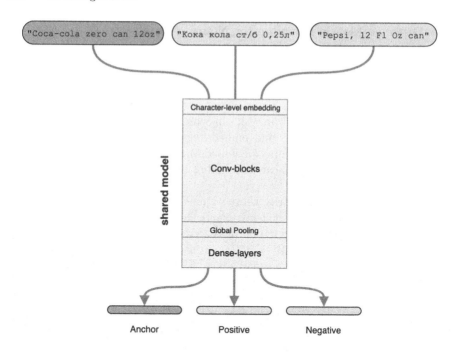

Fig. 3. CNN architecture with a triplet loss

We could see that there is a certain drop in the accuracy for predicting known brands, namely about 1%, after the addition of novel brands into the search space. These are the expected results, since we increased the amount of brands in the search space by 33%.

For the known brands SVMs and CNN have shown the best performance of 94.6%, outperforming LSTM by 0.3%. However, we can assume that LSTM could also achieve better results with better FastText embeddings. In our research, we have used FastText weights from Common Crawl [11], without finetuning it.

As expected, extensible kNN-based methods which have utilized deep embeddings, outperformed shallow TF-IDF embeddings because the former are directly trained to extract brand information. However, in our results we are still 5% below perfect accuracy. After analyzing a subset of the errors (exhaustive investigation of almost 100000 misclassifications would have taken too long), we have realized that in UHTT dataset, there are certain amounts of labeling mistakes or situations when titles do not contain anything related to a brand. For example, some "Coca-Cola" entries have contained the following product titles - "Soft drink 12 Fl Oz cans", without mentioning "Coca-Cola" brand at all.

Also, in Fig. 3 we visualized embeddings for product names which have not been seen during training. Visualization was done via t-SNE dimensionality reduction (Fig. 4).

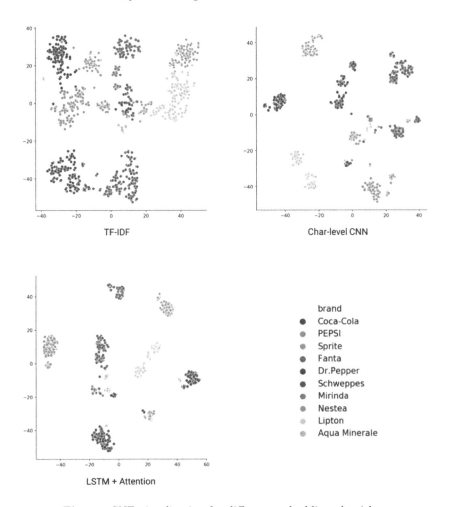

Fig. 4. t-SNE visualization for different embedding algorithms

Table 1. Results. All numbers represent classification accuracy

Method	Known brands	Known in global	Unknown in global
TF + SVM	$94.6 \pm 0.1\%$	–	–
TF-IDF + kNN	$88.5 \pm 0.2\%$	$87.0 \pm 0.2\%$	$86.8 \pm 0.7\%$
LSTM (with softmax)	$93.3 \pm 0.2\%$	–	–
LSTM (with triplet loss) + kNN	$94.3 \pm 0.1\%$	$93.4 \pm 0.1\%$	$91.1 \pm 0.6\%$
CNN (with softmax)	$94.4 \pm 0.3\%$	–	–
CNN (with triplet loss) + kNN	$94.6 \pm 0.2\%$	$\mathbf{93.9 \pm 0.3\%}$	$\mathbf{92.7 \pm 0.5\%}$

6   Conclusion

In this paper we present the comparison of multiple methods designed to extract brand names from product titles. During our evaluation, we have shown superiority of kNN classifier built atop deep embeddings extracted by character-level CNN. Moreover, we have also shown that this method meets the extensibility requirement - as it can work with new brands just by adding a few labeled instances.

Besides, we have established a state-of-the-art brand classification method on UHTT dataset, which supports active learning and few-shot learning. We hope that this method would serve as the basis for future research in the field of brand classification with the UHTT dataset.

Acknowledgments. We would like to thank Nikita Tarasov and Mikhail Bortnikov for their helpful insights, and Maya Stoyanova for the careful proofreading of this work.

References

1. More, A.: Attribute extraction from product titles in ecommerce. arXiv preprint arXiv:1608.04670 (2016)
2. Majumder, B.P., et al.: Deep Recurrent Neural Networks for Product Attribute Extraction in eCommerce. arXiv preprint arXiv:1803.11284 (2018)
3. Universe-HTT barcode reference. https://github.com/papyrussolution/UhttBarcodeReference
4. GS1 General Specification. https://www.gs1.org/standards/barcodes-epcrfid-id-keys/gs1-general-specifications
5. Hoffart, J., et al.: Robust disambiguation of named entities in text. In: Proceedings of the Conference on Empirical Methods in Natural Language Processing, pp. 782–792. Association for Computational Linguistics (2011)
6. Malkov, Y.A., Yashunin, D.A.: Efficient and robust approximate nearest neighbor search using hierarchical navigable small world graphs. IEEE Trans. Pattern Anal. Mach. Intell. (2018)
7. Zhang, X., Zhao, J., LeCun, Y.: Character-level convolutional networks for text classification. In: Advances in Neural Information Processing Systems (2015)
8. Tang, D., Qin, B., Liu, T.: Document modeling with gated recurrent neural network for sentiment classification. In: Proceedings of the 2015 Conference on Empirical Methods in Natural Language Processing (2015)
9. Lai, S., et al.: Recurrent convolutional neural networks for text classification. In: Twenty-Ninth AAAI Conference on Artificial Intelligence (2015)
10. Joulin, A., et al.: Bag of tricks for efficient text classification. arXiv preprint arXiv:1607.01759 (2016)
11. FastText pretrained word vectors. https://fasttext.cc/docs/en/english-vectors.html
12. Srivastava, N., et al.: Dropout: a simple way to prevent neural networks from overfitting. J. Mach. Learn. Res. **15**(1), 1929–1958 (2014)
13. Bahdanau, D., Cho, K., Bengio, Y.: Neural machine translation by jointly learning to align and translate. arXiv preprint arXiv:1409.0473 (2014)

14. Kingma, D.P., Ba, J.: Adam: a method for stochastic optimization. arXiv preprint arXiv:1412.6980 (2014)
15. Conneau, A., et al.: Very deep convolutional networks for text classification. arXiv preprint arXiv:1606.01781 (2016)
16. Ioffe, S., Szegedy, C.: Batch normalization: Accelerating deep network training by reducing internal covariate shift. arXiv preprint arXiv:1502.03167 (2015)
17. Schroff, F., Kalenichenko, D., Philbin, J.: FaceNet: a unified embedding for face recognition and clustering. In: Proceedings of the IEEE Conference on Computer Vision and Pattern Recognition (2015)

Wear the Right Head: Comparing Strategies for Encoding Sentences for Aspect Extraction

Valentin Malykh[1,2,3], Anton Alekseev[1(✉)], Elena Tutubalina[1,3],
Ilya Shenbin[1], and Sergey Nikolenko[1,4]

[1] Samsung-PDMI Joint AI Center, Steklov Mathematical
Institute at St. Petersburg, St. Petersburg, Russia
`anton.m.alexeyev@gmail.com`
[2] Moscow Institute of Physics and Technology, Moscow, Russia
[3] Kazan Federal University, Kazan, Russia
[4] Neuromation OU, 10111 Tallinn, Estonia

Abstract. In this work we investigate the impact of encoding mechanisms used in neural aspect extraction models on the quality of the resulting aspects. We concentrate on the neural attention-based aspect extraction (ABAE) model and evaluate five different types of encoding mechanisms: simple averaging, self-attention with and without positional encoding, recurrent, and convolutional architectures. Our experiments on four datasets of user reviews demonstrate that, in the family of ABAE-like architectures, all models with different encoding mechanisms show the similar results in terms of standard coherence metrics for English and Russian data. Our qualitative study shows that all models yield interpretable aspects as well, and the difference in quality is often very minor.

Keywords: Aspect-based sentiment analysis · Self-attention · User reviews · Neural network · Deep learning · Aspect extraction

1 Introduction

Recent advances in neural architectures have made it into a method of choice for modern natural language processing. Many recent works in opinion mining have applied deep learning methods; see a brief overview in Sect. 2 and in the survey [19]. Neural attention-based models have been successfully applied to aspect-based opinion mining, and it is currently well established that neural models are able to identify latent topical aspects in reviews in an unsupervised way. A topical aspect is defined as an attribute or feature of the product that has been commented upon in a review and can be clustered into coherent topics, or *aspects*; e.g., *guitar* and *violin* are part of the topic *musical instruments* for music.

© Springer Nature Switzerland AG 2019
W. M. P. van der Aalst et al. (Eds.) AIST 2019, LNCS 11832, pp. 166–178, 2019.
https://doi.org/10.1007/978-3-030-37334-4_15

Recently, He et al. [8] proposed an unsupervised neural attention-based aspect extraction (ABAE) approach that encodes word-occurrence statistics into word embeddings and applies an attention mechanism to remove irrelevant words, learning a set of aspect embeddings. The ABAE model is effective in discovering meaningful and coherent aspects [8], proving that attention mechanisms are effective for the aspect extraction task as well. Yet, the classical attention mechanism is learning the weights of a linear combination of input vectors and is thus ignoring word order, i.e., it basically converts the input into a bag of words.

. In this work, we seek to answer the following research question: *Does order information help to improve aspect extraction quality?* In order to answer this question, we evaluate the impact of several popular neural network architectures (that would not discard word order information by default) used as "heads" in the ABAE model. In particular, we use recurrent and convolutional architectures as well as simple averaging and self-attention to learn sentence representations for a collection of reviews.

The paper is structured as follows. Section 2 briefly surveys related work. In Sect. 3, we begin with the model description, describing attention mechanisms (Sect. 3.1) and the existing ABAE model (Sect. 3.2). In Sect. 3.3, we define the ABAE model variations that we have experimented with. The experimental setup and results, as well as a discussion and an answer to our research question, are presented in Sect. 4. We conclude with a summary of our results and possible future research directions in Sect. 5.

2   Related Work

The deep learning revolution has affected all fields of machine learning, and natural language processing is no exception. Different neural architectures define state of the art in many NLP tasks. In particular, recurrent neural networks (RNN) have been used for natural language processing for a long time, starting from at least [11], where RNNs were used for grammatical inference. After the deep learning revolution, Mikolov et al. [12] used RNNs for language modeling, and in general recurrent architectures have been a standard technique for NLP tasks such as sentiment analysis or short text (sentence) classification for the last several years. A survey of all recent progress is beyond the scope of this paper, but we must mention the work of Bahdanau et al. [2] that introduced the attention mechanism in NLP. It was introduced for RNNs but later adapted for other neural architectures.

As for other architectures, there is a long history of using dense neural networks for text processing. Collobert et al. [5] used simple dense neural networks to build a state of the art language model. Recently, Vaswani et al. [17] used dense neural networks to achieve state of the art results in machine translation. On top of the dense networks they used a modified attention mechanism. Convolutional neural networks are also a staple of NLP, starting from the work of Collobert and Weston [4], where CNNs were used for such tasks as part-of-speech tagging. More recently, Gehring et al. [7] achieved state of the art in machine translation with a convolutional architecture.

Over the last years, aspect-based sentiment analysis has been tightly linked with topic modeling and deep learning. Topic modeling has become the method of choice for a number of applications dealing with general text-level analysis; the most popular basic model for further variation is Latent Dirichlet Allocation (LDA) [3]. For a recent overview on deep learning for opinion mining see [19].

In this work, we concentrate on the ABAE model [8]. Since it was put forward in 2017, recent studies have utilized ABAE for various NLP tasks including rating prediction [15] and user profiling [13]. Researchers from the *Airbnb* team applied ABAE to a large corpus of accomodation reviews in order to generate review summaries and user profiles [13]. They evaluated ABAE across these two tasks. For the first task of extractive summarization, they used sentence-level aspects inferred by ABAE to select representative review sentences of an accommodation for a given aspect. Aspect embeddings in ABAE were initialized using k-means centroids. For the second task, the authors used sentence-level aspects to compute user profiles by grouping all reviews coming from a given user. Quantitative and qualitative analysis conducted in [13] showed that these user profiles are effective in reranking reviews and accommodations.

Another recent model, Aspect-based Rating Prediction (AspeRa), has been proposed in [15] for learning rating- and text-aware recommender systems based on neural attention-based aspect extraction produced by the ABAE model, metric learning, and autoencoder-enriched learning. The proposed model outperformed state of the art aspect-based recommender systems on several real-world datasets of user reviews. Moreover, aspects discovered by AspeRa as a side product of the rating prediction task proved to be readily interpretable and, when evaluated in terms of standard topic coherence metrics, showed quality similar to LDA.

3   Neural Architectures with Attention for NLP

3.1   Attention Mechanisms

While attention mechanisms initially appeared in computer vision problems, they were quickly adapted to recurrent architectures in NLP. Attention mechanisms were introduced to overcome a commonly known flaw of RNNs, the lack of long-term memory: without additional modifications, RNNs can quickly forget early timesteps [10]. Attention serves as a kind of recall mechanism, allowing the network to recall different parts of the input when necessary. The already classical approach was defined in [2]; a more recent and advanced version of attention, which has also already served as a basis for many extensions, was presented in [17].

The idea could be described as choosing the most "interesting" part of the input sequence to produce the current step of the output. A soft alignment model produces *attention weights* a_i that control how much each input word influences the word currently being produced. The score a_i indicates whether the network should be focusing on this specific word right now, and z_s is the text vector that summarizes all information from the words. Since attention is soft (a_i are real numbers), the gradients are able to flow through the entire network, and the

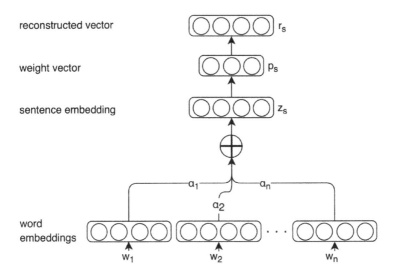

reconstructed vector r_s

weight vector p_s

sentence embedding z_s

word embeddings

w_1 w_2 w_n

a_1 a_2 a_n

Fig. 1. Architecture of the neural attention-based aspect extraction model (ABAE) [8].

model can be trained end-to-end. Soft attention drastically improves translation (see [2]) and other tasks, allowing recurrent architectures to operate with longer sentences than without it; it is now a standard approach.

More formally, the basic attention mechanism is defined as

$$a_i = \frac{\exp\left(\boldsymbol{w}_i^\top \boldsymbol{y}_s\right)}{\sum_{j=1}^{n} \exp\left(\boldsymbol{w}_j^\top \boldsymbol{y}_s\right)},$$
$$\boldsymbol{z}_s = \sum_{i=1}^{n} a_i \boldsymbol{w}_i, \tag{1}$$

where $\boldsymbol{y}_s$ is the *key vector* produced separately (we discuss it in the case of ABAE in the next section); intuitively, $\boldsymbol{y}_s$ represents the context, meaning that vectors which are closer to the current context one should have more weight; $\{\boldsymbol{w}_i\}_{i=1}^{n}$ are the *value vectors* from which we construct $\boldsymbol{z}_s$; in case of ABAE and many other NLP models, the value vectores are a sequence of word embeddings corresponding to words from the input text, and n is the number of words in the input.

3.2   Neural Attention-Based Aspect Extraction Model

The Neural Attention-Based Aspect Extraction Model [8] is a neural architecture intended to capture the topical content of input texts. Similar to classical topic modeling [3], the user chooses a finite number of topics (called *aspects* in this context), and the goal of ABAE is to learn the aspects themselves and the extent to which each document corresponds to each of the aspects.

In essence, the ABAE model is an autoencoder; the primary component of ABAE loss function is the reconstruction loss between the (weighted) sum of word embeddings used as the sentence representation and a linear combination of aspect embeddings. The sentence embedding is weighted by the so called *self-attention*, an attention mechanism where the values are embeddings of words in a sentence and the key is the mean embedding of the same words.

Figure 1 illustrates the ABAE model in more detail. The first step is to compute the embedding $z_s \in \mathbb{R}^d$ for a sentence s. In order to do this, we first compute attention weights a_i as a multiplicative self-attention model:

$$a_i = \frac{\exp\left(w_i^\top A y_s\right)}{\sum_{j=1}^n \exp\left(w_j^\top A y_s\right)}, \text{ where } y_s = \frac{1}{n}\sum_{i=1}^n w_i.$$

Here w_i is a word embedding for the word w_i, $w_i \in \mathbb{R}^d$, and $A \in \mathbb{R}^{d \times d}$ is a matrix to be learned during end-to-end training. Importantly, the attention mechanism in ABAE is slightly different from the one described above in Sect. 3.1; here, a simple dot product $w_i^\top y_s$ is replaced by a more complex bilinear transformation with a trained matrix $w_i^\top A y_s$. This modification does not change the dimension of the output vector and improves the model's expressiveness.

Once we have attention weights, we compute the text representation z_s as

$$z_s = \sum_{i=1}^n a_i w_i.$$

The second step is to compute the aspect-based sentence representation $r_s \in \mathbb{R}^d$ from an aspect embeddings matrix $T \in \mathbb{R}^{k \times d}$, where k is the number of aspects:

$$r_s = T^\top p_s, \text{ where } p_s = \text{softmax}(W z_s + b).$$

Here $p_s \in \mathbb{R}^k$ is the vector of probability weights over k aspect embeddings, and $W \in \mathbb{R}^{k \times d}$, $b \in \mathbb{R}^k$ are the parameters of a multi-class logistic regression model.

To train the model, ABAE defines the reconstruction error as the cosine distance between r_s and z_s with a contrastive max-margin objective function [18]. In addition, an orthogonality penalty term is added to the objective, which tries to learn the aspect embedding matrix T that would produce as diverse aspect embeddings as possible. The entire architecture is presented on Fig. 1.

3.3   Different Strategies for Sentence Embeddings

The part of the ABAE model described in Sect. 3.2 that we are studying in this work is how to produce sentence embeddings. ABAE employs self-attention, but it is by no means guaranteed to be best, and other options can be investigated. There are several common ways often used in literature to produce constant-size embeddings from ordered texts, including simple approaches based on averaging and more complex neural architectures of all kinds: recurrent, convolutional, or dense (see also Sect. 2 for an overview of existing approaches).

There are five model variations that we have compared in our experiments. All of them operate with word position in a sequence, but by different means. Besides, we experiment with the original self-attention model and the so called averaging model which are not using positional information. All variations are addressing only the mechanism of how to construct the sentence representation vector z_s. Next, we introduce and define each variation.

Averaging. The averaging model is one of two baseline models we are experimenting with. It is simply using the mean of input word vectors as the sentence representation z_s:

$$z_s = \frac{1}{n} \sum_{i=1}^{n} w_i.$$

Self-attention. The self-attention model is the original ABAE model, where the mean of input word vectors serves as a key vector for an attention mechanism. It has been described in detail in Sect. 3.2.

Positional Encodings. Following [17], we added positional encodings to enrich input word vectors for the attention mechanism. We use the sine and cosine functions introduced in [17] to produce a positional encoding and concatenate it to the word embedding. It is important to mention that we use positional encodings only to produce attention weights in the attention mechanism. After that, we use these attention weights in the standard fashion to produce the sentence embedding z_s.

$$a_i = \frac{\exp\left([w_i, p_i]^\top A y_s\right)}{\sum_{j=1}^{n} \exp\left([w_i, p_i]^\top A y_s\right)}, \quad \text{where } y_s = \frac{1}{n} \sum_{i=1}^{n} [w_i, p_i],$$

where $p_i \in \mathbb{R}^d$ is the positional encoding for position i. Thus, the matrix A is extended to $A \in \mathbb{R}^{2d \times 2d}$. The positional encoding p_i is computed as follows:

$$p_{(i,2j)} = sin(i/10000^{2j/d}), \quad p_{(i,2j+1)} = cos(i/10000^{2j/d}),$$

where j is an index in the vector p_i.

Recurrent Neural Network. For this variation, we preserve the information regarding word order by running a recurrent neural network (RNN) on the input. We process input word vectors with the RNN and use its outputs as input vectors for the attention mechanism. We use the last cell state as the key vector for the attention mechanism, obtaining attention weights from the attention mechanism and using them in the standard fashion.

$$y_s = c_n,$$
$$h_i = \text{RNN}(w_i, c_{i-1}),$$
$$a_i = \frac{\exp\left(h_i^\top A y_s\right)}{\sum_{j=1}^{n} \exp\left(h_i^\top A y_s\right)},$$

where c_n is the RNN memory state after reading the entire input sequence; h_i is the RNN output produced using w_i (a word embedding) and c_{i-1} (memory state from the previous timestep) as inputs.

Bidirectional RNN. The setup for bidirectional RNN is following the previously described setup for RNN. We use two identical RNNs in forward and backward directions and concatenate their outputs and cell states respectively.

Convolutional Neural Network. Another common approach to handling input sequences where order matters is to use convolutional neural networks (CNNs). We use a standard convolutional layer over the matrix E composed of (possibly padded) word embeddings for the input text. It is followed with a global max-pooling layer to produce the text embedding:

$$b_{ij} = \sum \sum k \odot E_{[i-s,i+s]},$$
$$z_s = \max_i B,$$

where $k \in \mathbb{R}^{v \times d}$ is a kernel matrix, v is the width of a kernel; $B \in \mathbb{R}^{(n-v) \times d}$ is a matrix composed of elements b_{ij}. The j axis is computed using different parallel kernels. The max operation is applied alongside the i axis.

4   Experimental Evaluation

We have used several corpora in different languages for our experiments: *City-Search* in English and *Automobiles*, *Drugs*, and *Hospitals* in Russian. The *City-Search* corpus is described in detail in [6]; it contains 50,000 reviews for New York City restaurants. The *Automobiles* corpus is described in detail in [16]; it contains 134,359 reviews for cars in Russian language. The *Drugs* and *Hospitals* datasets are an extended collection of reviews described in [1]; both corpora are available via GitHub[1]. The *Drugs* and *Hospitals* corpora contains 215,780 drug reviews and 56,441 reviews of hospitals in Russian language, respectively.

4.1   Experimental Setup

Following [8], we set the aspects matrix ortho-regularization coefficient equal to 0.1. Since this model utilizes an aspect embedding matrix to approximate aspect words in the vocabulary, initialization of aspect embeddings is crucial. The work [8] used *k-means* clustering-based initialization, where the aspect embedding matrix is initialized with centroids of the resulting clusters of word embeddings. Word embeddings length set to be 100 in our experiments.

All model variations share the basic set of hyper-parameters described above, but some variations also have additional hyper-parameters. We chose the positional encoding size to be equal to the word embedding size, so that A is a

---

[1] https://github.com/dartrevan/health_data

200×200 matrix. We use the LSTM cell as described in [9] as the basic RNN unit, using 256 LTSM units in one layer. We also utilized the LSTM unit in the bidirectional RNN, again using 256 units in the forward layer and 256 units in the backward layer.

4.2 Evaluation Metrics

For evaluation, we used *word2vec*-based metrics introduced in [14]; these metrics join the semantic properties of *word2vec* (and broadly speaking, dense word vector representations) and topic modeling in the form of top words for the topics, and they have been shown to more closely match human evaluation than other metrics. We adopt the following topic quality measures in order to evaluate the aspects discovered by our models:

(1) d_{L_2}: mean averaged L_2 distance between *word2vec* vectors corresponding to all pairs of word vectors $(\boldsymbol{w}_i, \boldsymbol{w}_j)$ for the words representing a topic t:

$$d_{L_2}(t) = \frac{1}{N(N-1)} \sum_{i \neq j \in W_t} \|\boldsymbol{w}_i - \boldsymbol{w}_j\|, \text{ where } \|\boldsymbol{v}\| = \sum_{i=1}^{d} v_i^2,$$

where N is the number of most probable words in a topic (20 in our case);
(2) d_{L_1}: similar to d_{L_2} but based on the L_1-distance:

$$\|\boldsymbol{v}\| = \sum_{i=1}^{d} |v_i|;$$

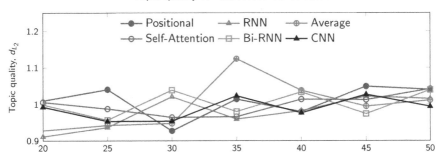

Topic quality for the Automobiles dataset

#	Pos.	RNN	Avg.	Self	BiRNN	CNN	#	Pos.	RNN	Avg.	Self	BiRNN	CNN
20	1.0095	0.9111	0.9280	1.0056	0.9987	0.9926	40	0.9798	0.9824	1.0379	1.0137	1.0335	0.9765
25	1.0403	0.9369	0.9427	0.9876	0.9569	0.9535	45	1.0487	1.0219	0.9945	1.0122	0.9740	1.0264
30	0.9274	1.0205	0.9479	0.9644	1.0386	0.9541	50	1.0398	1.0150	1.0109	1.0418	1.0391	0.9938
35	1.0138	0.9594	1.1244	0.9660	0.9800	1.0229	**Avg**	**1.0085**	**0.9782**	**0.9980**	**0.9987**	**1.0030**	**0.9886**

Fig. 2. Topic quality scores d_{L_2} on the Automobiles dataset.

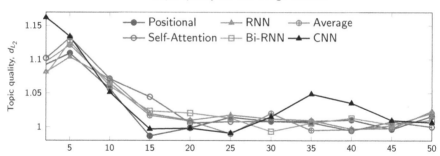

Topic quality for the Drugs dataset

#	Pos.	RNN	Avg.	Self	BiRNN	CNN	#	Pos.	RNN	Avg.	Self	BiRNN	CNN
2	1.0925	1.0815	1.0969	1.1018	1.0779	1.1618	30	1.0083	1.0120	1.0201	1.0082	0.9934	1.0151
5	1.1091	1.1036	1.1214	1.1322	1.1265	1.1337	35	1.0085	1.0098	0.9952	1.0065	1.0058	1.0487
10	1.0567	1.0709	1.0651	1.0714	1.0603	1.0513	40	1.0104	0.9971	0.9962	0.9945	1.0131	1.0352
15	0.9867	1.0196	1.0170	1.0444	1.0236	0.9970	45	0.9966	0.9979	1.0019	1.0078	1.0032	1.0096
20	0.9986	1.0099	1.0089	1.0058	1.0206	0.9982	50	1.0160	1.0233	1.0206	1.0001	1.0103	1.0066
25	1.0144	1.0177	0.9900	1.0079	1.0131	0.9909	Avg	1.0271	1.0312	1.0303	1.0346	1.0316	1.0407

Fig. 3. Topic quality scores d_{L_2} on the Drugs dataset.

(3) $d_{\cos}$: similar to d_{L_2} but based on the cosine distance:

$$\text{dist}(\boldsymbol{u}, \boldsymbol{v}) = 1 - \frac{\boldsymbol{u}^\top \boldsymbol{v}}{\|\boldsymbol{u}\|\|\boldsymbol{v}\|}, \text{ where } \|\boldsymbol{v}\| \text{ is the Euclidean distance.}$$

4.3   Results

In the first part of our experiments, we have found that all metrics are very well correlated with each other. Tables 1 and 2 show sample results for different metrics that show that the order of different model variations remains virtually the same for all metrics. Therefore, we concentrate on the L_2-based metric as being best suited for *word2vec* representations.

The experiments for the *Automobiles* corpus across different numbers of aspects are shown in Fig. 2. The experiments for the *Drugs* and *Hospitals* corpora are shown in Figs. 3 and 4, respectively. Experimental results for the *CitySearch* dataset are presented in Fig. 5. One can see that there is basically no evidence in the data that word order matters for the aspect extraction task on this corpus. For the *CitySearch* corpus the best results on average are given by simple averaging and the self-attention unit. Interestingly, for the *Automobiles* corpus the best model variation is RNN, with CNN being a close second, and the rest of the models trailing a bit behind in the average metrics. Note that for practical use, the CNN "head" might be preferable since its running time is only about 20% of the RNN running time (see Tables 1 and 2).

Next, we performed qualitative analysis of the actual aspects extracted by our models. Tables 3 and 4 show sample aspects extracted from the *CitySearch* dataset, while Tables 5 and 6 show the same kind of information for the *Automobiles* dataset. We see that the aspects are quite easy to interpret and show that

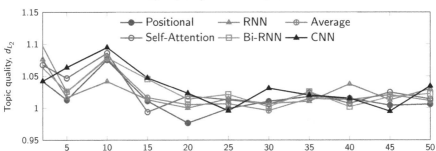

Fig. 4. Topic quality scores d_{L_2} on the Hospitals dataset.

#	Pos.	RNN	Avg.	Self	BiRNN	CNN	#	Pos.	RNN	Avg.	Self	BiRNN	CNN
2	1.0420	1.0764	1.0968	1.0676	1.0631	1.0422	30	1.0104	1.0069	0.9959	1.0059	1.0019	1.0308
5	1.0125	1.0162	1.0251	1.0468	1.0256	1.0635	35	1.0178	1.0104	1.0147	1.0237	1.0255	1.0200
10	1.0752	1.0416	1.0774	1.0788	1.0946	40		1.0149	1.0374	1.0130	1.0072	1.0020	1.0151
15	1.0107	1.0128	1.0162	0.9936	1.0454	1.0472	45	1.0041	1.0124	1.0205	1.0246	1.0174	0.9948
20	0.9765	1.0000	1.0052	1.0185	1.0137	1.0229	50	1.0058	1.0297	1.0112	1.0139	1.0222	1.0343
25	0.9991	1.0141	1.0065	1.0127	1.0211	0.9955	**Avg**	**1.0154**	**1.0235**	**1.0257**	**1.0273**	**1.0288**	**1.0328**

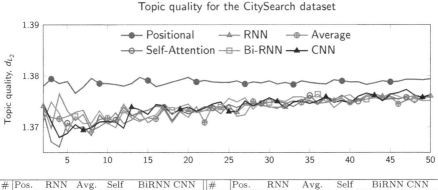

Fig. 5. Topic quality scores d_{L_2} on the CitySearch dataset.

#	Pos.	RNN	Avg.	Self	BiRNN	CNN	#	Pos.	RNN	Avg.	Self	BiRNN	CNN
5	1.3789	1.3732	1.3721	1.3708	1.3692	1.3688	30	1.3790	1.3739	1.3740	1.3748	1.3748	1.3757
10	1.3785	1.3711	1.3718	1.3707	1.3708	1.3711	35	1.3787	1.3752	1.3751	1.3762	1.3755	1.3746
15	1.3791	1.3709	1.3718	1.3733	1.3719	1.3725	40	1.3788	1.3745	1.3748	1.3752	1.3763	1.3746
20	1.3797	1.3733	1.3719	1.3726	1.3745	1.3739	45	1.3788	1.3760	1.3759	1.3755	1.3765	1.3764
25	1.3787	1.3737	1.3744	1.3744	1.3756	1.3730	50	1.3794	1.3761	1.3763	1.3757	1.3752	1.3763
26	1.3790	1.3729	1.3721	1.3738	1.3749	1.3730	**Avg**	**1.3787**	**1.3738**	**1.3736**	**1.3736**	**1.3740**	**1.3738**

the ABAE model works well for aspect extraction with basically every "head" that we studied. Note how the aspects are actually relevant for the problem at hand: e.g., some aspects about restaurants in Table 3 show different kinds of food, others are sentimental aspects (consisting of sentiment words, e.g., "fantastic", "incredible", and "fabulous"), and yet others correspond to specific aspects important for the evaluation of a restaurant (e.g., price).

Table 1. Topic quality scores on the CitySearch dataset for 8 aspects.

	Self-attention	Positional	Average	RNN	BiRNN	CNN
L_1	15.7240	15.7243	15.7296	15.7503	15.7170	15.8711
L_2	1.3684	1.3693	1.3699	1.3701	1.3708	1.3795
Cosine	0.9388	0.9396	0.9398	0.9407	0.9376	0.9527
Running time, s	406.99	520.38	327.28	2095.43	3793.22	441.99

Table 2. Topic quality scores on the Automobiles dataset for 9 aspects.

	Positional	RNN	Average	Self-attention	BiRNN	CNN
L_1	15.6838	15.7257	15.7785	15.7532	15.7195	15.8707
L_2	1.3656	1.3696	1.3734	1.3712	1.3688	1.3795
Cosine	0.9339	0.9391	0.9444	0.9415	0.9382	0.9527
Running time, s	446.03	4335.11	272.23	278.81	7963.07	438.38

Table 3. Sample topics discovered in the CitySearch dataset by ABAE with self-attention.

broiled fish grilled meat rubbery bone sardine beef overcooked snapper fatty broth
cheesecake cooky gelato banana brownie strawberry muffin vanilla tiramisu sorbet
fantastic incredible fabulous outstanding excellent terrific amazing superb wonderful
celebration celebrate celebrated celebrating anniversary daughter yesterday holiday
considering overpriced ridiculously quantity expensive tad relatively fairly

Table 4. Sample topics discovered in the CitySearch datasets by ABAE with positional encoding.

blood marg kiwi prickly pomegranate ale passionfruit yea mushi hibiscus gelee flaming
tiramisu cooky espresso cheesecake gelato brownie chocolate strawberry mousse sorbet
average lousy sub certainly okay overpriced poor unfortunately par expensive
scallop tuna salmon seared bass tartar ceviche halibut lamb duck yellowtail lobster
nite afternoon thursday monday wednesday tuesday thanksgiving sunday saturday

Table 5. Sample topics discovered in the Automobiles dataset by ABAE with self-attention.

перевозить сумка вещь отделение полагать коляска багажный холодильник влезать
расходник поломка замена гарантия ГРМ меняться менять ремонт вложение поменять
сильный заводиться зима улица тепло мороз сугроб холод справляться снег страшный
поездка путешествие дача отдых лес рыбалка местность природа сельский выезд гриб
стойка бампер пружина крыло порог амортизатор дтп уплотнительный глушитель защита

Table 6. Sample topics discovered in the Automobiles dataset by ABAE with positional encoding.

мороз заводиться зима холод теплый замерзать жара печка тепло прогреваться холодный
глушитель болезнь крыло пружина радиатор пыльник шарнир резинка стойка подшипник
дорого дешевый цена запчасть стоимость доступный дорогой дороговатый дешево
педаль скорость светофор газ разгон ускорение ускоряться разгоняться секунда старт
помещаться ребенок взрослый влезать трое двое перевозить возить умещаться коляска

5 Conclusion

We have conducted results for four corpora in different languages. We found that the word order indeed did not seem to matter for aspect extraction. For the English language corpus the simplest model won the race. At the same time, for Russian language corpora the winners were a recurrent architecture and a positional encoder that take the word order into account, albeit it improved the final results only slightly on average. The latter fact draws a question of the actual importance of the word ordering in the results.

This discrepancy brings even more questions that we pose as future research directions: is it the property of a language that could make word order important? How well did the neural architectures that we considered actually capture word order and could they do better? Besides, all corpora that we have considered in this work consist of short texts, so another natural question arises: how does the length of the input text affect word order importance? The same questions could be posed for different languages, and the results might correlate with some important differences in their linguistic structure.

Acknowledgements. Work on problem definition and model development was carried out at the Samsung-PDMI Joint AI Center at PDMI RAS and supported by Samsung Research. Work on experiments with Russian-language datasets was carried out by S.N. and E.T. and supported by the Russian Science Foundation grant no. 18-11-00284.

References

1. Alimova, I., Tutubalina, E., Alferova, J., Gafiyatullina, G.: A machine learning approach to classification of drug reviews in Russian. In: 2017 Ivannikov ISPRAS Open Conference (ISPRAS), pp. 64–69. IEEE (2017)
2. Bahdanau, D., Cho, K., Bengio, Y., Aharoni, R.: Neural machine translation by jointly learning to align and translate. In: Proceedings of International Conference of Learning Representation (2014)
3. Blei, D.M., Ng, A.Y., Jordan, M.I.: Latent Dirichlet allocation. J. Mach. Learn. Res. **3**(4–5), 993–1022 (2003)
4. Collobert, R., Weston, J.: A unified architecture for natural language processing: deep neural networks with multitask learning. In: Proceedings of the 25th International Conference on Machine Learning, pp. 160–167. ACM (2008)

5. Collobert, R., Weston, J., Bottou, L., Karlen, M., Kavukcuoglu, K., Kuksa, P.: Natural language processing (almost) from scratch. J. Mach. Learn. Res. **12**(Aug), 2493–2537 (2011)
6. Ganu, G., Elhadad, N., Marian, A.: Beyond the stars: improving rating predictions using review text content. In: WebDB, vol. 9, pp. 1–6. Citeseer (2009)
7. Gehring, J., Auli, M., Grangier, D., Yarats, D., Dauphin, Y.N.: Convolutional sequence to sequence learning. In: Proceedings of International Conference on Machine Learning (2017)
8. He, R., Lee, W.S., Ng, H.T., Dahlmeier, D.: An unsupervised neural attention model for aspect extraction. In: Proceedings of the 55th Annual Meeting of the Association for Computational Linguistics, vol. 1: Long Papers, pp. 388–397 (2017)
9. Hochreiter, S., Schmidhuber, J.: Long short-term memory. Neural Comput. **9**(8), 1735–1780 (1997). Based on TR FKI-207-95, TUM (1995)
10. Kirkpatrick, J., et al.: Overcoming catastrophic forgetting in neural networks. Proc. Natl. Acad. Sci. **114**(13), 3521–3526 (2017)
11. Lawrence, S., Giles, C.L., Fong, S.: Natural language grammatical inference with recurrent neural networks. IEEE Trans. Knowl. Data Eng. **12**(1), 126–140 (2000)
12. Mikolov, T., Karafiát, M., Burget, L., Černocký, J., Khudanpur, S.: Recurrent neural network based language model. In: Eleventh Annual Conference of the International Speech Communication Association (2010)
13. Mitcheltree, C., Wharton, V., Saluja, A.: Using aspect extraction approaches to generate review summaries and user profiles. In: Proceedings of the 2018 Conference of the North American Chapter of the Association for Computational Linguistics: Human Language Technologies, vol. 3 (Industry Papers), pp. 68–75 (2018)
14. Nikolenko, S.I.: Topic quality metrics based on distributed word representations. In: Proceedings of the 39th International ACM SIGIR Conference on Research and Development in Information Retrieval, pp. 1029–1032. ACM (2016)
15. Nikolenko, S.I., Tutubalina, E., Malykh, V., Shenbin, I., Alekseev, A.: AspeRa: aspect-based rating prediction model. In: Azzopardi, L., Stein, B., Fuhr, N., Mayr, P., Hauff, C., Hiemstra, D. (eds.) ECIR 2019. LNCS, vol. 11438, pp. 163–171. Springer, Cham (2019). https://doi.org/10.1007/978-3-030-15719-7_21
16. Tutubalina, E.: Identifying product failures from reviews in noisy data by distant supervision. In: Ngonga Ngomo, A.-C., Křemen, P. (eds.) KESW 2016. CCIS, vol. 649, pp. 142–156. Springer, Cham (2016). https://doi.org/10.1007/978-3-319-45880-9_12
17. Vaswani, A., et al.: Attention is all you need. In: Advances in Neural Information Processing Systems, pp. 5998–6008 (2017)
18. Weston, J., Bengio, S., Usunier, N.: WSABIE: scaling up to large vocabulary image annotation. In: Walsh, T. (ed.) IJCAI, pp. 2764–2770. IJCAI/AAAI (2011). http://dblp.uni-trier.de/db/conf/ijcai/ijcai2011.html#WestonBU11
19. Zhang, L., Wang, S., Liu, B.: Deep learning for sentiment analysis: a survey. Wiley Interdisc. Rev. Data Min. Knowl. Discov. **8**(4), e1253 (2018)

Combined Advertising Sign Classifier

Valentin Malykh[1,3,4]([⊠]) and Aleksei Samarin[2,4]

[1] Neural System and Deep Learning Laboratory,
Moscow Institute of Physics and Technology, Dolgoprudny, Russia
`valentin.malykh@phystech.edu`
[2] Saint-Petersburg State University, Saint-Petersburg, Russia
[3] Kazan (Volga Region) Federal University, Kazan, Russia
[4] VK Research, Saint-Petersburg, Russia

Abstract. The article describes the problem of classifying photographs of advertising signs of commercial establishments according to the type of services provided. The proposed solution is based on the *sharing of textual and visual features*. We provide a *composite model* that includes a *text recognition module* and an extractor of *visual characteristics* to improve classification accuracy. We achieve F_1 of 0.24 exceeding strong baseline quality for 10%.

Keywords: Sharing of textual and visual features · Composite model · Optical character recognition · Visual characteristics

1 Introduction

At the present time in pattern recognition and its marketing applications, the problem of recognizing advertising signs is urgent [1–3]. The signs recognition problem is extremely difficult since the signage fonts are often unique and their size can vary arbitrarily. But at the same time, there is a narrower problem of classifying advertising signs, which can be solved partly by means of optical text recognition, partly by means of image classification.

Currently, there are a lot of methods for classifying images that have demonstrated their effectiveness using open data sets [4–6]. It should also be noted that dissimilar objects images classifying is significantly different from the problem of classifying advertising signs by the nature of the objects represented in the image. The first significant difference is the apparent external similarity between all the signs, that makes separating them by subject matter much more difficult than classifying objects from general datasets [7,8]. This circumstance determines the low results of the models for the classification of heterogeneous images for the problem considered in this article (Sect. 4). The second essential feature of signage as visual objects is a presence of textual data, in some cases containing the most useful information for classifying an object. Today, there are many approaches for semantic classification of photographs of documents based only on optical text recognition with subsequent analysis of text data, based

© Springer Nature Switzerland AG 2019
W. M. P. van der Aalst et al. (Eds.) AIST 2019, LNCS 11832, pp. 179–185, 2019.
https://doi.org/10.1007/978-3-030-37334-4_16

on tools such as [9–11]. These methods have proven themselves efficient in solving problems associated with dividing photographs of documents by type, but that approaches do not demonstrate high efficiency in solving problems of advertising signs classification (Sect. 4). The reasons for this are such factors as the possible lack of sufficient information in the text of the advertising sign and the difficulty of solving the problem of optical text recognition with varying angles, fonts, styles of signage and lighting (Fig. 3), that is typical for photographs of advertising signs.

Our paper describes a neural network method based on the extraction of persistent visual characteristics and characteristics derived from textual information presented on an advertising sign. Thus, this method shows better efficiency than methods that use only visual information or based only on the analysis of the text recognized during photo processing (see in Sect. 4).

2   Problem Statement

In the current article, we consider the problem of advertising signs classification by the type of advertised institution or the scope of services provided. That problem can be formulated as follows. Image containing advertising sign Q should be assigned to one of the classes $C = C_i$, where $i \in [0, N]$.

It should be noted that in addition to the formal statement of the classification problem, to address the requirements imposed resistance to noise and various shooting conditions (varying the angle and lighting).

We use the following assumptions. Images are taken using a camera fixed on a car, following along the roadway, therefore: (a) may contain visual defects - sun glare, defocus areas, noise, including those that greatly impede optical text recognition; (b) angle, framing, lighting and colour balance are unknown; (c) the scale and placement of advertising signs in the frame can vary greatly.

3   Proposed Method

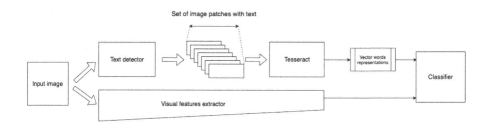

Fig. 1. Architectural diagram of the proposed solution.

Since the most progressive and effective methods for classifying images at the present time are based on convolutional neural networks, that by a large margin surpassed all other approaches on open data sets for image classification [4–6], one of our classifier modules is based on the model [6]. This model is a sequence of convolutional, fully connected layers and residual connections [6,12]. As shown in Sect. 4 it is possible to improve the classification of photographs of advertising signs by adding textual information analysis. For that purpose, a special module is added to the overall architecture of the classifier, which is a combination of a text detector [10], OCR engine and text model (Fig. 1). The output of the visual features extractor and the text model output are combined by concatenation, then the resulting vector is projected onto a space of dimension 4 (according to the number of classes). The SoftMax function is superimposed on the projected vector. The values obtained are interpreted as the probabilities of the classes.

3.1   Visual Feature Extractor

As an extractor of visual signs, several modern architectures were chosen, which proved themselves (at different times) to be the best on the task of image classification. There were selected: MobileNet, described in Howard's work [6]; Inception, described in [4]; ResNet, described in [5].

Additionally, experiments were performed with the excluded module for the selection of visual signs.

3.2   Text Recognition Module

According to the above, in addition to the purely visual features extractor, the proposed model includes a text information analyzer. The process of analyzing textual information consists of two main stages. The first stage is an optical recognition procedure, the result of which is applied to the original image is the text that could be recognized in the input photo of the poster. Recognition occurs in two stages. The first step is to localize image areas containing textual information, for which the EAST detector, described in [10], is used. It is a universal neural network text detector, resistant to a varying angle. Its architecture is depicted in Fig. 2. The advantages of this detector are the use of a multi-channel neural network to highlight the characteristic signs of text and then localize the text position of the scene space.

Fig. 2. EAST architecture [10].

Further, the parts of the image in which a text was found are transmitted to the input of the module of optical recognition, in the role of which Tesseract is used [11]. An OCR module output, i.e. character sequences, are transmitted to the text model input.

3.3   Text Embedding Module

As a text processing model, several approaches were used: word2vec, RoVe, and without any processing. The text processing module receives a sequence of characters from the text recognition module, splits this sequence into spaces and inputs the models of the vector representation of words to the input.

No Embedding. In this approach, extracted textual data is completely ignored and only visual characteristics are used.

Word2vec. This approach is based on the word2vec vector representation of the words described in the work of Mikolov 2013 [13]. Vector representations are taken for all words that are in the dictionary of this model, then they are averaged, and the resulting vector is used as a feature vector in the classifier.

RoVe. This approach is based on the vector representation of the words described in [14]. This model possesses an designed resistance to noise, which includes the construction of a special character-based representation of a word. Similar to the approach word2vec, the vector representations of words derived from the model are averaged and used as a feature vector.

For this method, there is a significant hyper-parameter affecting the quality of the model performance on the data. It stands for the level of noise in the training data. The noise level is defined as the probability of applying a replacement symbol in the input text. In this paper, two noise levels are considered: 0.0 (corresponding to the "clean" text) and 0.05 (which is a characteristic of natural text according to [15]).

4   Experiments

Metric which was used as the quality metric in all experiments is F_1. It is formulated as follows:

$$Precision = \frac{TP}{TP + FP}, \tag{1}$$

$$Recall = \frac{TP}{TP + FN}, \tag{2}$$

$$F_1 = \frac{2 \cdot Precision \cdot Recall}{Precision + Recall}, \tag{3}$$

where TP is a number of objects which are known to belong to the class in question and are marked up by a model to belong to this class; FP is a number of objects which are known to be from some other class but are marked up by a model to belong to the class in question; and FN is a number of objects which are known to belong to the class in question but are marked up by a model to belong to some other class. The F_1 score is formulated to measure quality for one class only. To measure a multi-class classification the F_1 scores for individual classed are averaged.

Visual Features. For developing the architecture described above, we used methods based on following neural network architectures: Inception [4], ResNet [5], MobileNet [6]. All of the models for visual feature extraction were trained on ImageNet dataset presented in [8].

Text Embedding. The text embedding module presented by word2vec, which was trained on Google News dataset[1], and RoVe architecture. The latter architecture has hyper-parameter of a noise level used during training process. We trained two models for RoVe architecture: with 0.0 and 0.05 levels. To train these models we have used Reuters dataset [16].

Dataset. All of the above versions of the sign classifier model were trained on a dataset based on an open set of photographs of advertising signs of establishments [17]. The photos in this dataset were obtained under different lighting conditions and camera angles. They were divided into 4 categories according to the type of services provided (hotels, shops, restaurants, "other"). In the category of "other", there are signs that can not be clearly attributed to any of previous three categories. Examples for each class are presented on Fig. 3.

The dataset consists of 350 images (approximately 88 samples per class). These images were split to a train set of 280 images, a validation set of 35 images, and a test set of 35 images. The class ratio in the all three subsets was explicitly set for matching.

Fig. 3. Examples of images from various classes of the considered data set: (a) image of a hotel signboard; (b) image of a store sign; (c) image of a restaurant's signboard; (d) image of an advertising sign from the category "other".

Results. The results of the listed model architectures are presented in Table 1. We could draw several conclusions from the table:

- Inception architecture for obtaining visual characteristics is non-optimal for this task, the other two examined architectures show similar results;

---

[1] We use publicly available pre-trained model which could be accessed here: https://drive.google.com/file/d/0B7XkCwpI5KDYNlNUTTlSS21pQmM/edit?usp=sharing.

– accounting text information significantly increases the quality of the classification of advertising images;
– the textual information itself is sufficient for classification, although the visual features improve the classification quality.

The best result was shown by the system using RoVe vector representations with a noise level close to the natural one and visual features extracted by MobileNet and ResNet architectures.

Table 1. F-measure for classification systems on a test set.

Visual	Textual			
	-	Word2vec	RoVe (0.0)	RoVe (0.05)
-	0.15	0.21	0.20	0.22
Inception	0.05	0.15	0.15	0.15
MobileNet	0.14	0.22	0.21	**0.24**
ResNet	0.15	0.20	0.22	**0.24**

5   Conclusion

In this paper we show the effectiveness of a model for classification of photographs of advertising signs based on composite neural network architecture. The proposed model has demonstrated better efficiency in comparison to methods based on the use of only visual or only textual features. It should be noted that we created a model allows us to retrieve noise-resistant features from images of advertising signs, and thus allows us to build a classifier of images of advertising posters according to the type of services provided, superior to efficiency classifiers based only on visual or textual information, according to Sect. 4. According to obtained results, the further development of the proposed approach can be focused on the study of the properties of the model on other classes of recognizable posters, and an improvement its characteristics due to the use of more expressive image descriptors, and the model performance optimization.

References

1. Intasuwan, T., Kaewthong, J., Vittayakorn, S.: Text and object detection on billboards. In: 2018 10th International Conference on Information Technology and Electrical Engineering (ICITEE), pp. 6–11 (July 2018)
2. Zhou, J., McGuinness, K., O'Connor, N.E.: A text recognition and retrieval system for e-business image management. In: Schoeffmann, K., et al. (eds.) MMM 2018. LNCS, vol. 10705, pp. 23–35. Springer, Cham (2018). https://doi.org/10.1007/978-3-319-73600-6_3
3. Watve, A., Sural, S.: Soccer video processing for the detection of advertisement billboards. Pattern Recogn. Lett. **29**(7), 994–1006 (2008)

4. Szegedy, C., et al.: Going deeper with convolutions. In: 2015 IEEE Conference on Computer Vision and Pattern Recognition (CVPR), pp. 1–9 (June 2015)
5. He, K., Zhang, X., Ren, S., Sun, J.: Deep residual learning for image recognition. In: 2016 IEEE Conference on Computer Vision and Pattern Recognition (CVPR), pp. 770–778 (June 2016)
6. Howard, A.G., et al.: MobileNets: efficient convolutional neural networks for mobile vision applications (2017)
7. Lin, T.-Y., et al.: Microsoft COCO: common objects in context. In: Fleet, D., Pajdla, T., Schiele, B., Tuytelaars, T. (eds.) ECCV 2014. LNCS, vol. 8693, pp. 740–755. Springer, Cham (2014). https://doi.org/10.1007/978-3-319-10602-1_48
8. Deng, J., Dong, W., Socher, R., Li, L., Li, K., Fei-Fei, L.: ImageNet: a large-scale hierarchical image database. In: 2009 IEEE Conference on Computer Vision and Pattern Recognition, pp. 248–255 (June 2009)
9. Tian, Z., Huang, W., He, T., He, P., Qiao, Y.: Detecting text in natural image with connectionist text proposal network. In: Leibe, B., Matas, J., Sebe, N., Welling, M. (eds.) ECCV 2016. LNCS, vol. 9912, pp. 56–72. Springer, Cham (2016). https://doi.org/10.1007/978-3-319-46484-8_4
10. Zhou, X., et al.: East: an efficient and accurate scene text detector. In: 2017 IEEE Conference on Computer Vision and Pattern Recognition (CVPR), pp. 2642–2651 (July 2017)
11. Smith, R.: An overview of the Tesseract OCR engine. In: Ninth International Conference on Document Analysis and Recognition (ICDAR 2007), vol. 2, pp. 629–633 (September 2007)
12. Liu, T., Fang, S., Zhao, Y., Wang, P., Zhang, J.: Implementation of training convolutional neural networks. CoRR. arXiv:1506.01195 (2015)
13. Mikolov, T., Sutskever, I., Chen, K., Corrado, G.S., Dean, J.: Distributed representations of words and phrases and their compositionality, vol. 26, p. 10 (2013)
14. Malykh, V.: Robust word vectors for Russian language. In: Proceedings of Artificial Intelligence and Natural Language AINL FRUCT 2016 Conference, Saint-Petersburg, Russia, pp. 10–12 (2016)
15. Cucerzan, S., Brill, E.: Spelling correction as an iterative process that exploits the collective knowledge of web users, vol. 4, pp. 293–300 (2004)
16. Lang, K.: NewsWeeder: Learning to filter netnews. In: Proceedings of the Twelfth International Conference on Machine Learning, pp. 331–339 (1995)
17. Wang, K., Babenko, B., Belongie, S.: End-to-end scene text recognition. In: Proceedings of the 2011 International Conference on Computer Vision, ICCV 2011, pp. 1457–1464. IEEE Computer Society, Washington (2011)

A Comparison of Algorithms for Detection of "Figurativeness" in Metaphor, Irony and Puns

Elena Mikhalkova[1]($\boxtimes$) [ID], Nadezhda Ganzherli[1] [ID], Vladislav Maraev[2] [ID], Anna Glazkova[1] [ID], and Dmitriy Grigoriev[3]

[1] University of Tyumen, Tyumen, Russia
{e.v.mikhalkova,n.v.ganzherli,a.v.glazkova}@utmn.ru
[2] University of Gothenburg, Gothenburg, Sweden
vladislav.maraev@gu.se
[3] OOO ITSK, Tyumen, Russia

Abstract. Figurative speech is an umbrella term for metaphor, irony, sarcasm, puns and some other speech genres and figures of speech. In research and competitions like SemEval, each of them is usually processed separately with a task-specific model. However, being altogether called "figurative speech", they should share some property: "figurativeness". If such a property exists, figurative speech can be processed simultaneously by one and the same algorithm. The present research compares performance of several NLP methods that were designed to detect one type of figurative speech (either metaphor, or irony, or puns) on short texts containing a combination of these types. The study shows that, despite being task-specific, state-of-the-art algorithms are able to process different types of figurative speech fairly well, and some of them are good even at cross-detection when the training set contains one type and the test set another.

Keywords: Figurative speech · Figurativeness · Metaphor · Irony · Pun · Cross-detection

1 Introduction

The word *figurative* collocated with *speech, language, meaning, utterance,* and *speech act* usually means *metaphoric* [3], *poetic, uncommon* [17], *creative* [14] and is opposed to *literal, clear, precise* [17]. In [17], the authors note that, although it is hard to draw a strict line between figurative and literal speech, different types of figurative speech share communicative goals. From where we assume that there can as well be other properties that unite all types of figurative speech and, hence, a software that can process them similarly well. In further research, its architecture can tell us what this property is.

The reported study was funded by RFBR according to the research project No. 18-37-00272.

In the present study, we consider three types of figurative speech: puns, irony and metaphor, to evaluate algorithms in cross-detection of these phenomena. By **cross-detection** we mean such a process in which these algorithms designed for one particular type are used for another or a combination, thereof. In machine learning, it would also imply that the training set should contain one type of figurative speech (or a mixture of types) and the test set another.

2   Related Work

There have been previous attempts to simultaneously process several types of figurative speech. First, in [16] the authors use the same algorithm for processing irony and humor. However, they end up distinguishing humour from irony, and humour from other topic-specific sentences (politics, etc.). The authors of [11] suggest to use their algorithm of pun detection (PunFields) for irony detection in Twitter. However, its performance is much lower on irony than on puns. Next, in [7] the authors note that their program designed for sarcasm detection incorrectly labels metaphors in a non-sarcastic text as sarcasm, which may indicate that their system is good at figurative speech detection. Finally, while this article was prepared, another research was published [18] describing a novel task for evaluation of humorousness and metaphor novelty of a text, the latter feature pointing at difference between a creative text and a text containing dead metaphors.

3   Experiment Setup

In this work, we consider figurative speech detection as a binary classification task on datasets with different types of figurative speech. Each algorithm is tested on a trivial combination of the same class (e.g. *irony/None* in the training and *irony/None* in the test set) and then on a mixed combination that we earlier called cross-detection (e.g. *irony/None* in the training and *puns/None* in the test set).

3.1   Dataset

Our dataset includes four classes of short texts: irony, puns, metaphor, and None. The sentences with irony were taken from SemEval-2018 [20]; puns – from SemEval-2017 [12]. Due to scarcity of extended *creative* metaphors in the existing corpora, our metaphor collection is composed of sentences from several corpora and sites: 673 sentences from [2]; 100 from 2018 VUA [8]; 82 from [13][1]; the rest are manually selected metaphors from NLTK corpora [1] (Reuters, Gutenberg), blogs and sites[2].

---

[1] Only sentences with metaphoricity index over 0.7.
[2] The full list of sources is available at https://github.com/evrog/FS_Datasets_AIST_2019.

The sentences in the None-class were harder to obtain. As they should not contain any figurativeness, we manually searched for them in several sources: non-ironic tweets from SemEval-2018 [20]; sentences without puns from SemEval-2017 [12]; paraphrased sentences from VUA [8], and literal sentences from NLTK corpora [1]. The final corpus of None-class includes 1,200 samples.

With texts of the None-class collected, we shuffled the three other datasets, cut their size to 1,200 samples and split into two sub-sets of 1,000 as the training set and 200 as the test set. For algorithms including neural networks, we took 100 samples from the training set and used them as a validation set. Table 1 describes the final size of datasets.

Table 1. Distribution of texts across datasets. Train-1 is used for algorithms without neural networks; Train-2 plus Validation – with neural networks.

Set	Metaphor	Irony	Puns	None
Train-1	1,000	1,000	1,000	1,000
Train-2	900	900	900	900
Val.	100	100	100	100
Test	200	200	200	200

3.2   Baseline Algorithms

In our analysis, we use three task-agnostic algorithms. The first is an ML classifier suggested in [19] to process irony in Twitter. It uses a TF-IDF smoothed Bag-of-Words algorithm with a Linear SVM classifier implemented with the help of the Scikit-learn library [15].

As deep learning models generally demonstrate good results in figurative speech processing [4,7–9,21], the two other baselines are standard deep learning models that learn from tokens, including special characters (punctuation, emoji, etc.). Hence, the text preprocessing is minimal: converting to lower case and tokenization. Our RNN-based model architecture combines three layers: Embedding, LSTM and Dense. The Convolutional neural network includes an Embedding layer, 1D Convolutional layer+ReLu, a Max-pooling layer, another 1D Convolutional layer+ReLu, a Max-over-time pooling layer, and a Dense layer[3].

3.3   State-of-the-Art Algorithms

As follows from Sect. 2, detection of several types of figurative speech by means of one and the same algorithm is quite rare. As for detection of one particular

---

[3] Hyperparameters for CNN: $d = 100, h_1 = h_2 = 7, k_1 = k_2 = 40$. Hyperparameters for RNN: $d = 50, n = 20, \epsilon = 0.1$. The scripts for NNs are available at https://github.com/evrog/Incongruity.

type (e.g. irony, sarcasm, metaphor), such a task is very common at the international NLP competitions [5,8,12,20]. There is also standalone research that often compares its result to existing achievements.

In our experiment, we chose several algorithms that showed efficient performance at the mentioned competitions or surpassed earlier results. However, we were able to run or reproduce only some of them. The short list of algorithms that we finally used is as follows and further referred to as state-of-the-art:

1. PunFields at SemEval-2017 [10] and SemEval-2018 [11]: an SVM-based classifier using Roget's Thesaurus to define semantic classes of words in a text.
2. THU-NGN for Irony Detection [21]: a neural network using Glove vectors.
3. A system [4] for Metaphor Detection: a neural network with embeddings trained on the authors' own corpus.

As concerns [4], due to Allennlp installation issues, we use the baseline version of the system that we tested by running the file "baseline/lexicalBaseline.py". Initially its result is $F_1 = 0.51$ and $A = 0.76^4$ for VUA corpus [8], $F_1 = 0.63$ and $A = 0.71$ for [2], $F_1 = 0.31$ and $A = 0.44$ for [13].

The system of [4] is designed for detecting metaphorical usage of verbs. While figurative speech is not necessarily expressed via verbs, they are often a part of a phrase that should be treated in two meanings simultaneously. To evaluate the system's performance on our dataset, we lemmatized each text using NLTK [1], selected verbs and ran the system on each verb. If the text referred to class None, the correct answer was no verbs used metaphorically. As for types of figurative speech, finding at least one metaphorical usage of a verb in a text was considered to be a correct answer.

4   System Performance and Error Analysis

Performance of the tested algorithms is reflected in Table 2. As concerns the usual I/I, P/P and M/M combinations, results of baseline algorithms (the first three columns) are much higher than those of the state-of-the-art systems. At SemEval-2018 [20], the Linear SVM baseline was better than many of the competing systems. Whereas in that competition it had to classify between ironic and non-ironic *tweets*, presently its performance on irony is even higher, as the None class contains fewer tweets and many topic-specific sentences from news reports and romantic novels. As our neural networks, CNN and RNN, were better tuned to detect these differences, their result on the Irony dataset is unbeatable. And, as could be expected, neural networks mainly erred attributing non-ironic tweets from None to irony (*home finally ! what a way to start the christmas break ! ! !*). At the same time, in most cases of cross-detection, i.e. when training and test sets contain different classes, these algorithms show very low results, with the exception of CNN that outperformed all other systems in M/P, as concerns accuracy. However, currently, we are not ready to conclude about its grounds.

---

4 A=accuracy.

Table 2. Cross-detection of figurative speech: F_1—F-score, acc.—accuracy; I—irony, P—puns, M—metaphor. Lin. SVM and PunFields were tested with 5-fold cross-validation; the result given is the average of 5 tests. The result in bold is the best result in a column, i.e. the best for each system. The underlined result is the best result in a row, i.e. the best result among all the systems for a given combination, e.g. for I/I.

Dataset ↓	Lin. SVM	CNN	RNN	PunFields	THU-NGN	Gao [4]
Measure →	F_1, acc.	F_1, acc.	F_1, acc.	F_1, acc.	F_1, acc.	F_1, acc.
I/I	**0.81 0.83**	**0.89 0.89**	**0.89 0.89**	0.59 0.60	**0.77 0.76**	0.19 0.51
I/P	0.18 0.51	0.29 0.54	0.26 0.53	0.52 0.56	0.22 0.51	0.18 0.52
I/M	0.10 0.49	0.31 0.55	0.32 0.55	0.68 0.68	0.45 0.61	0.18 0.52
P/I	0.33 0.54	0.36 0.56	0.41 0.58	0.44 0.51	0.57 0.59	0.53 0.59
P/P	0.69 0.72	0.86 0.86	0.86 0.86	0.63 0.64	0.68 0.68	0.53 0.60
P/M	0.39 0.57	0.52 0.64	0.49 0.62	0.45 0.51	0.58 0.61	**0.57** 0.65
M/I	0.37 0.52	0.35 0.54	0.27 0.53	0.38 0.54	0.43 0.59	0.48 0.53
M/P	0.50 0.59	0.55 0.64	0.46 0.61	0.41 0.54	0.56 0.56	0.48 0.54
M/M	0.73 0.73	0.82 0.82	0.82 0.83	**0.70 0.72**	0.69 0.65	0.56 **0.67**

As for achievements in cross-detection, the state-of-the-art systems show an incomparably better performance. If the baselines have 10 best results (underlined in Table 2) in combinations of I/I, P/P and M/M, the state-of-the-art systems have the same number in cross-detection.

PunFields, that was designed for puns, presently shows its best performance in detecting metaphors. Also, it is the only algorithm that processes a combination of different types (I/M) better than it processes a single type (I/I) and the type it was designed for (P/P). Processing I/M, PunFields mainly erred in cases of short texts (*the flowers moved in the wind*), fiction (*the misfortune of Harriet's cold had been pretty well gone through before her arrival*) and non-ironic tweets some of which, probably, contain metaphor (*Main issue with the walking dead- you forget to breathe when youre watching*).

As could be expected, THU-NGN, a neural network learning from pre-trained embeddings, does its best in single-class classification in the combination of I/I. However, to our surprise, designed for irony, THU-NGN gives a fairly even top performance in cross-detection, when trained on puns and metaphors. Regarding its errors, beside shortness, wrongly classified texts do not reveal any obvious regularity, which could also speak in favor of the system as being insensitive to the mentioned topic specificity of non-figurative texts.

The result of Gao [4] is very preliminary due to the mentioned technical issues. However, we can trace that it fails to process irony and is usually better at processing puns than metaphors.

As concerns column-wise results, all the systems but PunFields and Gao are best at detection of irony. We tend to attribute these results to the mentioned topic specificity of None-class, as all these algorithms use word embeddings or

Bag-of-Words. As for PunFields, its best performance on M/M dataset is very close to that of Linear SVM baseline which PunFields also uses. However, the features it uses are unusual in NLP[5], and it could be that its performance on puns and irony is only good because they contain metaphors.

5 Conclusion and Future Work

The main focus of our study was to compare performance of different algorithms in detection of three types of figurative speech: metaphor, irony and puns. We tested two types of systems: baseline and state-of-the-art. As for systems that perform best on each dataset, in cases, when the training and test set contained the same type of figurative speech, the CNN and RNN baselines are the best. The same stands for state-of-the-art systems in cross-detection (combinations of I/P, I/M, etc.). Surprisingly, in our experiment state-of-the-art systems perform better on types they were not designed for originally. Beside that, the performance of some algorithms (PunFields for irony, THU-NGN for puns) in cross-detection (e.g. I/P, I/M) approaches that in single-class detection (I/I). Also, all algorithms but one were best in detecting irony in I/I combination, probably due to the specificity of tweets.

At this stage, we find it too early to conclude about the reasons and suggest any improvements. We plan to continue our research with a study on combinations of features, associated with figurative speech in different works (e.g. ambiguity, polarity, unexpectedness in [16], incongruity in [6], etc.) and their influence on performance. We would also like to elaborate on the datasets to select purely one-class texts, i.e. such texts that contain only one type of figurative speech, which requires a survey of human respondents.

References

1. Bird, S., Klein, E., Loper, E.: Natural Language Processing with Python: Analyzing Text with the Natural Language Toolkit. O'Reilly Media, Inc., Sebastopol (2009)
2. Birke, J.: A clustering approach for the unsupervised recognition of nonliteral language. Ph.D. thesis, School of Computing Science-Simon Fraser University (2005)
3. Cohen, T.: Figurative speech and figurative acts. J. Philos. **72**(19), 669–684 (1976)
4. Gao, G., Choi, E., Choi, Y., Zettlemoyer, L.: Neural metaphor detection in context. arXiv preprint: arXiv:1808.09653 (2018)
5. Ghosh, A., et al.: SemEval-2015 task 11: sentiment analysis of figurative language in Twitter. In: Proceedings of the 9th International Workshop on Semantic Evaluation (SemEval 2015), pp. 470–478 (2015)
6. Hempelmann, C.F., Attardo, S.: Resolutions and their incongruities: further thoughts on logical mechanisms. Humor Int. J. Humor Res. **24**(2), 125–149 (2011). https://doi.org/10.1515/HUMR.2011.008

---

[5] We do not know of any other systems that use Roget's Thesaurus in detection of figurative speech.

7. Joshi, A., Tripathi, V., Patel, K., Bhattacharyya, P., Carman, M.: Are word embedding-based features useful for sarcasm detection? arXiv preprint: arXiv:1610.00883 (2016)
8. Leong, C.W.B., Klebanov, B.B., Shutova, E.: A report on the 2018 VUA metaphor detection shared task. In: Proceedings of the Workshop on Figurative Language Processing, pp. 56–66 (2018)
9. Mihalcea, R., Strapparava, C., Pulman, S.: Computational models for incongruity detection in humour. In: Gelbukh, A. (ed.) CICLing 2010. LNCS, vol. 6008, pp. 364–374. Springer, Heidelberg (2010). https://doi.org/10.1007/978-3-642-12116-6_30
10. Mikhalkova, E., Karyakin, Y.: PunFields at SemEval-2017 task 7: employing Roget's thesaurus in automatic pun recognition and interpretation. In: Proceedings of the 11th International Workshop on Semantic Evaluation (SemEval-2017), pp. 426–431 (2017)
11. Mikhalkova, E., Karyakin, Y., Voronov, A., Grigoriev, D., Leoznov, A.: PunFields at SemEval-2018 task 3: detecting irony by tools of humor analysis. In: Proceedings of the 12th International Workshop on Semantic Evaluation, pp. 541–545 (2018)
12. Miller, T., Hempelmann, C., Gurevych, I.: SemEval-2017 task 7: detection and interpretation of English puns. In: Proceedings of the 11th International Workshop on Semantic Evaluation (SemEval-2017), pp. 58–68 (2017)
13. Mohammad, S., Shutova, E., Turney, P.: Metaphor as a medium for emotion: an empirical study. In: Proceedings of the Fifth Joint Conference on Lexical and Computational Semantics, pp. 23–33 (2016)
14. Moreno, R.E.V.: Creativity and Convention: The Pragmatics of Everyday Figurative Speech, vol. 156. John Benjamins Publishing, Amsterdam (2007)
15. Pedregosa, F., et al.: Scikit-learn: machine learning in Python. J. Mach. Learn. Res. **12**, 2825–2830 (2011)
16. Reyes, A., Rosso, P., Buscaldi, D.: From humor recognition to irony detection: the figurative language of social media. Data Knowl. Eng. **74**, 1–12 (2012)
17. Roberts, R.M., Kreuz, R.J.: Why do people use figurative language? Psychol. Sci. **5**(3), 159–163 (1994)
18. Simpson, E., Do Dinh, E.L., Miller, T., Gurevych, I.: Predicting humorousness and metaphor novelty with Gaussian process preference learning. In: Proceedings of the 57th Annual Meeting of the Association for Computational Linguistics (2019)
19. Van Hee, C.: Can machines sense irony?: Exploring automatic irony detection on social media. Ph.D. thesis, Ghent University (2017)
20. Van Hee, C., Lefever, E., Hoste, V.: SemEval-2018 task 3: irony detection in English tweets. In: Proceedings of the 12th International Workshop on Semantic Evaluation, pp. 39–50 (2018)
21. Wu, C., Wu, F., Wu, S., Liu, J., Yuan, Z., Huang, Y.: Thu_ngn at SemEval-2018 task 3: tweet irony detection with densely connected LSTM and multi-task learning. In: Proceedings of the 12th International Workshop on Semantic Evaluation, pp. 51–56 (2018)

Authorship Attribution in Russian with New High-Performing and Fully Interpretable Morpho-Syntactic Features

Elena Pimonova(ID), Oleg Durandin(ID), and Alexey Malafeev$^{(\boxtimes)}$(ID)

National Research University Higher School of Economics,
Nizhny Novgorod, Russia
hpimonova@gmail.com, oleg.durandin@gmail.com,
amalafeev@yandex.ru

Abstract. This work tackles the problem of modeling author style in Russian. In particular, we solve the task of authorship attribution using the collected dataset of 30 authors, 1506 texts written in the period of 18th–21st century. We apply various approaches to solving the attribution problem: Random Forest, Logistic Regression, SVM Classifier. In terms of text representation, we use seven models in three language levels: lexis, morphology, and syntax. Most importantly, we propose our own set of morpho-syntactic features that perform on about the same level as doc2vec, but are fully interpretable. The conducted experiments show the effectiveness of their standalone use, as well as the increase in the quality of classification when using these attributes along with the classic doc2vec-based approach. All code, including feature extraction, is made freely available. Additionally, we analyze the performance of individual features as style markers. Finally, we study classification errors in order to identify the patterns in the misattribution of specific authors.

Keywords: Authorship attribution · Author style · Text classification · Text representation · Morpho-syntactic features · Language feature engineering · Machine learning · Natural language processing

1 Introduction

The task of authorship attribution, i.e. identifying the author of a given text while having access to a pool of various other texts written by known authors, has existed for decades. Yet the problem has not been completely solved. Nowadays there are many approaches to authorship attribution in the English language, whereas there are not so many algorithms or methods for authorship attribution in Russian. Most existing methods are based on simple statistical models of text representation (for example, in the form of character bigrams or trigrams). Although these models show sufficiently high accuracy in clustering and classification problems, the results are difficult or impossible to interpret. We assume that more linguistically sound models for text representation could help determine which language components are actual style markers. In other words, character n-grams and such do help identify authors of texts, but do they tell us anything about what author style is? With this in mind, we set a goal

© Springer Nature Switzerland AG 2019
W. M. P. van der Aalst et al. (Eds.) AIST 2019, LNCS 11832, pp. 193–204, 2019.
https://doi.org/10.1007/978-3-030-37334-4_18

to increase the interpretability of text representation models in solving the attribution problem in order to determine by which language means the author style is expressed. In achieving this goal, we also strive to obtain classification accuracy comparable to existing models.

2   Related Work

One of the first algorithms for authorship attribution in Russian was proposed by Dmitriy Khmelev [5] and was based on character bigrams. Given a set of texts with known authorship, a transition matrix of character bigram frequencies was created. It served as a probability estimate for texts of unknown authorship. A large corpus of texts was used (1000 works, 82 authors of Russian literature). With the help of such a relatively simple model, Khmelev managed to achieve 82% accuracy. A similar study was conducted by Leonid Borisov et al. [2]. They compared bi-, tri- and fourgram-based text representation models. They also studied the dependence of clustering and classification accuracy on text length and the homogeneity of texts written by the same author. The authors concluded that the classification problem is best solved by using the model of character trigrams (on the dataset of 300 texts by 30 authors, the authorship was determined unmistakably). At the same time, the clustering problem was best solved by using the character bigram model ($\sim 85\%$ accuracy).

A slightly different approach was presented by Vasily Poddubny et al. [13]. They represented texts as word frequency matrices, which could help identify words and phrases that are most relevant in terms of style differences. However, they found out that such an approach shows a tendency to separating texts according to their functional speech style (e.g. business style, academic style, etc.). Thus, word frequency analysis is not always sufficient for highlighting the style of a particular author.

Another approach to the problem of authorship attribution is morpho-syntactic analysis. This approach was proposed by the group of creators of the information and analytical system "SMALT" (Alexander Rogov et al.) [14]. They used morphological and syntactic criteria, namely frequency analysis of POS, morphological characteristics, syntactic units and syntactic relations. The authors gave preference to syntactic analysis, as did Harald Baayen et al., who leveraged syntactic annotation to improve authorship attribution accuracy [1]. From the texts annotated with the CCPP and TOSCA parsers they extracted relations based on a constituency grammar. These relations comprise a phrase model of constituent units (for example, NP $\rightarrow$ DTP + N, where NP is Noun Phrase, DTP is Determiner Phrase, N is Noun) with function labels (for example, SU for Subject) and attribute labels (for example, sing for singular). As a result, there were 4194 types of tokens. Their frequency in the text was used as attributes to determine the author of the text. The researchers compared the accuracy of word-based and syntax-based approaches and came to the conclusion that the syntactic method shows as good results as the word-based one. Thus, it may be a new useful technique for authorship attribution.

In addition to the methods used to solve the problem of authorship attribution, it is useful to consider models of text representation in the adjacent problem of author profiling. In [8], the authors suggest using the following parameters as attributes:

(1) type-token ratio (calculated by V/C formula, where V is the dictionary size, and C is the corpus size), (2) word length, (3) readability (value obtained by averaging the length of words and sentences), (4) POS distribution, (5) function word frequency, (6) chunks (frequent combinations). The authors also note the important role of syntactic analysis arguing that the author does not consciously control the syntactic characteristics and therefore they may form reliable markers of style. More complex linguistic characteristics were used by Polina Panicheva and Yanina Ledovaya [11]. In the Russian Facebook texts, they identified people belonging to the so-called Dark Triad, i.e. people with a pronounced tendency to Machiavellianism, Narcissism and Psychopathy. For this purpose, they performed morphological and semantic analysis using PyMorphy library and word2vec model. Following this approach, they revealed the words and morphological characteristics that are most typical of this category of people.

The authors of the article "Cross-lingual syntactic variation over age and gender" (Anders Johannsen, Dirk Hovy, and Anders Søgaard) were engaged in determining the gender and age of the author on the basis of six languages, namely English, French, German, Italian, Spanish and Swedish [4]. To accomplish such a large-scale task, they used the Universal Dependencies markup format. After receiving POS and syntactic relations tag-sets, they extracted subtrees consisting of three tokens called treelets. The authors decided not to make a distinction between right and left relations going from the root word, although they made the separation of treelets consisted of one, two, and three tokens. A treelet of one token is part of speech, a treelet of two tokens is a relationship between the main and dependent word (for example, VERB $\to$ (nsubj) NOUN). A treelet of three words can be one of the two kinds: (1) when two words come from one main word (for example, NOUN $\leftarrow$ VERB $\to$ NOUN), (2) when words are lined up in a chain of successive subordination (for example, VERB $\to$ NOUN $\to$ PRON). The combination of all possible tokens served as attributes for determining author's age and gender. According to the results of the study, syntax features make it possible to identify some gender and age differences that are consistently manifested in different languages.

Thus, there are many different approaches to authorship attribution. In general, one can observe a tendency to unite linguistic and statistical/mathematical models of author style. In this work we combine several approaches, namely lexical, morphological, syntactic and semantic, to compare the effectiveness of different text representation models and find out whether the results can become more interpretable without losing classification accuracy. In the following sections we describe the methodology adopted in the present study.

3   Tools

The processing part of our work was implemented in Python using some external libraries. In particular, for natural language processing we used the spaCy library. It was chosen because it provides a convenient pipeline with NLP tools for at all basic levels of natural language processing (word and sentence tokenization, morphological and syntactic analysis, support for vector representations of words). The official website

[17] provides language models for seven languages, but, unfortunately none for Russian. However, a Russian language model is currently being developed by enthusiasts [15] and is freely available, which allowed us to conduct research using this latest tool. This model was tested on SyntagRus corpus [3] and at present it has an accuracy of UAS = 90.467, LAS = 87.227. The accuracy of POS-tagging is 93.796.

The analysis in spaCy was done using the Universal Dependencies (UD) markup format [9]. The UD project is aimed at developing a universal markup format for various languages. The UD format is relevant for this study for a number of reasons. Firstly, the theoretical framework for UD is the dependency relation method, which is best suited for Russian. Secondly, the UD creators work not only on quantity, but also on quality, reducing the percentage of markup errors associated with transferring the outlined packages into a universal format to a minimum. In UD the morphological markup is also carried out in addition to syntactic markup. However, we do not use it in all models of text representation because it does not contain some parts of speech that are Russian-specific (e.g., the so-called 'state words' like *dosadno* that is approximately equivalent to *it is a shame*). Considering this, we also use the PyMorphy2 library [6] for part of speech tagging. In addition, we use such well-known toolkits for machine learning and statistical processing as Scikit-Learn [12], NumPy and Pandas.

4 Data Collection

The material for the study was 215 works of Russian literature, created by 30 authors spanning XVIII–XXI centuries. We used works by Andreev, Astafiev, Bianchi, Bulgakov, Bulychev, Bunin, Vasilyev, Gogol, Goncharov, Gorky, Dostoyevsky, Zhitkov, Zamyatin, Karamzin, Lermontov, Lukyanenko, Nabokov, Nosov, Platonov, Prishvin, Pushkin, Rasputin, Skrebitskiy, Solzhenitsyn, Sologub, Tolstoy, Turgenev, Chernyshevsky, Chekhov, Sholokhov. The principles of selecting material were as follows:

1. The selected authors have played a significant role in Russian literature. They are recognized by the Russian and international community and write in a rich, literary language.
2. The analyzed works were created from the end of the 18th to the beginning of the 21st century, that is, in a broad sense, texts written in modern Russian.
3. Each author's works were selected in such a way that they covered only one approximate period of the writer's creative life. This was done to minimize the changes in the same writer's style throughout his life.

The length of the works by different authors varied greatly. In order to equalize the volume of texts per each author, texts were divided into blocks of 350 sentences (in some cases, texts were taken entirely if they consisted of a smaller number of sentences). Thus, we obtained 1506 blocks of text.

5   Text Representation Models

Since our goal was to increase the interpretability of the results, we used linguistic models of representation, which differed in complexity and emphasis on different levels of language. In total, we had 7 models. Let us consider each of them in more detail. At this stage, each text is 350 sentences or less. For each sentence in the text, we performed syntactic analysis with the Russian spaCy model. As output, we obtained parse trees where each word corresponded to a part of speech. Syntactic relations between words were also established.

5.1   The Simple Morphology and Simple Syntax Models

In the simple morphology model, we obtained relative frequencies for each part of speech in the text. There were 17 of them (e.g., ADJ for an adjective). Similarly, the relative frequencies of syntactic relations were also calculated and presented as attributes of the simple syntax model. Their number was 35 (e.g., obj for a direct object).

5.2   The Complex Morphology Model

In order to increase interpretability, we decided to use more complex features. This concerned both morphology and syntax. In our opinion, for morphology, an increase in interpretability was more necessary. Since syntax itself is a more complex level of language, including all subordinate units and relations, its interpretability should be higher than that of morphology. To improve the morphological model, we decided to develop new criteria for the morphological markup of the text. It seemed to us that using semantic features of words is more effective than using just their general categorical meaning. To this end, we chose the OpenCorpora [10] markup used in PyMorphy, since it contains all Russian parts of speech, while in the Universal Dependencies tagset used in spaCy, some of them are not represented because of its compatibility with a large number of languages.

Thus, we obtained POS-tagging for all words with PyMorphy and classified them into groups according to their semantic properties. For example, adjectives in the full and concise forms, as well as the comparative degree, were combined into one group that we called "attribute". The most ambiguous part of speech in terms of determining semantic features appeared to be the noun. It is known that the noun can have various semantic characteristics (e.g., object, process, state, etc.), so we decided to determine for each noun in the text its semantic feature. We used Shvedova's Russian semantic dictionary [16] to achieve this goal. We split all nouns into the following groups: concrete, abstract, attribute, process and state nouns. The nouns that were not found in the semantic dictionary were added to a special group and categorized manually.

Thus, we obtained 13 groups of words with common semantic features (aside from those that characterize nouns, there were such groups as pronoun, attribute, process, number, action descriptor, state words and function words). The main drawback of our separation is the unsolved problem of ambiguity. A polysemantic word in its various senses may possess different semantic features, and, therefore, a certain error is present in the separation. However, theoretically, even with some inaccuracies, the usage of this semantics-based grouping should give better results that can be interpreted.

Then we developed the criteria for lexico-morphological analysis of the text. For example, the so-called 'action descriptiveness' was calculated as follows: the number of words belonging to the action descriptor group that consists of adverbs was divided by the number of words that belong to the process group (verbs, infinitives, participles and verbal participles). In total, we got 16 criteria: abstractness, objectivity, pronominal replacement, action feature, generalized action, descriptiveness, action descriptiveness, number, dynamism, state, real modality, passive, present tense, past tense, future tense and action completeness. They were used as features in the classification task.

5.3   The Complex Syntax Model

Similarly, a more complex syntactic model was developed. It was initially intended to represent attributes on three levels of syntactic unit complexity: phrase, simple sentence and complex sentence. Later, we decided to combine the latter two levels into one due to the peculiarities of spaCy library: relations between clauses in a complex sentence are not singled out separately, but are considered within the same inter-word relations. At the phrase level, we developed criteria for combining phrases into groups if they have the same: (1) communication type (coordination, agreement, regimen, contiguity), (2) structural type (complex phrase, simple phrase), (3) degree of phrase components unity (syntactically free and non-free phrase), (4) lexico-grammatical type (nominal phrase, verbal phrase, adverbial phrase). The values for these features were calculated as normalized frequencies of relations representing each group (e.g., coordination is represented by such relations as nsubj, csubj, expl, csubj: pass and nsubj: pass).

At the sentence level, we considered contracted and uncontracted sentences. We considered as contracted all sentences with an interjection or a particle as the root word. The vocatives (the root word is a proper name) and the genitive sentences (the root word is a noun in the genitive or partitive case) were also counted. We also made a division into two-member and one-member sentences (among them there were definitely personal, indefinitely personal, infinitive, impersonal and nominative). A number of complex structures were also considered. Among them were parenthetic and epenthetic constructions, interjections, appeals, appositions, homogeneous constructions, adjectival and participial constructions. As a result, 28 attributes were developed in accordance with a complex syntactic representation model.

5.4   The Treelet Bigram and Trigram Models

The idea of this text representation model was taken from [4]. According to this model, text is represented in the form of treelets, that is, typed relationships between tokens. Treelet bigrams represent the relationship between two tokens, namely between the main and dependent word (for example, nsubj relation between VERB and NOUN). Treelet trigrams can be of two types: with two dependent words and one main word (for example, NOUN $\leftarrow$ VERB $\rightarrow$ ADV) and with consecutive subordination of words (for example, VERB $\rightarrow$ NOUN $\rightarrow$ ADJ). The set of all possible treelets serves as attributes for this representation model.

5.5 The Doc2vec Model

For comparison, we chose one of the modern embedding-based text representation methods, namely doc2vec [7], as a widely used baseline for text classification. A distinctive feature of this method is the linking of words to each other in context, that is, predicting a word based on nearby words, or vice versa, predicting a set of words based on one specific word. Identifying the set of semantically close words for each author, that is, the individual lexicon, in theory, could serve as a good model for solving the authorship attribution problem. In addition, the doc2vec method uses artificial neural networks as classification algorithms. Such an approach seems more promising than a simple frequency analysis of word forms in the text.

6 Experiments

We solve the text classification problem using the types of text representation mentioned above. In our study, three classification algorithms were considered, namely Random Forest (with 20 base estimators and default parameters from the scikit-learn framework), Logistic Regression («One-VS-Rest» multiclassification type, l1-regularization) and SVM with a linear kernel (Linear SVC from the scikit-learn framework). To evaluate classification quality, we used stratified 3-fold cross-validation and the standard metrics of accuracy, precision, recall and F1-score. When presenting the results, we will focus on accuracy as the main metric as the sample was balanced. The best classifier turned out to be the Logistic Regression algorithm, and the second in efficiency for the authorship attribution task was Linear SVC (see Fig. 1). The difference between them is insignificant. Sometimes Linear SVC surpassed Logistic Regression (for example, with the model of simple and complex morphology and treelet bigrams). Random Forest showed worse results. Still, we used it in order to find out which of the features appear to be the most important for particular types of representation.

The best representation model turned out to be the doc2vec model. It showed 93% accuracy with the Logistic Regression algorithm. After we tuned the doc2vec parameters, accuracy increased to 98% (vector dimension = 100, window size = 5). This result was expected because doc2vec is the state-of-the-art model. The second most effective model was simple syntax. Its accuracy with Logistic Regression was 89%. The simple morphology model, which is methodologically similar to simple syntax, performed worse (on average, 10% worse than syntax). The same trend is observed with complex syntax and morphology (on average, syntax is 14% better). This allows us to draw the first conclusion that syntax-based models are more relevant for solving the authorship attribution problem than morphological ones. The reason for this may be that with the complexity of the language level individual features become more visible. Morphology conveys features of the Russian language as a whole, and syntax, being more complex, shows how the authors combine lower-level language units in accordance with their individual preferences. The second observation we made was that simpler models consistently show better accuracy than complex ones. This applies to both morphology and syntax, as well as treelet bigrams and trigrams. Although in the case of treelets, the difference is not so noticeable (e.g., the accuracy with the Logistic

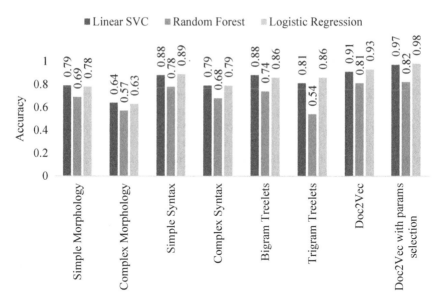

Fig. 1. Accuracy of single text representation models in solving the classification task.

Regression algorithm is the same, 86%). In general, the treelet models showed quite a high result. In comparison with all models of text representation, treelet trigrams ranked third in terms of efficiency.

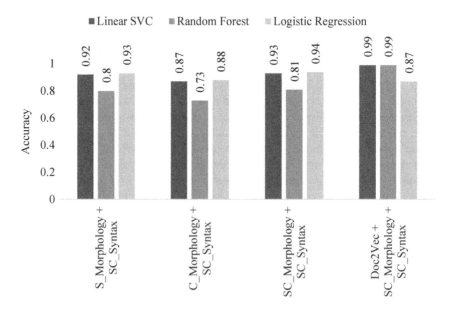

Fig. 2. Classification accuracy with combined text representation models.

After we obtained classification results for all representation models, we decided to also test model combinations. We made four: (1) simple morphology with syntactic models, (2) complex morphology with syntactic models, (3) all morphologic and syntactic models, (4) all morphologic and syntactic models with doc2vec model. In general, combining different models of textual representation led to increased classification accuracy (see Fig. 2). It is noteworthy that the combination of simple morphology and syntax in the Logistic Regression led to the accuracy that was originally in the best doc2vec model (93%). The combination of all morphological and syntactic models showed even better result (94%). Their combination with the doc2vec model resulted in the highest accuracy: 99%. This proves that text representation as a set of morphological and syntactic features is an efficient method for the authorship attribution task. These features improved the quality of the best doc2vec model, and even their standalone use led to quite good accuracy in comparison with other models. Most importantly, unlike doc2vec, they have the property of interpretability. This is confirmed by the analysis of important features provided by the Random Forest algorithm.

7 Important Feature Analysis

The analysis of important features was carried out for all morpho-syntactic models of representation. Among the important were, for example:

1. voice as the only grammatical category selected as a style marker,
2. phrase link type (coordination, agreement, contiguity),
3. various complex structures (adjectival, participial, epenthetic constructions and appositions) (Table 1).

A certain regularity may be observed in the selection of feature significance. Elements and relations at a simple level are part of a more complex level and continue to be assessed as important. Thus, we can conclude that these features stand out not by chance but naturally and, therefore, can really be regarded as markers of style. In addition, as they are interpretable, it is possible to find explanation why this or that feature stood out as significant.

For example, of function words, the particle and conjunction fell into the category of important ones. Contrarily, the preposition, being a function word, turned out to be at the very end of the significance distribution list. This may be due to the fact that prepositions are almost never chosen freely. They are used to designate relations between words, and, therefore, their presence in the texts is of more deterministic character, so they can hardly serve as markers of style. Particles and conjunctions, by contrast, vary greatly in texts authored by different people. Their usage is more optional, which can really serve as an indicator of author preferences.

Of content words, nouns identified as important features in the simple morphology model are reflected in the complex morphology in the objectivity coefficient. This coefficient can tell whether the author reasons in his works more about the objective world or about the world of abstract phenomena. At the same time, nouns are also reflected in the model of simple and complex syntax. In the simple syntax model, we have identified the subject as an important attribute. In most cases, it is exactly the noun

Table 1. The relationship of important features in different models of representation.

Simple morphology	Complex morphology	Simple syntax	Complex syntax
particle	–	discourse (emotional evaluation components)	–
conjunction	–	conj (relationship between homogeneous members)	homogeneous members as a complicator of the sentence
noun	objectivity (used in the text to state facts)	nsubj (connection between subject and predicate)	coordination and agreement
adverb	action feature and action descriptiveness	admod and advcl (relationship between the main word and modifier)	contiguity

that acts in the role of subject. In the complex syntax model, nouns are associated with coordination and agreement. Coordination, being a connection between the subject and the predicate, once again confirms the importance of this feature. Agreement as a typical type of connection between nouns and adjectives, can show whether the author is inclined to characterize objects. It is also noteworthy that, in addition to coordination and agreement, contiguity was also highlighted as an important type of communication.

In the complex syntax model, a number of complex structures were considered significant. Among them were adjectival, participial, epenthetic constructions and appositions. The identification of these features as significant is easy to explain. The use of complicators largely depends on the preferences of the author, since the basic information in the sentence can be conveyed without them.

8 Error Analysis

Another stage of our work was error analysis. We assumed that classification algorithms can naturally assign texts of one author to another. In this case, there is reason to believe that styles of these writers, who cannot be distinguished from each other, are similar. We analyzed confusion matrices in all representation models. We did not take individual errors into account, since it is very likely that this classification inaccuracy arose by chance or was due to the sample or algorithm bias. Under our attention were the repeated mistakes of the classifier when it mistakenly assigned texts of one author to another multiple times, namely 2–4 times (occasional misattribution) or always (consistent misattribution).

There were not so many errors of the second type. In almost every case, these mistakes were made by one model and not by others. As far as occasional misattribution is concerned, based on the number of such errors for each author, we ranked all writers into three groups: (1) 0–3 errors, (2) 4–6 errors, (3) 7+ such errors. It can be assumed that writers of the first group have a unique style that is very difficult to confuse with another (e.g., Sholokhov, Andreev, Gorky, Karamzin, Solzhenitsyn,

Tolstoy, among others). The writers of the third group, on the contrary, have an ambiguous style that is more difficult to identify (e.g., Vasilyev, Pushkin, Prishvin, Nosov, Gogol, Bulgakov). The authors of the second group are between these two extremes (e.g., Nabokov, Chernyshevsky, Goncharov, Lukyanenko, etc.).

It is noteworthy that the majority of the authors are in the first group. Some authors regularly had errors in different models of text representation. On this basis, we decided to determine which authors had such errors and whether it is possible to find some justification for this. There were 17 cases of such regular errors (when texts of one author were assigned to another one more than once). One of the most interesting examples is the misclassification of texts by Bulychev and Nosov in both directions (Bulychev $\rightarrow$ Nosov: 3, Nosov $\rightarrow$ 3 Bulychev: 4). This mistake can be explained by the fact that these writers created works in one genre, namely child fiction. Thematically, their books are also similar. They describe the adventures of primary school age children. Another example is the misattribution of Pushkin's texts to Karamzin. This phenomenon may be explained by the fact that Pushkin was a follower of Karamzin and learned from the latter's works, possibly adopting style elements from his predecessor. Thus, error analysis can help identify patterns in the misattribution of specific authors.

9   Conclusion and Future Work

In this study, we solved the task of authorship attribution using various text representation models. The best single model turned out to be the embedding-based doc2vec. With tuned parameters, its accuracy with the Logistic Regression algorithm reached 98%. We also proposed our own morpho-syntactic text representation models, whose standalone use also yields a comparable result (94%). When combined with doc2vec, these models make it possible to further improve the quality: 99%. Most importantly, they are fully interpretable, which makes it possible to determine linguistic markers of style. The code, including feature extraction, is freely available[1]. Finally, we also carried out an error analysis and identified the possible reasons for author misattribution.

In the future, we plan to test our models on other tasks related to stylometry (e.g. author profiling), as well as plagiarism detection tasks. Furthermore, an interesting area of research is the cross-lingual aspect and the identification of universal markers of style. Another possible direction for future work is testing the scalability of the proposed approach.

---

[1] https://github.com/OlegDurandin/AuthorStyle.

References

1. Baayen, R., van Halteren, H., Tweedie, F.: Outside the cave of shadows: using syntactic annotation to enhance authorship attribution. Lit. Linguist. Comput. **11**(3), 121–132 (1996)
2. Borisov, L., Orlov, Y., Osminin, K.: Authorship attribution by the distribution of letter combination frequencies. 27th edn. Institute of Applied Mathematics named after M. Keldysh of the Russian Academy of Sciences, Moscow (2013)
3. Dyachenko, P., et al.: The current state of the deeply annotated corpus of Russian language texts (SinTagRus). In: Proceedings of the Institute of Russian Language named after V.V. Vinogradov, vol. 6, pp. 272–299 (2015)
4. Johannsen, A., Hovy, D., Søgaard, A.: Cross-lingual syntactic variation over age and gender. In: Proceedings of the Nineteenth Conference on Computational Natural Language Learning: CoNLL, pp. 103–112. Association for Computational Linguistics, Beijing (2015)
5. Khmelev, D.: Recognition of the text author using the Markov chains. MSU Bull. **9**(2), 115–126 (2000)
6. Korobov, M.: Morphological analyzer and generator for Russian and Ukrainian languages. In: Khachay, M.Y., Konstantinova, N., Panchenko, A., Ignatov, D.I., Labunets, V.G. (eds.) AIST 2015. CCIS, vol. 542, pp. 320–332. Springer, Cham (2015). https://doi.org/10.1007/978-3-319-26123-2_31
7. Le, Q., Mikolov, T.: Distributed representations of sentences and documents. In: ICML 2014 Proceedings of the 31st International Conference on Machine Learning, pp. 1188–1196. JMLR, Beijing (2014)
8. Luyckx, K., Daelemans, W., Vanhoutte, E.: Stylogenetics: clustering-based stylistic analysis of literary corpora. In: Proceedings of LREC 2006: The 5th International Language Resources and Evaluation Conference, Workshop Towards Computational Models of Literary Analysis, pp. 30–35. ILC, Genova (2006)
9. de Marneffe, M.-C., et al.: Universal Stanford dependencies: a cross-linguistic typology. In: Proceedings of the 9th International Conference on Language Resources and Evaluation (LREC), pp. 4585–4592. European Language Resources Association (ELRA), Reykjavik (2014)
10. OpenCorpora. http://opencorpora.org/. Accessed 30 Apr 2019
11. Panicheva, P., Ledovaya, Y., Bogolyubova, O.: Lexical, morphological and semantic correlates of the dark triad personality traits in Russian facebook Texts. In: Proceedings of the AINL FRUCT 2016 Conference, pp. 72–79. Institute of Electrical and Electronics Engineers Inc., St. Petersburg (2017)
12. Pedregosa, F., et al.: Scikit-learn: machine learning in Python. JMLR **12**, 2825–2830 (2011)
13. Poddubny, V., Shevelev, O., Kravtsova, A., Fatykhov, A.: Vocabulary and analytical block of the Style Analyzer. In: 14th Russian Scientific and Practical Conference, pp. 138–140. Tomsk University, Tomsk (2010)
14. Rogov, A., Sidorov, U., Solopova, A., Surovtsova, T.: The information-analytical system "SMALT". In: International Conference "Dialogue 2007", pp. 470–474. Petrozavodsk State University, Bekasovo (2007)
15. Russian language models for spaCy. https://github.com/buriy/spacy-ru. Accessed 21 Apr 2019
16. Shvedova, N.: Russian Semantic Dictionary. Explanatory Dictionary, Systematized by Classes of Words and Meanings, 3rd edn. Azbukovnik, Moscow (2003)
17. spaCy. https://spacy.io/. Accessed 21 Apr 2019

Evaluation of Sentence Embedding Models for Natural Language Understanding Problems in Russian

Dmitry Popov, Alexander Pugachev, Polina Svyatokum,
Elizaveta Svitanko[(✉)], and Ekaterina Artemova

National Research University Higher School of Economics, Moscow, Russia
{dgpopov,avpugachev,posvyatokum,eisvitanko}@edu.hse.ru,
echernyak@hse.ru

Abstract. We investigate the performance of sentence embeddings models on several tasks for the Russian language. In our comparison, we include such tasks as multiple choice question answering, next sentence prediction, and paraphrase identification. We employ FastText embeddings as a baseline and compare it to ELMo and BERT embeddings. We conduct two series of experiments, using both unsupervised (i.e., based on similarity measure only) and supervised approaches for the tasks. Finally, we present datasets for multiple choice question answering and next sentence prediction in Russian.

Keywords: Multiple choice question answering · Next sentence prediction · Paraphrase identification · Sentence embedding

1 Introduction

With word embeddings have been in the focus of researchers for several decades, sentence embeddings recently started to gain more and more attention. A word (sentence) embedding is a projection in a vector space of relatively small dimensionality, that can capture word (sentence) meaning by making the embeddings of two words (sentences) that are similar to get closer in this vector space. With no doubts, the usage of the properly trained word embeddings boosted the quality of the majority of natural language processing (NLP), information extraction (IE), neural machine translation (NMT) tasks. However, it seems that word embedding models are facing their limits when it comes to polysemy and ambiguity. A natural solution to these problems lies in sentence embedding models too, as they allow to capture the context–dependent meaning of any part of the sentence.

For the last two years, the amount of projects and papers on sentence embeddings has increased dramatically. Several research groups show that a complex pre–trained language model may serve both as an input to another architecture and as a standalone sentence embedding model. The most famous models of this type, ELMo, and BERT, named after characters of Sesame Street show, can be

© Springer Nature Switzerland AG 2019
W. M. P. van der Aalst et al. (Eds.) AIST 2019, LNCS 11832, pp. 205–217, 2019.
https://doi.org/10.1007/978-3-030-37334-4_19

treated as "black boxes", that read a sentence in and output a vector representation of the sentence. The efficiency of these models for the English language is well studied already not only in several natural language understanding (NLU) tasks but also for language modeling and machine translation. However little has been done to explore the quality of sentence embeddings models for other languages, including Russian, probably due to the absence of NLU datasets.

The contribution of this paper is two–fold. **First**, we create two novel NLU datasets for the Russian language: (a) multiple choice question answering dataset, which consists of open domain questions on various topics; (b) next sentence prediction dataset, that can be treated as a kind of multiple choice question answering. Given a sentence, one needs to choose between four possible next sentences. Correct answers are present in both datasets by design, which makes supervised training possible. **Second**, we evaluate the quality of several sentence embedding models for three NLU tasks: multiple choice question answering, next sentence prediction and paraphrase identification.

The results confirm, that in the tasks under consideration, sentence–level representations perform better than the word–level ones as in many other tasks.

The remainder is structured as follows. Section 2 presents an overview of related works; Sect. 3 introduces the datasets, Sect. 4 describes the methods we used to tackle the tasks. The results of the experiments are presented in Sect. 5. Section 6 concludes.

2   Related Work

A word embedding is a dense vector representation of a word that allows modeling some sort of semantic (or functional) word similarity. Two words are considered similar if a similarity measure, such as cosine function, for example, between corresponding vectors is high enough. As this definition is rather vague, there are two main approaches to evaluating the quality of a word embedding model. Intrinsic evaluation is based on conducting on standard word pairs and analogies datasets, such as Word–353[1] or Simlex–999[2]. External evaluation requires an external machine learning tasks, such as sentiment classification or news clustering, which can be evaluated by a quality measure, such as accuracy or Rank index. All factors of machine learning algorithms are held equals so that these quality measures are affected by the word embeddings model only.

The methodology of creation and evaluation of sentence embeddings is less developed, when compared to the zoo of word embedding models. The evaluation of sentence embeddings models is usually conducted of natural language understanding tasks, such as semantic textual similarity, natural language inference, question answering, etc. Further, we overview the basic and more advanced models of sentence embeddings.

---

[1] http://www.cs.technion.ac.il/~gabr/resources/data/wordsim353/.
[2] https://fh295.github.io/simlex.html.

2.1   Unsupervised Sentence Embeddings

The simplest way of obtaining the sentence embedding is by taking the average of the word embeddings in the sentence. The averaging can treat words equally, rely on $tf - idf$ weights [1], take the power mean of concatenated word embeddings [17] and employ other weighting and averaging techniques.

Another approach of unsupervised training of sentence embeddings is Doc2Vec [12], which succeeds after Word2Vec and FastText. Word2Vec [13] by Mikolov et al. is a predictive embedding model and has two main neural network architectures: continuous Bag–of–Words (CBoW) and continuous skip–gram [14]. Given a central word and its context (i.e., k words to the left and k words to the right), CBoW tries to predict the context words based on the central one, while skip–gram on tries to predict the context words based on the central one.

Joulin et al. [9] suggest an approach called FastText, which is built on Word2Vec by learning embeddings for each subword (i.e., character n–gram, where n can vary between some bounds and is a hyperparameter to the model). To achieve the desired word embeddings, the subword embeddings are averaged into one vector at each training step. While this adds a lot of additional computation to training, nevertheless it enables word embeddings to encode sub–word information, which appears to be crucial for morphologically rich languages, the Russian language being one of them.

The goal of the aforementioned Doc2Vec approach, introduced by Mikolov et al. [12], is to create an embedding of a document, regardless of its length.

All the studies as mentioned earlier work properly on the English language and some of them release pre–trained Russian embeddings, too. Russian–specific RusVectores [11] pre–trained model, which was trained on Russian National Corpora, Russian Wikipedia and other Web corpora possesses several pre–trained word embeddings models and allows to conduct a meaningful comparison between the models and their hyperparameters.

The Skip–Thought model follows the skip–gram approach: given a central sentence, the context (i.e., the previous and the next) sentences are predicted. The architecture of Skip–Thought consists of a single encoder, which encodes the central sentence and two decoders, that decode the context sentences. All three parts are based on recurrent neural networks. Thus their training is rather difficult and time–consuming.

2.2   Supervised Sentence Embeddings

In recent years, several sources say the unsupervised efforts to obtain embeddings for larger chunks of text, such as sentences, are not as successful as the supervised methods. Conneau et al. [5] introduced the universal sentence representation method, which works better than, for instance, Skip–Thought [10] on a wide range of transfer tasks. It's model architecture with BiLSTM network and max–pooling is the best current universal sentence encoding method. The paper on Universal Sentence Encoder [3] discovers that transfer learning using sentence embeddings, which tends to outperform the word level transfer. As an advantage, it needs a small amount of data to be trained in a supervised fashion.

2.3   Language Models

One of the recently introduced and efficient methods are embeddings from Language Models (ELMo) [16] that models both complex characteristics of word use, and how it is different across various linguistic contexts and can also be applied to the whole sentence instead of the words. Bidirectional Encoder Representations from Transformers (BERT) [6] has recently presented state–of–the–art results in a wide variety of NLP tasks including Natural Language Inference, Question Answering, and others. The application of an attention model called Transformer allows a language model to have a better understanding of the language context. In comparison to the single–direction language models, this one uses another technique namely Masked LM (MLM) for a bidirectional training.

2.4   Evaluation of Sentence Embedding Models

While different embedding methods are previously discussed, the most suitable evaluation metric is also a challenge. Quality metrics are largely overviewed in RepEval[3] proceedings. In [2] both widely–used and experimental methods are described. SentEval is used to measure the quality of sentence representations for the tasks of natural language inference or sentence similarity [4]. The General Language Understanding Evaluation (GLUE) benchmark[4] is one of the popular benchmarks for evaluation of natural understanding systems. The top solutions, according to the GLUE leadership, exploit some sort of sentence embedding frameworks. GLUE allows testing any model in nine sentence or sentence–pair tasks, such as natural language inference (NLI), semantic textual similarity (paraphrase identification, STS) or question answering (QA).

3   Datasets

3.1   Multiple Choice Question Answering (MCQA)

The portal geetest.ru provides tests on many subjects such as history, biology, economics, math, etc., with most of the tasks being simple wh–questions. These tests were downloaded to create a multiple choice question answering dataset.

Every test is a set of questions in a specific area of a certain subject. For the final dataset we handpicked tests from the following subjects: Medicine, Biology, History, Geography, Economics, Pedagogy, Informatics, Social Studies.

The selection was based on two criteria. Firstly, questions should be answerable without knowing the topic of the test. For example, some questions in test could not be answered correctly without presenting a context of a specific legal

---

[3] https://aclweb.org/anthology/events/repeval-2017/.
[4] https://gluebenchmark.com.

system. Secondly, questions should test factual knowledge and not skills. For example, almost any math test will require to perform computations, and such type of task is not suitable for this dataset.

What is more, we have collected questions on history and geography from ege.sdamgia.ru. The selection of questions was similar to described above. As a result, the total number of questions is around 11k with three subjects being larger than other. These subjects are: medicine, (4k of questions), history (3k of questions), biology (2k of questions). The resulting dataset is somewhat similar to Trivia QA dataset, however the domains are different [8].

3.2 Multiple Choice Next Sentence Prediction (NSP)

We have collected a new dataset with 54k multiple choice questions where the objective is to predict the correct continuation for a given context sentence from four possible answer choices. The dataset was produced using the corpora of news articles of "Lenta.ru"[5]. To sample correct and incorrect answer choices we chose a trigram and a context sentence which ends on this trigram. The correct answer choice was the continuation of the context sentence, and the incorrect choices were the sentences following the trigram in other sentences in the corpora. Labels of correct answers were sampled uniformly. So, the random or constant predictions results in an accuracy score of 0.25.

3.3 Paraphrase Identification (PI)

For a paraphrase detection task, we used Russian language paraphrase dataset collected from news titles[6]. The dataset contains 7k pairs of titles which are the same, close and different by meaning. Constant prediction on this dataset gives us an accuracy score of 0.64. In a sense, Microsoft Research Paraphrase Corpora [7] is similar to this dataset. Both of them were collected from news titles.

3.4 Dataset Statistics

Frequency distribution of top 25 most frequent tokens, the number of unique tokens and the total number of tokens in the datasets can be found in Fig. 1. Sentence length distribution, average and median sentence length can be found in Fig. 2.

4 Methods

There are two types of problems we were considered for the comparison of sentence embeddings:

- **Multiple choice questions.** Datasets for this type are MCQA and NSP ones. The objective of the problem is to predict the correct answer for a given question/context and four answer choices (Tables 1 and 2).

---

[5] https://github.com/yutkin/Lenta.Ru-News-Dataset.
[6] http://paraphraser.ru.

– **Paraphrase identification.** The goal for such a problem is to predict if two given sentences are a paraphrase or not (Table 3).

Table 1. Examples from MCQA dataset. The correct answer is bolded.

Question
Какие из указанных симптомов характерны для фарингита?
Answer choices
1. резкая боль в горле
2. першение и дискомфорт в горле
3. затруднение проглатывания слюны
4. субфебрильная температура

Table 2. An example from NSP dataset. The correct answer is bolded.

Context
Мартин Скорсезе намеревается приступить к съемкам экранизации романа «Молчание» японского писателя Сюсаку Эндо в 2014 году,
Answer choices
1. был оснащен одиннадцатиметровым стеклянным полом шириной два метра, сообщается в полученном «Домом» пресс-релизе компании «Сен-Гобен».
2. когда он провозил мак для одной из продуктовых баз
3. сообщает Deadline. Финансированием проекта займутся компании Emmett/Furla Films и Corsan Films
4. а производить трубы там начали уже спустя два года. В числе поставщиков «Газпрома» ЗТЗ появился в 2017 году

Table 3. Examples from the PI dataset.

text 1	text 2	label
Мэрилин Мэнсон передумал выступать в России.	Мэрилин Мэнсон отменил тур по России.	1
Бывший чемпион мира по боксу умер в 48 лет.	Как судей судили за их решения.	0

To compare different methods of obtaining sentence embeddings, we have explored supervised and unsupervised scenarios of using such embeddings.

4.1 Unsupervised Approach

In unsupervised methods, we are interested in a similarity between sentence embeddings in terms of cosine similarity.

Multiple Choice Questions. For such type of problem for a given set $\{q, a_1, a_2, a_3, a_4\}$, where q is an embedding of either a question or a context sentence and a_i is an embedding of i–th answer choice, the predicted answer choice is the choice which embedding is the most similar to q.

Paraphrase Identification. Let for a given pair of sentences t_1 and t_2 are sentence embeddings of this pair respectively. First of all, we split the dataset into training and test sets, after that searching for a threshold t on a training set such that pairs with $sim(t_1, t_2) > t$ will be labeled as a paraphrase. Finally, we will evaluate results on a test set.

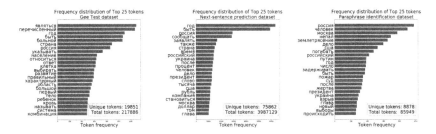

Fig. 1. Frequency distribution of top 25 tokens for MCQA, NSP, PI datasets.

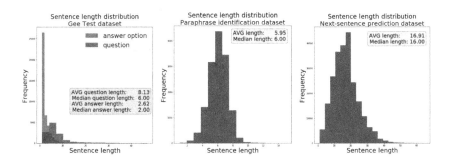

Fig. 2. Sentence length distribution for MCQA, NSP, PI datasets.

4.2 Supervised Approach

Text vector representations are often inputs to machine learning models. In this approach, we are aiming to figure out which methods of obtaining vector representations are better as inputs into linear models such as the logistic regression and which methods are better as inputs to a gradient boosting models such as the CatBoost [15].

Multiple Choice Questions. Since we cannot just run a multiclass classification as the correct answers numbers are not related to questions we will make a binary classification model which predicts a probability for a given question–answer (context–answer) pair to be correct, i.e. the answer is the correct answer choice for the question/context. Then for a given question/context and four answers, the predicted answer is the answer such that the model gives the highest probability.

Paraphrase Identification. For this problem, we just build a binary classifier on a concatenation of sentence embeddings.

5   Experiments and Results

5.1   FastText

As it can be seen from Table 4, FastText model accomplishes differently depending on tasks and methods. The best quality on each task was reached by Cat-Boost method according to Accuracy metrics. According to the F1 score, Cat-Boost also outperforms other methods in all of the tasks except Paraphrase identification where the best score with a large margin was achieved by the unsupervised method. There can also be seen a big quality difference between CatBoost and other methods in next sentence prediction task. Logistic regression shows the worst results in each task. The overall performance of methods based on FastText model is far from perfect. The results achieved using FastText can be considered as a baseline for our investigation.

Table 4. FastText embeddings results.

Method	MCQA		NSP		PI	
	Accuracy	F1	Accuracy	F1	Accuracy	F1
Unsupervised	0.305	0.304	0.337	0.337	0.719	0.806
Logistic regression	0.287	0.287	0.248	0.248	0.704	0.612
CatBoost	0.318	0.317	0.496	0.496	0.762	0.715

5.2   ELMo

We have used three ELMo models[7] pre–trained on Russian Wikipedia, "Lenta.Ru" news articles and Russian tweets corpora, respectively. So, one of the main results obtained from our experiments is how different domains of pre–trained models affects final results.

The results achieved by three methods based on three different pre–trained ELMo models are presented in Table 5. The performance of these methods is quite sensitive to the source of training data for ELMo model. For example, regarding the MCQA task, models trained on the News and Twitter corpora perform better than the model trained on Wikipedia, especially when logistic

---

[7] http://docs.deeppavlov.ai/en/master/intro/pretrained_vectors.html.

Table 5. ELMo unsupervised experiments results.

Method	Domain	MCQA		NSP		PI	
		Accuracy	F1	Accuracy	F1	Accuracy	F1
Unsupervised	Wikipedia	0.300	0.300	0.645	0.645	0.807	0.867
	News	0.293	0.293	0.691	0.691	0.807	0.866
	Twitter	0.291	0.291	0.559	0.559	0.803	0.863
Logistic regression	Wikipedia	0.301	0.300	0.249	0.249	0.684	0.652
	News	0.318	0.318	0.249	0.248	0.702	0.668
	Twitter	0.317	0.316	0.251	0.250	0.705	0.674
CatBoost	Wikipedia	0.314	0.314	0.647	0.647	0.773	0.729
	News	0.310	0.310	0.669	0.669	0.797	0.758
	Twitter	0.314	0.314	0.631	0.631	0.779	0.741

regression is used. In the unsupervised setting, the quality of next sentence prediction task highly depends on the source of training data ELMo model, too. However, in most cases there is no significant difference in results between three ELMo models.

One can notice that logistic regression in both cases (FastText model and all three ELMo models) shows the worst results in the majority of the tasks. Regarding the next sentence prediction task, the performance of logistic regression has not gone far away from random choice. However, it shows better results than other models in the MCQA task. The unsupervised method achieved the best results for the paraphrase identification task. We can claim that the use of ELMo model contributed to better results in next sentence and paraphrase identification tasks, as there was observed significant improvement in accuracy and F1 scores. Speaking of MCQA task, the results are comparable with the previously obtained.

5.3 BERT

The results of different methods based on BERT embeddings are shown in Tables 6, 7, 8, 9, 10 and 11. All the results were obtained using BERT–Base Multilingual Cased model[8]. There were considered different BERT model layers and combinations of layers. For each method and task, there is presented the best result achieved with layer indication. From Table 6 we can see that using BERT embeddings we can significantly improve results in MCQA task. The CatBoost method based on BERT model noticeably outperforms FastText and ELMo models within this task. According to Table 7, we can claim that BERT model could not get better results compared to ELMo. Even the performance of logistic regression remained the same. However, as we can notice

---

[8] https://github.com/google-research/bert/blob/master/multilingual.md

Table 6. BERT embeddings results on MCQA dataset.

Method	Best score			Average score		Worst score	
	Accuracy	F1	Layer	Accuracy	F1	Accuracy	F1
Unsupervised	0.303	0.302	Concatenation of layers from 1 to 6	0.292	0.292	0.274	0.274
Logistic regression	0.336	0.335	Layer number 1	0.324	0.323	0.310	0.310
CatBoost	0.346	0.346	Max pooling of layers from 4 to 6	0.326	0.326	0.312	0.311

Table 7. BERT embeddings results on NSP dataset.

Method	Best score			Average score		Worst score	
	Accuracy	F1	Layer	Accuracy	F1	Accuracy	F1
Unsupervised	0.508	0.508	Layer number 12	0.457	0.457	0.429	0.429
Logistic regression	0.255	0.255	Max pooling of layers from 7 to 9	0.249	0.248	0.244	0.244
CatBoost	0.514	0.514	Average pooling of layers from 7 to 12	0.479	0.479	0.414	0.414

Table 8. BERT embeddings results on PI dataset.

Method	Best score			Average score		Worst score	
	Accuracy	F1	Layer	Accuracy	F1	Accuracy	F1
Unsupervised	0.801	0.857	Average pooling of layers from 1 to 3	0.787	0.851	0.775	0.843
Logistic regression	0.715	0.676	Average pooling of layers from 7 to 12	0.693	0.651	0.670	0.615
CatBoost	0.778	0.732	Layer number 3	0.763	0.713	0.749	0.694

from Table 8, the performance of BERT model regarding paraphrase identification task is comparable to the ELMo results. In both cases, the best score was achieved by the unsupervised method and the worst by logistic regression. We suppose that in many cases BERT could not outperform ELMo model because it was not trained for these tasks and the best way for BERT is to fine–tune it by ourselves (Table 12).

Table 9. BERT embeddings results on MCQA dataset (1st and 12th layer).

Method	1st layer		12th layer		Average pooling	
	Accuracy	F1	Accuracy	F1	Accuracy	F1
Unsupervised	0.298	0.298	0.296	0.296	0.296	0.296
Logistic regression	0.336	0.335	0.321	0.321	0.326	0.326
CatBoost	0.318	0.318	0.318	0.318	0.329	0.330

Table 10. BERT embeddings results on NSP dataset (1st and 12th layer).

Method	1st layer		12th layer		Average pooling	
	Accuracy	F1	Accuracy	F1	Accuracy	F1
Unsupervised	0.429	0.429	0.508	0.508	0.484	0.484
Logistic regression	0.244	0.244	0.248	0.248	0.247	0.246
CatBoost	0.414	0.414	0.497	0.497	0.514	0.514

Table 11. BERT embeddings results on PI dataset (1st and 12th layer).

Method	1st layer		12th layer		Average pooling	
	Accuracy	F1	Accuracy	F1	Accuracy	F1
Unsupervised	0.795	0.853	0.781	0.850	0.789	0.853
Logistic regression	0.670	0.615	0.704	0.664	0.703	0.662
CatBoost	0.762	0.708	0.758	0.706	0.764	0.718

Table 12. Final results.

Model	MCQA		NSP		PI	
	Accuracy	F1	Accuracy	F1	Accuracy	F1
FastText	0.318	0.317	0.496	0.496	0.762	0.806
ELMo	0.318	0.318	0.691	0.691	0.807	0.867
BERT	0.346	0.346	0.514	0.514	0.801	0.857

6 Conclusion

We tested three sentence embedding models: (a) FastText averaged over words in the sentence, (b) three pre–trained on various sources ELMo models, (c) BERT model in three NLU tasks for the Russian language. These tasks are (a) multiple choice question answering, (b) next sentence prediction, (c) paraphrase identification. For the first two tasks, we presented our own new datasets. These datasets are designed as multiple choice questions: given a question/a sentence one need to choose a correct option from four possible answers/continuations.

Our experiments show that the MCQA dataset is much more complicated than the other two datasets. The quality of the results for this task is not as high as for two others. All models perform somewhat similar in next sentence prediction and paraphrase identification tasks. The paraphrase identification dataset is highly unbalanced, and the positive examples are in the minority, which may affect the quality if the results.

Overall, we can claim that we started to evaluate the popular sentence embeddings frameworks in GLUE–like fashion for the Russian language. So far we can state that (1) the word–level embeddings are outperformed by the sentence–level embeddings, (2) the pre–trained models available online with no doubts can attempt some of the NLU tasks with little or almost no fine tuning. The directions of the future work may include probing of embedding models for Russian rich morphology and free word order. The code of all experiments is available on GitHub[9].

Acknowledgements. This project was supported by the framework of the HSE University Basic Research Program and Russian Academic Excellence Project "5–100".

References

1. Arroyo-Fernández, I., Méndez-Cruz, C.F., Sierra, G., Torres-Moreno, J.M., Sidorov, G.: Unsupervised sentence representations as word information series: revisiting TF-IDF. Comput. Speech Lang. **56**, 107–129 (2019)
2. Bakarov, A.: A survey of word embeddings evaluation methods (2018)
3. Cer, D., et al.: Universal sentence encoder. arXiv preprint: arXiv:1803.11175 (2018)
4. Conneau, A., Kiela, D.: SentEval: an evaluation toolkit for universal sentence representations (2018)
5. Conneau, A., Kiela, D., Schwenk, H., Barrault, L., Bordes, A.: Supervised learning of universal sentence representations from natural language inference data. In: Proceedings of the 2017 Conference on Empirical Methods in Natural Language Processing, Copenhagen, Denmark, pp. 670–680. Association for Computational Linguistics, September 2017. https://www.aclweb.org/anthology/D17-1070
6. Devlin, J., Chang, M.W., Lee, K., Toutanova, K.: Bert: pre-training of deep bidirectional transformers for language understanding (2018)
7. Dolan, W., Quirk, C., Brockett, C.: Unsupervised construction of large paraphrase corpora: exploiting massively parallel news sources. https://www.microsoft.com/en-us/download/details.aspx?id=52398
8. Joshi, M., Choi, E., Weld, D.S., Zettlemoyer, L.: TriviaQA: a large scale distantly supervised challenge dataset for reading comprehension. arXiv preprint: arXiv:1705.03551 (2018)
9. Joulin, A., Grave, E., Bojanowski, P., Mikolov, T.: Bag of tricks for efficient text classification. CoRR abs/1607.01759 (2016). http://arxiv.org/abs/1607.01759
10. Kiros, R., et al.: Skip-thought vectors (2015)

---

[9] https://github.com/foksly/aist-sentence-embeddings.

11. Kutuzov, A., Kuzmenko, E.: WebVectors: a toolkit for building web interfaces for vector semantic models. In: Ignatov, D.I., et al. (eds.) AIST 2016. CCIS, vol. 661, pp. 155–161. Springer, Cham (2017). https://doi.org/10.1007/978-3-319-52920-2_15

12. Le, Q.V., Mikolov, T.: Distributed representations of sentences and documents (2014)

13. Mikolov, T., Chen, K., Corrado, G., Dean, J.: Efficient estimation of word representations in vector space. CoRR abs/1301.3781 (2013). http://arxiv.org/abs/1301.3781

14. Mikolov, T., Sutskever, I., Chen, K., Corrado, G., Dean, J.: Distributed representations of words and phrases and their compositionality. CoRR abs/1310.4546 (2013). http://arxiv.org/abs/1310.4546

15. Ostroumova, L., Gusev, G., Vorobev, A., Dorogush, A.V., Gulin, A.: CatBoost: unbiased boosting with categorical features. arXiv preprint: arXiv:1706.09516 (2018)

16. Peters, M., et al.: Deep contextualized word representations. In: Proceedings of the 2018 Conference of the North American Chapter of the Association for Computational Linguistics: Human Language Technologies (2018). https://doi.org/10.18653/v1/n18-1202

17. Rücklé, A., Eger, S., Peyrard, M., Gurevych, I.: Concatenated power mean word embeddings as universal cross-lingual sentence representations. arXiv preprint: arXiv:1803.01400 (2018)

Noun Compositionality Detection Using Distributional Semantics for the Russian Language

Dmitry Puzyrev[1], Artem Shelmanov[2], Alexander Panchenko[2],

and Ekaterina Artemova[1(✉)]

[1] National Research University Higher School of Economics, Moscow, Russia
dapuzyrev@edu.hse.ru, echernyak@hse.ru
[2] Skolkovo Institute of Science and Technology, Moscow, Russia
{a.shelmanov,a.panchenko}@skoltech.ru

Abstract. In this paper, we present the first gold-standard corpus of Russian noun compounds annotated with compositionality information. We used Universal Dependency treebanks to collect noun compounds according to part of speech patterns, such as ADJ-NOUN or NOUN-NOUN and annotated them according to the following schema: a phrase can be either compositional, non-compositional, or ambiguous (i.e., depending on the context it can be interpreted both as compositional or non-compositional). Next, we conduct a series of experiments to evaluate both unsupervised and supervised methods for predicting compositionality. To expand this manually annotated dataset with more non-compositional compounds and streamline the annotation process we use active learning. We show that not only the methods, previously proposed for English, are easily adapted for Russian, but also can be exploited in active learning paradigm, that increases the efficiency of the annotation process.

1 Introduction

A phrase is *compositional* if its meaning can be derived from the meaning of its parts, like in "green tree" or "tall building", otherwise it is *non-compositional*, like in "red herring" or "hot dog". Automatic detection of compositionality can be of great use for various natural language processing applications and this problem was extensively studied for English and some other languages where gold standard datasets are available to enable reproducible scientific research (see Sect. 6 for a survey of prior works in this field). To the best of our knowledge, there have been a few attempts to study idioms in the Russian language [1,17], but in general, no systematic studies on compositionality in the Russian language are available. Our study makes the first step towards the solution of this issue providing a dataset and a comparison of several methods for compositionality detection that showed state-of-the-art results for English.

The quality of natural language processing applications is heavily dependent on the quality of vector representations of text elements. The streamline NLP research encompasses many works on building various distributional semantic models (DSMs), and on methods for combining vector representations of atomic elements like words into representations of bigger fragments: phrases, sentences, texts. The simple but strong

© Springer Nature Switzerland AG 2019
W. M. P. van der Aalst et al. (Eds.) AIST 2019, LNCS 11832, pp. 218–229, 2019.
https://doi.org/10.1007/978-3-030-37334-4_20

baseline for this task suggests averaging word embeddings of a text fragment (sometimes weighted, e.g., according to TF-IDF). Although the result vector representation is rough compared to results, which could be achieved by more elaborate neural network encoding methods, it was shown that this baseline has high performance in many tasks [2, 14, 15, 29]. The main advantages of such methods are computational efficiency and an ability to use them in an unsupervised setting, while neural encoders would commonly require heavy computational power, labeled datasets, and substantial time for training. However, simple averaging of word embeddings is often too naïve. Idiomatic noun phrases are one of the cases where the averaging of the phrase parts would yield a wrong result since the meaning of such phrases is metaphorical and could not be directly "summed up" from meanings of its components. Therefore, it would be beneficial to have a DSM that tackles this problem by having a distinct phrase embedding. The ability to detect compositionality for noun compounds is considered beneficial for many tasks including machine translation, semantic parsing, as well as word sense disambiguation.

Namely, the **contribution** of this paper is three-fold:

1. we present the first *gold-standard dataset* for Russian annotated with compositionality information of noun compounds;[1]
2. we provide an *experimental evaluation* of models and methods for predicting compositionality of noun compounds: The methods from the previous work trained on the our dataset achieve comparable performance with results on English data.
3. we show the usefulness of *active learning* for creation of compositionality datasets.

The remainder of this paper is structured as follows. Section 2 describes the developed linguistic resource. In Sect. 3, we present our methodology for detection of noncompositional compounds. Section 4 describes our approach to dataset expansion using active learning. Section 5 discusses the conducted experiments on the created datasets and reports our results. Section 6 presents related work on predicting compositionality of noun compounds. Finally, Sect. 7 concludes the paper and outlines promising direction for the future work.

2   Noun Compound Dataset

2.1   Data Collection

The compound phrases are collected from the Russian Universal Dependency (UD) treebanks[2] according to part of speech patterns, such as adjective (ADJ) + noun (NOUN) or noun + noun, based on gold-standard UD annotations, which guarantees that not only no preprocessing but also no POS tagging and disambiguation is required. We use all Russian treebanks available in the UD project. They consist of texts from the following genre: news, nonfiction, fiction. To extract nominal compounds, we loop over all nouns and select only those, which has noun or adjective dependent (i.e., are "heads" of other noun phrases). We filter out non-frequent compounds, and from the list

---

[1] The dataset and the code: https://github.com/slangtech/ru-comps.
[2] https://universaldependencies.org.

of frequent compounds, we randomly select 1,000 compounds to be annotated. Note, that this procedure is coarse and does not rely on more precise compound definition such as the exact type of the dependency between the head and dependant tokens.

Each compound is lowercased and lemmatized. Accented characters are omitted. The head noun is provided in the nominal case and in singular number (if it exists), and the dependant adjectives are put in grammatical agreement with the head noun in case and gender, while dependant nouns remain unchanged.

2.2   Annotation Setup and Agreement

Each compound phrase in the selected list was annotated by two experts according to the following schema: (0) the phrase is non-compositional, such as "hot dog"; (1) the phrase is compositional, such as "green tree"; (2) the phrase is ambiguous, which means that exact compositionality of the phrase is dependant on the corresponding context, such as "melting pot" (see Table 2).

Next, annotators' answers were reviewed by a moderator. Out of 1,000 randomly selected compounds, the moderator sampled 220 pairs and resolved the ambiguity left from the first two annotators. Inter-annotator agreement metrics of the first two annotators on the dataset of 1,000 compounds are presented in Table 1. They show that annotators achieved a substantial agreement. The typical problematic cases that are hard to annotate are compounds, which meaning tends to be compositional in a metaphorical way, e.g., *"otkritoe more"* [open sea] and compounds that contain polysemic words: *"hod dela"* [justicement or the course of business].

Table 1. Inter-annotator agreement metrics for our dataset.

Agreement metric	Value
Pearson's correlation	0.541
Cronbach's alpha	0.700

Table 2. Examples of compositional (1), non-compositional (0), and ambiguous (2) compounds.

Type of compound	Compound samples
Non-compositional (0)	*goryachaya tochka* [trouble spot], *zheleznyi zanaves* [iron curtain], *kamennyi vek* [the Stone Age], *tsar gory* [king of the hill], *novaya volna* [new wave]
Compositional (1)	*aviatsiannaya bomba* [aircraft bomb], *gimn strany* [national anthem], *gornolyzhnyi kurort* [ski resort], *dno okeana* [ocean bed], *federalnyi zakon* [federal law]
Ambiguous (2)	*novyi god* [New year celebration or new year], *krupnaya set'* [big net or big network], *ogromnaya massa* [big mass of or big amount of], *pozitsiya kompanii* [company place or company position], *drevnyaya professiya* [ancient profession or prostitution]

Table 3. The number of compositional and non-compositional compounds in our dataset.

	Adjective-Noun	Noun-Noun	Total
Non-compositional (0)	23	10	33
Compositional (1)	71	96	167
Ambiguous (2)	9	11	20
Total	103	117	220

Под воздействием этого поля ядра	**атомов водорода**	в теле исследуемого , каждый со своим слабым магнитным полем , ориентируются определенным образом относительно сильного поля магнита .
Прозрачная жидкость , в которой на два	**атома водорода**	приходится один атом кислорода , может быть водой , а может быть и смесью жидких водорода и кислорода
Нам удалось сложить кучку из восьми атомов - двух атомов углерода и шести	**атомов водорода**	, изображенную на рисунке .
С чего начинать : сдвинуть два атома углерода или приставить	**атом водорода**	к атому углерода ?
Китайский	**Новый год**	и другие праздники , отмечаемые тайскими китайцами , отличаются в обоих случаях , так как они рассчитываются по китайскому календарю .
Перед самым	**Новым годом**	отключили поселок Никольское .
Речь , конечно же , идет об очередной заморозке до	**нового года**	цен на бензин .
В нашем рейтинге лучших подарков мужчине под	**Новый год**	пневматическая винтовка с ночным прицелом твердо заняла первое место .
Нынешнее заседание Госсовета - первое в	**новом году**	и последнее , на котором Владимир Путин выступит как президент страны .
Но это же был единственный русский фильм на	**Новый год**	, у него были все шансы на успех " .
А у нас политик	**второго эшелона**	ниже этого эшелона не опустится " , - говорит эксперт .
Несмотря на озабоченность Минобрнауки бесконтрольным размножением экономистов и недоверие солидных работодателей к дипломам вузов	**второго эшелона**	, молодой экономист сегодня вряд ли останется на обочине жизни
Пока потребители	**второго эшелонов**	дожидаются сезона распродаж или приобретают подержанные вещи , лидеры консюмеризма переходят к следующей фазе потребления .
Опускаясь по стратификационной лестнице , они опережают по статусу тех , кто находится во	**втором эшелоне**	, то есть в предшествующей фазе потребительской гонки .

Fig. 1. Compounds and their contexts. In this example, *atom vodoroda* [hydrogen atom] is compositional (1), *novyi god* [New year celebration or new year] is ambiguous (2), and *vtoroyi eshelon* [second tier] is non-compositional (0).

2.3   Dataset Description

The resulting dataset consists of 220 compound phrases with several full-sentence contexts collected from source texts. The number of contexts is not fixed. So far, the contexts are not annotated. A few examples of compounds are presented in Table 2. Table 3 presents the cross-tabulation of compound patterns and compositionality classes. Each compound is provided with a sentence context. The number of contexts is not fixed as we extract all contexts that contain the compound from the UD treebanks. The contexts so far are not used in the experiments. However, one of the possible directions for the future work would be compound disambiguation. For instance, the compound "belyi dom" in Russian could be interpreted as a reference to the specific parliament building or simply mean a random house of white color depending on the context. The same goes, for instance, for the phrase "melting pot" in English. Examples of the compound contexts are presented in Fig. 1.

3   Compositionality Detection Methods

We investigate various unsupervised and supervised methods for compositionality detection stems from the previous work for English. We train a DSM model that includes embeddings not just for single words but also for compounds. This is achieved by replacing in the training corpora all occurrences of compounds from the proposed resource with single tokens composed of their parts.

The use of hyperbolic embeddings [16] for noun compositionality detection was explored by Jana et al. [11]. Similar to our work, both supervised and unsupervised settings were reported. The authors found that in some settings hyperbolic embeddings outperform Euclidean embeddings, such as fastText and word2vec, on the task of noun compositionality prediction. However, for training such embeddings large amounts of hypernymy relations are needed. We leave the testing of such alternative embeddings for future work.

3.1   Unsupervised Methods

The unsupervised methods investigated in this work are adopted from [7]. They rely solely on a similarity between a compound embedding and an embedding composed from its parts using an additive function. Other functions like subtraction or element-wise multiplication may be used to obtain a compound embedding as well. The value of the similarity should correlate with annotators' judgments in the created resource.

Consider w_1, w_2 are words of a given compound and a function $v(\cdot)$ yielding vector representation of a word/compound. Then the similarity $sim(w_1, w_2)$ metric equals to:

$$sim(w_1, w_2) = cos(v(w_1 w_2), v(w_1 + w_2)) = \frac{v(w_1 w_2) \cdot v(w_1 + w_2)}{\|v(w_1 w_2)\| \, \|v(w_1 + w_2)\|}, \quad (1)$$

where $v(w_1 + w_2)$ is the sum of normalized vectors:

$$v(w_1 + w_2) = \frac{v(w_1)}{\|v(w_1)\|} + \frac{v(w_2)}{\|v(w_2)\|}. \quad (2)$$

In addition to cosine, we use similarity measures based on distance metrics between embeddings: Chebyshev distance (L_∞-norm), Manhattan distance (L_1-norm), and Euclidean distance (L_2-norm). When using these distances, instead of normalized sum, we use a simple averaging:

$$v(w_1 + w_2) = \frac{1}{2}(w_1 + w_2). \quad (3)$$

3.2   Supervised Methods

Supervised methods consider compositionality detection as a binary classification task. In this case, we simply train various supervised machine learning methods on vector representations of a compound and its parts. We use the following classification algorithms: linear support vector machine (LSVC) [18], three-layer perceptron (MLP) [10], decision tree (DT) [6], Naïve Bayes (NB) [30]. For feature representation, we use a concatenation of a compound embedding with embeddings of compound parts acquired from the DSMs. We also apply standard scaling to the generated vectors.

4 Expanding the Noun Compound Dataset with Active Learning

Non-compositional compounds, such as "hot dog" are rarer as compared to compositional compounds, such as "green tree". In our case, there are approximately only 3–4% of such compounds in the original treebank corpora used to collect the gold standard. This fact makes the straightforward approach to the annotation process quite inefficient since annotators have to look through and label many pairs of the dominant compositional type, which are not very useful for training a detection model. Non-compositional compounds are much more informative objects for supervised machine learning methods and, hence, finding and labeling them is more important.

We address the problem of the scarcity of non-compositional compounds using an active learning framework [12, 27]. Active learning incorporates the feedback from an acquisition machine learning model into the annotation process. This feedback is used in a query strategy that samples from large unannotated dataset objects that if labeled, potentially would be most informative for model training. It has been shown in many works, that active learning can significantly reduce the amount of manual labor required to achieve the specified performance level of machine learning models (in some cases by several times) [25].

In this work, we apply active learning to perform biased sampling of word pairs to be annotated and give a priority to non-compositional compounds. To implement active learning annotation, we use an active learning toolbox – the program framework that provides several query strategies and a widget for interactive annotation directly in a Jupyter environment [26][3] (Fig. 2).

We take half of the created gold-standard dataset of 220 pairs (with classes 0 and 1) to make a seed labeled set, which is used for initial training of the binary classifier that can detect compositionality of the unseen compounds (LSVC is used at this point). The rest half of the gold standard is left for validation. We prepare a large unlabeled dataset of 24,348 word pairs and train a fastText DSM that contains all compounds from this dataset and from the training half (300 dimensions, 2 minimum frequency, 5 epochs). The DSM is used for feature generation as in supervised methods for compositionality detection. For the selection of objects for annotation, we use "maximum error" query strategy [26]. This strategy is mainly aimed at noise reduction. It considers unannotated objects as negative samples (in our case as compositional compounds) trains a classifier on the labeled and unlabeled datasets and suggests for annotation objects that are assigned the largest negative margin by the trained model. The annotators could label 40 objects at maximum in each iteration before retraining of the acquisition model.

Two annotators were involved in the active learning annotation. We acquired 235 answers from the first annotator and 183 answers from the second annotator. We merged the answers excluding the pairs that received contradicting labels. The resulting dataset contains 357 labeled word pairs, 87 of which are non-compositional. In the experimental section, we show that using the dataset obtained with active learning we can significantly improve the performance of the supervised compositionality detection models.

---

[3] https://github.com/IINemo/active_learning_toolbox.

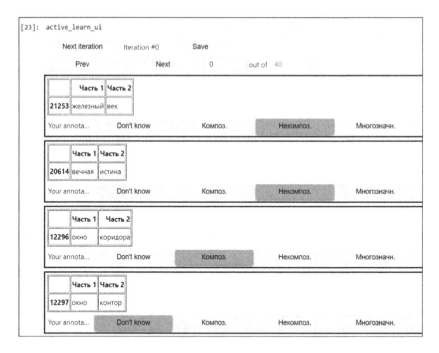

Fig. 2. The Jupyter widget for noun compound compositionality annotation making use of active learning framework proposed by Suvorov and Shelmanov [26].

5   Experiments

5.1   Experimental Setup

We train DSMs using fastText [5] and word2vec [14] algorithms with CBOW architecture implemented in the *gensim* library [21][4]. A dump of Russian Wikipedia is used as a training corpus[5], with Universal Dependencies raw texts as an enrichment, which helps to deal with cases of missing compounds. Both Wikipedia articles and compounds are lemmatized with MyStem [24][6].

We performed experiments on several sets of hyperparameters (dimensionality and amount of training epochs). We found that dimensionality of 300 and five epochs give good or the best results across all considered settings, therefore, we report results only for this set of hyperparameters. To simplify the task, in the experimental evaluation, we do not consider contextual information of compounds. It means that no ambiguity is under consideration and only phrases with compositionality classes of 1 and 0 are qualified for evaluation, which leaves 200 compounds. For three of them, the DSMs lack an embedding, which leaves 197 phrases for experiments: 167 are compositional and 33 are non-compositional, which approximately corresponds to 0.83 to 0.17 ratio (see Table 3).

---

[4] https://radimrehurek.com/gensim.

[5] We used a Wikipedia dump from 02.05.2019, which consists of 1,542,621 articles.

[6] http://github.com/nlpub/pymystem3.

The experiments with unsupervised methods follow the evaluation pipeline presented in [7]. It is based on Spearman's rank correlation (ρ) between the value of the unsupervised metrics and the compositionality class in the annotated dataset.

For experiments with supervised methods, we train an additional DSM that does not have any modifications (it does not contain embeddings for compounds). In this setup, embeddings of compound parts are obtained from this unmodified model. We record the average performance of the classifiers on 25 stratified randomized splits of the selected dataset (75% for training and 25% for testing). We use the following hyperparameters: For the LSVC model: $C = 1$; for the MLP model: $\alpha = 1$, LBFGS solver, three layers featuring respectively 200, 20, and 20 neurons; for the DT model: maximum depth of 10, the maximum number of features is 20. We calculate the Spearman's correlation with the target compositionality class from annotation, as well as precision, recall, and F_1-score.

We also measure the impact of the additional data annotated with active learning on the performance of the supervised models. In this experiment, the first set of classifiers is trained on the half of the original gold standard that was used for training acquisition models, the second set of classifiers is trained on this half joined with the merged dataset annotated using active learning. For testing, we use the other half of the original gold standard that was not used for acquisition model training. We do not make randomized splits in this experiment to make the comparison fair. The features for the supervised classifiers are generated by the DSM used during the active learning annotation. We measure the changes of the F1-score for non-compositional class and the changes of the Spearman's correlation.

Table 4. Spearman's correlation (ρ) of the unsupervised metrics with annotators' judgments.

Metric\Model	fastText	word2vec
cos (norm.)	**0.42**	0.37
L_∞ (avg.)	0.33	0.09
L_1 (avg.)	0.33	0.14
L_2 (avg.)	0.33	0.14

Table 5. The performance of the supervised classifiers. The evaluation metrics are presented for the non-compositional class (0).

Supervised model	Spearman's ρ	Precision	Recall	F-measure
Liner Support Vector classifier (LSVC)	**0.47**	0.37	**0.78**	0.48
Multi-layer Perceptron (MLP)	**0.46**	0.32	**0.82**	0.44
Decision Tree (DT)	0.18	0.31	0.36	0.31
Naïve Bayer (NB)	**0.43**	**0.55**	0.52	**0.52**

Table 6. The performance of the classifiers trained on the half of the gold standard and on the extended dataset with annotations acquired with active learning (AL). The evaluation metrics are presented for the non-compositional class (0). The delta is between "before" and "after" AL.

Supervised model		Spearman's ρ	F-measure
Liner Support Vector classifier (LSVC)	Before AL	0.42	0.50
	After AL	0.61	0.67
	Delta	+0.19	+0.17
Multi-layer Perceptron (MLP)	Before AL	0.42	0.36
	After AL	**0.71**	**0.74**
	Delta	**+0.35**	**+0.32**
Decision Tree (DT)	Before AL	0.26	0.39
	After AL	−0.05	0.17
	Delta	−0.31	−0.22
Naïve Bayer (NB)	Before AL	0.28	0.37
	After AL	0.04	0.23
	Delta	−0.24	−0.14

5.2   Results and Discussion

The results of the experimental evaluation of the unsupervised methods are presented in Table 4, the evaluation results of the supervised methods are presented in Table 5.

We see that all unsupervised metrics, L_1, L_2, and L_∞ present substantial negative correlation. That can be explained by the nature of embedding vectors. The bigger the distance value, the further compound is from its components in a semantic sense. If the sense of the compound substantially differs from corresponding senses of its components, it is deemed as non-compositional. To be comparable with previous papers, we present a positive correlation bringing minus of a distance instead.

Taking this into consideration, all metrics perform comparably on the gold-standard dataset. We can see a not strong, yet stable and substantial correlation between similarity and the compositionality class. As non-compositional compounds are in the minority in this dataset, and detecting idiomatic phrases provides more interest practice-wise, we report on zero-class quality metrics to access algorithm performance.

Supervised models LSVC, MLP, and NB present higher ρ than the unsupervised counterparts. LSVC and MLP also give relatively high recall on non-compositional examples. Overall, linear SVC and multi-layer perceptron perform better than the other models across all metrics.

As we can see from Table 6, annotating additional data with active learning significantly improves the performance of LSVC and MLP classifiers. The biggest improvement is achieved for MLP with 32% points of F1-score and 35% points of Spearman's correlation. Due to active learning, we managed to quickly detect in the large unlabeled dataset a set of informative word pairs containing an increased ratio of non-compositional compounds that according to the results appeared to be very beneficial for supervised classification models. We consider that degradation of the performance of NB and DT classifiers is mainly related to overfitting and a number of noise samples in the dataset annotated with active learning.

6 Related Work

The construction of compositionality datasets can be traced back to as early 2000s when Baldwin and Villavicencio [4] proposed chunk-based extraction methods for English verb-prepositional combinations and gave some binary judgments on the subject of considering them as phrasal verbs. In the follow-up paper Baldwin et al. [3], the same framework is used to retrieve 1,710 NOUN-NOUN compounds from the Wall Street Journal corpus. The authors use LSA to calculate the similarity between a phrase and its components as one of the early compositionality prediction attempts.

McCarthy et al. [13] evaluate 116 candidates of English phrasal verbs using three annotators' predictions on a scale from 0 to 10. Venkatapathy and Joshi [28] use 800 verb-object collocations obtained from British National Corpus to give annotations from 1 to 6 where one stands for total non-compositionality and 6 for complete compositionality.

Reddy et al. [20] contains 90 English noun compounds and uses an average of 30 judgments to give each phrase compositionality scores. This work provided compositionality assessments for both the phrase and its constituents enabling the use of various operations with corresponding embeddings of a compound and its distinctive parts in the context of linking human validations with measurements of semantic distance.

Ramisch et al. [19] extended this dataset to 180 phrases presenting two parallel sets for French and Portuguese languages. English NOUN-NOUN compounds were mapped with NOUN-PREP-NOUN and NOUN-ADJ constructions according to the grammar equivalents. Farahmand et al. [8] issue considerably larger dataset, which has 1042 NOUN-NOUN compounds annotated with the help of 4 experts.

We also should note some works on compositionality detection datasets for non-English languages. Gurrutxaga and Iñaki [9] studies 1,200 Basque NOUN-VERB collocations and resolves classification task into three classes: idiom, collocation, and free combination.

Roller et al. [22] provide 244 German compounds with compositionality scores assigned from 1 to 7 as an average from 30 validations. PARSEME project [23] is devoted to the multilingual annotation of multiword expressions (MWE) of arbitrary length and syntactical structure. By design, PARSEME is more suited for MWE extraction tasks rather than compositionality evaluation. This dataset includes annotated verbal MWEs for several Slavic languages.

However, to the best of our knowledge, prior to our work there was no dataset for compositionality detection task for any Slavic language structurally similar to the datasets of Reddy et al. [20] and Farahmand et al. [8].

7 Conclusion

In this paper, we presented the first Russian-language dataset of noun compounds with compositionality information. Each compound in the dataset follows the pattern

NOUN-NOUN or the pattern ADJ-NOUN and is labeled as non-compositional, compositional, or ambiguous. The latter type of compound can be either compositional or not depending on the context, e.g. "melting pot". Each compound is provided along with the sentence contexts. The inter-annotator agreement metrics show that human judgments on the compositionality classes agree well. We investigated the performance of various algorithms from the previous work and showed that the state-of-the-art models for English can be successfully used for a Slavic language.

Non-compositional compounds, such as "hot dog" are rarer than compositional ones, such as "green house". We used active learning framework to cope with this natural disbalance and quickly gather more non-compositional terms showing that these additional examples lead to substantial gains in classification performance.

We hope that the resource developed in this work will foster research in the area of compositionality detection for Russian and other Slavic languages.

Acknowledgements. Dmitry Puzyrev and Ekaterina Artemova were supported by the framework of the HSE University Basic Research Program and Russian Academic Excellence Project "5–100".

References

1. Aharodnik, K., Feldman, A., Peng, J.: Designing a Russian idiom-annotated corpus. In: Proceedings of the Eleventh International Conference on Language Resources and Evaluation (LREC 2018) (2018)
2. Anke, L.E., Schockaert, S.: Seven: augmenting word embeddings with unsupervised relation vectors. In: Proceedings of the 27th International Conference on Computational Linguistics, pp. 2653–2665 (2018)
3. Baldwin, T., Bannard, C., Tanaka, T., Widdows, D.: An empirical model of multiword expression decomposability. In: Proceedings of the ACL 2003 Workshop on Multiword Expressions: Analysis, Acquisition and Treatment, vol. 18, pp. 89–96 (2003)
4. Baldwin, T., Villavicencio, A.: Extracting the unextractable: a case study on verb-particles. In: Proceedings of CoNLL, pp. 1–7 (2002)
5. Bojanowski, P., Grave, E., Joulin, A., Mikolov, T.: Enriching word vectors with subword information (2016)
6. Breiman, L.: Classification and regression trees (2017)
7. Cordeiro, S., Ramisch, C., Idiart, M., Villavicencio, A.: Predicting the compositionality of nominal compounds: giving word embeddings a hard time. In: Proceedings of the 54th Annual Meeting of the Association for Computational Linguistics (Volume 1: Long Papers), vol. 1, pp. 1986–1997 (2016)
8. Farahmand, M., Smith, A., Nivre, J.: A multiword expression data set: annotating non-compositionality and conventionalization for English noun compounds. In: Proceedings of the 11th Workshop on Multiword Expressions, pp. 29–33 (2015)
9. Gurrutxaga, A., Alegria, I.: Combining different features of idiomaticity for the automatic classification of noun+verb expressions in Basque. In: Proceedings of the 9th Workshop on Multiword Expressions, pp. 116–125 (2013)
10. Hinton, G.E.: Connectionist learning procedures. Artif. Intell. **40**(1–3), 185–234 (1989)
11. Jana, A., Puzyrev, D., Panchenko, A., Goyal, P., Biemann, C., Mukherjee, A.: On the compositionality prediction of noun phrases using poincaré embeddings. In: The 57th Annual Meeting of the Association for Computational Linguistics (ACL) (2019)

12. Lewis, D.D., Gale, W.A.: A sequential algorithm for training text classifiers. In: SIGIR 1994, pp. 3–12 (1994)
13. McCarthy, D., Keller, B., Carroll, J.: Detecting a continuum of compositionality in phrasal verbs. In: Proceedings of the ACL 2003 Workshop on Multiword Expressions: Analysis, Acquisition and Treatment, MWE 2003, vol. 18, pp. 73–80 (2003)
14. Mikolov, T., Sutskever, I., Chen, K., Corrado, G.S., Dean, J.: Distributed representations of words and phrases and their compositionality. In: Advances in Neural Information Processing Systems, pp. 3111–3119 (2013)
15. Mitchell, J., Lapata, M.: Vector-based models of semantic composition. In: Proceedings of ACL 2008: HLT, pp. 236–244 (2008)
16. Nickel, M., Kiela, D.: Poincaré embeddings for learning hierarchical representations. In: Advances in Neural Information Processing Systems, pp. 6338–6347 (2017)
17. Peng, J., Feldman, A.: Automatic idiom recognition with word embeddings. In: Lossio-Ventura, J.A., Alatrista-Salas, H. (eds.) SIMBig 2015-2016. CCIS, vol. 656, pp. 17–29. Springer, Cham (2017). https://doi.org/10.1007/978-3-319-55209-5_2
18. Platt, J.C.: Probabilistic outputs for support vector machines and comparisons to regularized likelihood methods. In: Advances in Large Margin Classifiers, pp. 61–74 (1999)
19. Ramisch, C., Cordeiro, S., Zilio, L., Idiart, M., Villavicencio, A.: How naked is the naked truth? A multilingual lexicon of nominal compound compositionality. In: Proceedings of the 54th Annual Meeting of the Association for Computational Linguistics (Volume 2: Short Papers) (2016)
20. Reddy, S., McCarthy, D., Manandhar, S.: An empirical study on compositionality in compound nouns. In: Proceedings of the 5th International Joint Conference on Natural Language Processing, pp. 210–218 (2011)
21. Rehurek, R., Sojka, P.: Software framework for topic modelling with large corpora. In: Proceedings of the LREC 2010 Workshop on New Challenges for NLP frameworks, pp. 45–50 (2010)
22. Roller, S., Schulte im Walde, S., Scheible, S.: The (un)expected effects of applying standard cleansing models to human ratings on compositionality. In: Proceedings of the 9th Workshop on Multiword Expressions, pp. 32–41 (2013)
23. Savary, A., et al.: PARSEME - PARSing and Multiword Expressions within a European multilingual network. In: 7th Language & Technology Conference: Human Language Technologies as a Challenge for Computer Science and Linguistics (LTC 2015) (2015)
24. Segalovich, I.: A fast morphological algorithm with unknown word guessing induced by a dictionary for a web search engine. In: MLMTA (2003)
25. Settles, B.: Active learning literature survey. Technical report, University of Wisconsin-Madison Department of Computer Sciences (2009)
26. Suvorov, R., Shelmanov, A., Smirnov, I.: Active learning with adaptive density weighted sampling for information extraction from scientific papers. In: Filchenkov, A., Pivovarova, L., Žižka, J. (eds.) AINL 2017. CCIS, vol. 789, pp. 77–90. Springer, Cham (2018). https://doi.org/10.1007/978-3-319-71746-3_7
27. Tong, S., Koller, D.: Support vector machine active learning with applications to text classification. J. Mach. Learn. Res. **2**, 45–66 (2001)
28. Venkatapathy, S., Joshi, A.K.: Measuring the relative compositionality of verb-noun (V-N) collocations by integrating features. In: Proceedings of the Conference on Human Language Technology and Empirical Methods in Natural Language Processing, pp. 899–906 (2005)
29. Weston, J., Bordes, A., Yakhnenko, O., Usunier, N.: Connecting language and knowledge bases with embedding models for relation extraction. In: Proceedings of the 2013 Conference on Empirical Methods in Natural Language Processing, pp. 1366–1371 (2013)
30. Zhang, H.: The optimality of naive bayes. In: Proceedings of the Seventeenth International Florida Artificial Intelligence Research Society Conference, FLAIRS 2004, vol. 2 (2004)

Deep JEDi: Deep Joint Entity Disambiguation to Wikipedia for Russian

Andrey Sysoev[1(✉)] and Irina Nikishina[1,2] 🆔

[1] Ivannikov Institute for System Programming of the Russian Academy of Sciences, Moscow, Russia
sysoev.msu@gmail.com, nia@ispras.ru
[2] Higher School of Economics, Moscow, Russia

Abstract. Over the past few years there has been a leap forward in both Entity Disambiguation and Entity Linking tasks. Meanwhile, Entity Disambiguation for Russian still lags behind advanced neural approaches developed for other languages. This paper introduces Deep JEDi— purely neural architecture, intended to identify the correct meaning for each mention in text. Combining sequence translation and sequence labeling approaches, our model achieves promising results on the Russian Wikipedia dataset. Significant improvement of its performance is attained by specific decoder that incorporates information about target mention position into attention mechanism. Additionally, we compare different approaches for learning distributed representations for tokens and entities and prove the importance of enriching joint embeddings with information about knowledge base structure.

Keywords: Disambiguation to Wikipedia · Wikification for Russian · Encoder-decoder neural network architecture · Joint embeddings · Sequence labeling

1 Introduction

Wikipedia is a widely known online encyclopedia, a great warehouse of human knowledge. Due to its size and quality, Wikipedia has become a good source of data for various NLP research including, but not limited to [14,20,26].

In the current work we concentrate on Disambiguation to Wikipedia task. Given some text and a list of mentions $\{m_1, m_2, ..., m_p\}$, the goal is to assign the correct Wikipedia article—hereinafter entity—to each m_i. The knowledge base, constructed from Wikipedia, provides a set of possible meanings for each mention, thus, the developed algorithm does not have to scan through all entities in search for the correct one. For instance[1], given the text[2]:

> Все джедаи называли Йоду своим учителем.
>
> *Vse dzhedai nazyvali Jodu svoim uchitelem.*
> (en. `All the Jedi called Yoda their teacher.`)

---

[1] Hereinafter examples are taken from Wikipedia.
[2] `Typewriter` font is used for Russian example, *italics*—for transliteration, *`both`*—for English version.

© Springer Nature Switzerland AG 2019
W. M. P. van der Aalst et al. (Eds.): AIST 2019, LNCS 11832, pp. 230–241, 2019.
https://doi.org/10.1007/978-3-030-37334-4_21

the system should assign the entity, corresponding to **Йода** (en. *Yoda*), to the underlined mention. The entity **Йод** (en. *Iodine*), even being the most common sense for this mention, is to be ignored.

In this work we introduce Deep JEDi—Deep Joint Entity Disambiguation to Wikipedia—purely neural architecture, intended to identify the correct meaning for each given mention in text. Our method is inspired by sequence labeling and sequence-to-sequence approaches. It attains high performance with specific decoder that incorporates information about target mention position into attention mechanism. Additionally, we compare different ways of learning distributed representations for tokens and entities and prove the importance of exploiting joint embeddings—when token and entity vectors are resided in the same space.

2   Related Work

Deep JEDi's motto conveys the following idea: developing a sequence-to-sequence model for Entity Disambiguation using appropriate embeddings. Each component of the task has its own background to be described in the next few sections.

2.1   Entity Linking and Disambiguation

Early methods devoted to Entity Linking and Disambiguation are based on hand-crafted features and supervised machine learning algorithms [16,19,22]. However, modern approaches usually exploit neural networks to score possible mention meanings with respect to local context and co-occurrent entities. Computed scores are then used in ranking [31], CRF [5,12], entity-mention graph regularization [11] or other kinds of models [4]. In contrast to current approaches, we do not restrict our neural network to local contexts: Deep JEDi is shown the whole document and it is allowed to decide which parts of the text are crucial for disambiguation.

Considering Entity Disambiguation for Russian, there is much room for improvement. Pioneering works [27,28] point out the significance of the task, providing further researchers with commonly available dataset.

2.2   Embeddings

Token vector representations (word2vec [17], GloVe [21], fastText [2]) are used nowadays in every other NLP research. Moreover, embeddings have always been quite popular for representing entities from knowledge bases [3,24]. In connection with disambiguation task, we spot [27], who implement entity embeddings in order to generate context features.

Meanwhile, combination of word and entity vector spaces has proven its effectiveness over separate for both Entity Linking and Entity Disambiguation. Approach, presented in [31], places token and entity embeddings into the same vector space by extending word skip-gram and knowledge base graph models with anchor model: it predicts context tokens for inner Wikipedia links. In [5] entity embeddings are gained from pretrained word2vec model for tokens by predicting words from corresponding Wikipedia pages and anchor contexts; entity co-occurrences are ignored. In contrast to the above methods, Deep JEDi treats tokens and entities equally. It uses skip-gram model to predict context elements of both types for tokens and entities. Additionally, it enriches joint distributed representations with graph information from knowledge base using node2vec [8].

2.3   Sequence-to-Sequence Architecture

Our neural architecture for disambiguating entities is inspired by sequence-to-sequence approach [25], vastly used in machine translation, text summarization and dialogue systems. It includes two main parts: encoder and decoder, traditionally constituted by Long Short-Term Memory (LSTM) [7,10]. Alternatively, (Bi)LSTM models may be substituted by convolutional neural network (CNN) [13] in both encoder and decoder [6]. The efficiency of CNN in both parts of sequence-to-sequence architecture is unstable and controversial, therefore, in the current paper we stack LSTM over CNN architecture, likewise in [9,29]. In order to ameliorate LSTM decoder performance, [1,15] introduce attention mechanism, which captures relevant information from encoder and helps to align input sequence to the decoder output. We also elaborate on attention mechanism enhanced with special mention alignment, discussed in Sect. 3.3.

3   Approach

In this section we introduce Deep JEDi—Deep Joint Entity Disambiguation to Wikipedia—purely neural network built for Disambiguation to Wikipedia task. The network accepts input token embeddings and vectors, specifying mention positions, and generates entity embeddings for each mention.

The model is a combination of sequence translation and sequence labeling approaches. As in sequence translation, it comprises an encoder and a decoder. Encoder role is to convert each input token embedding into some intermediate representation; at the same time it maps the whole token sequence into certain fixed-size structure. The decoder aims to unwrap it together with intermediate token representations into a sequence of target entity embeddings. At the same time, our model is to some extent similar to sequence labeling: the decoder additionally consumes vectors, denoting target mention positions.

Thus, the whole model is based on three cornerstones—token and entity embeddings (Sect. 3.1), encoder (Sect. 3.2) and decoder (Sect. 3.3). Some extra information about model implementation is presented in Sect. 3.4.

Yoda is a fictional character in the Star Wars franchise, wisest jedi

Fig. 1. Context generation for joint word and entity embeddings.

3.1   Embeddings

Separate Vector Spaces for Tokens and Entities. The simplest and the most obvious way to learn distributed representations for both words and Wikipedia entities is to place them into separate vector spaces, likewise in [27]. In the current study we also implement separate token and entity embeddings and consider them a baseline for complex joint embeddings. For learning word vector representations we implement fastText model [2], for entities—word2vec [18].

Joint Embeddings. The idea of learning embeddings jointly proves its efficiency in several works mentioned in Sect. 2.2. However, almost none of those methods takes entity relative positions in text into account. That is why in the current study we implement our approach that is mostly consistent with [30, 32] and relies on entity co-occurence.

While learning vector representations jointly, we meet the following constraints:

1. if two tokens have similar contexts they should have similar embeddings;
2. if a word and an entity have similar contexts they should have similar embeddings;
3. if two entities have similar contexts they should have similar embeddings.

Thus, while generating context for learning joint embedding space we include all possible combinations (see Fig. 1). On each step the closest entity mentions are taken, even though they may be considerably distant from the current token/entity.

Enriching Joint Embeddings with Wikipedia Graph Structure. Most studies about Entity Linking and Entity Disambiguation rely on knowledge base graph structure, as it allows to capture relations between entities and their categories.

Wikipedia knowledge base may be regarded as a graph, where articles and links between them are considered nodes and edges respectively. In order to incorporate graph data into joint vector representations we implement node2vec's

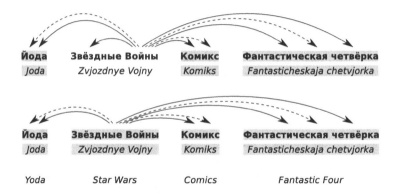

Fig. 2. Context generation for word and entity embeddings from Wikipedia graph.

sampling strategy [8]. A fixed size random walk from each node simulates sequences of Wikipedia entities consistently linked to one another. Those sequences (see Fig. 2) are regarded as texts and fed into embeddings computation model along with text corpus.

Model. For the current study we implement skip-gram model with negative sampling [18], intended to predict context given the token/entity. The context comprises a number of elements of each type located on both sides of the current token or entity. Moreover, we employ information about token n-grams as in [2] to take Russian morphology into account. As input we provide a multi-hot vector for token or a one-hot vector for entity and expect its embedding to be close to a positive example from the generated context. Moreover, we randomly select some words and entities that do not appear in the context and consider them being negative examples.

Comparison. Sample results for various embedding computations are presented in Table 1. For each type we provide top three closest tokens and entities for the word джедай (*dzhedaj*, en. `Jedi`) and entity Джедаи (*Dzhedai*, en. `Jedi`). Considering separate embeddings, relations between tokens and entities cannot be computed, as they are placed in different vector spaces. Comparing semantic associates for joint and joint with graph data variants we may see that the latter embeddings are slightly more accurate in this case.

3.2 Encoder

The encoder converts input token embeddings $\{e_1, e_2, ..., e_n\}$ into intermediate representations $\{h_1, h_2, ..., h_n\}$, at the same time trying to squeeze information about the whole sequence into a fixed-size vector pair (c, h).

LSTM cell has proven its usefulness as part of the encoder in various NLP tasks. What is more important for this work, LSTM-based encoder performs well for Disambiguation to Wikipedia [27]. Thus, as the first option we just implement the same architecture: input token embeddings are passed through BiLSTM with dropout.

Table 1. Top similar tokens and entities for word **джедай** (en. *Jedi*) and entity **Джедаи** (en. *Jedi*).

separate	joint	joint with graph data
	tokens for token джедай (en. Jedi)	
джеддак / *jeddak*	джедаи / *Jedi*	джедаи / *Jedi*
джедаи / *Jedi*	джеддак / *jeddak*	джедая / *Jedi*
джед / *djed*	джед / *djed*	джедаю / *Jedi*
	entities for token джедай (en. Jedi)	
	Ниндзя / *Ninja*	Оби-Ван Кеноби / *Obi-Wan Kenobi*
—	Джедаи / *Jedi*	Джедаи / *Jedi*
	Дарт Вейдер / *Darth Vader*	Дарт Вейдер / *Darth Vader*
	tokens for entity Джедаи (en. Jedi)	
	R2-D2 / *R2-D2*	джедаи / *Jedi*
—	Скайуокер / *Skywalker*	джедая / *Jedi*
	Оби-Ван / *Obi-Wan*	джедай / *Jedi*
	entities for entity Джедаи (en. Jedi)	
Оби-Ван Кеноби / *Obi-Wan Kenobi*	Ситхи / *Sith*	Оби-Ван Кеноби / *Obi-Wan Kenobi*
Люк Скайуокер / *Luke Skywalker*	Сила (Звёздные войны) / *The Force*	Дарт Вейдер / *Darth Vader*
Хан Соло / *Han Solo*	Оби-Ван Кеноби / *Obi-Wan Kenobi*	Ситхи / *Sith*

Another approach for building encoder is based on convolutional neural network. However, this option is more controversial. Convolutions have proven being efficient as part of the architecture in machine translation [6], but they do not manage to capture long-term dependencies, that are extremely vital for disambiguation. Nevertheless, applying convolutional layers—one or several—can be viewed as constructing local context-aware intermediate representations. As a result, they tend to increase the whole neural network capacity.

Keeping in mind described intuitions, we construct a neural network from convolutional gated units [6] with dropout, followed by BiLSTM encoder. All layers are bypassed with residual connections to facilitate training. In case input and output have different dimensionality, input vectors, before summation with output, are multiplied by a trainable matrix to make the operation feasible.

3.3 Decoder

The decoder task is to unwrap encoder output into a series of target entity embeddings. In this work the decoder is based on LSTM cell with attention [1]. Our main contribution is to explicitly incorporate information about target mention positions into attention computation. The general decoder architecture is presented in Fig. 3.

Encoder-computed vectors c and h are converted into initial decoder state by passing them through distinct dense layers. LSTM cell sequentially transmutes its current state and concatenation of previous step attention with output into vector v and next state. Cell outcome v and intermediate token representations $\{h_1, h_2, ..., h_n\}$ are used to compute Bahdanau [1] alignment $\alpha = \{\alpha_1, \alpha_2, ..., \alpha_n\}$, where each $\alpha_i \in [0, 1]$ and $\sum_i \alpha_i = 1$. Alignments act as coefficients in weighted average for attention computation: $u = \sum_{i=1}^{n} \alpha_i h_i$.

In this work instead of directly using α_i coefficients, we employ $\widetilde{\alpha}_i = 0.5(\alpha_i + \beta_i)$, where $\beta = \{\beta_1, \beta_2, ..., \beta_n\}$ is mention-encoding alignment. It is a multi-hot vector with non-zero weights corresponding to mention intersecting tokens (see Table 2 for an example). Finally, LSTM output v and attention vector u are concatenated and passed through linear dense layer. Its output is treated as a target entity embedding z.

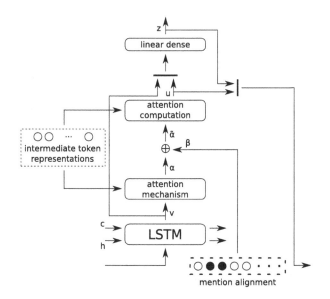

Fig. 3. Decoder.

Table 2. Sample mention alignment.

Оби	Ван	Кеноби	–	падаван	Квай-Гона	.
Obi	*Van*	*Kenobi*	–	*padavan*	*Kvaj-Gona*	.
(en. *Obi-Wan Kenobi is Qui-Gon's Padawan.*)						
0.33	**0.33**	**0.33**	0	0	0	0
0	0	0	0	1	0	0

3.4 Implementation Details

In this part we describe various implementation details, regarding our approach.

Throughout the whole work, token and entity embeddings of size 100 are used; they are L2-normalized before being passed into the model.

In experiments described below each LSTM cell size is 1000 (within decoder and BiLSTM encoder). In advanced version of encoder we stack three convolutional gated units before BiLSTM. Each convolution has 1000 filters of size 5. Dropout probability is kept constant at 0.3.

Squared Euclidean distance loss with negative examples is used to train the neural network:

$$Loss(z, z^+, z_{1...k}^-) = \omega\|z - z^+\|^2 + (1 - \omega)\frac{1}{k}\sum_{i=1}^{k}\| - z - z_i^-\|^2,$$

where $z_1^-, ..., z_k^-$ are vectors of the wrong meanings of the mention, closest to the correct entity embedding z^+. Parameter ω (set to 0.7) controls the balance between positive and negative parts of the loss function; k—number of wrong meanings—is 3.

The neural network is trained in a distributed fashion on three GPUs with gradients averaging using Horovod [23]: each worker batch size is 32, initial learning rate—10^{-4}; effectively, these values are multiplied by the number of workers.

Finally, learning rate ϵ is annealed over 50 epochs (starting from 0) according to the formula:

$$\epsilon(epoch) = \epsilon^{initial} max(0.05, e^{-0.1 epoch}).$$

4 Evaluation

To explore the usefulness of our propositions we employ the Disambiguation to Wikipedia dataset from [27]. It is constructed from good and featured Wikipedia articles: the former are used as development set (2968 texts; 190.1 tokens and 16.3 entity mentions per document); the latter constitute test set (1056 texts; 249.1 tokens and 21.3 entity mentions per document). We exclude such pages from Wikipedia dump when building training data for our models. In accordance with [27] macro-averaged accuracy is the target metric in comparisons.

To experiment with various embedding types the simplest neural network architecture—with BiLSTM encoder—is employed. We consider three ways of learning embeddings described in Sect. 3.1: separate, joint and joint enriched with graph data. The line graph (see Fig. 4) compares their performance on development and test sets every five epochs. Overall, the quality rises dramatically during the first ten epochs and then slightly increases or fluctuates until the end of training period. However, models with graph and joint vector representations steadily outperform the one with separate embeddings by more than one percentage point. In contrast to this, the gap between joint and joint with graph data types remains small, but steady in favor of the latter through all 50 epochs. Thus, we apply them for further experiments.

As it was expected, prepending convolutional gated units to BiLSTM encoder significantly improves efficiency of the overall architecture (see Table 3).

The final results are presented in Table 4. Deep JEDi outperforms most common sense and GLOW [22] adaptations for the Russian language, presented in [27]. Being a barely neural network approach, Deep JEDi is only a small step behind the much more complicated hybrid solution, which incorporates smart network-generated context into more classical machine learning algorithm.

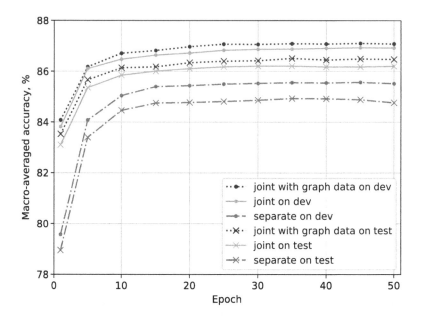

Fig. 4. Embeddings evaluation.

Table 3. Evaluation results for encoders.

Encoder	Accuracy, %	
	Development set	Test set
BiLSTM	87.08	86.47
BiLSTM + convolutions	**88.70**	**88.21**

Table 4. Final evaluation results.

Approach	Accuracy on test set, %
Most common sense	83.01
GLOW	87.80
GLOW$_{no\_linker}$	88.01
GLOW$_{no\_linker}$ + smart context features (BiLSTM)	**88.30**
Deep JEDi	88.21

Finally, we illustrate Deep JEDi's eligibility with an example. Consider a sentence with several mentions:

Люк Скайуокер отправляется в систему Дагоба с целью найти Йоду и пройти обучение джедая, как ему сказал дух Оби-Вана Кеноби.

Ljuk Skajuoker otpravljaetsja v sistemu Dagoba s cel'ju najti Jodu i projti obuchenie dzhedaja, kak emu skazal duh Obi-Vana Kenobi.

(en. `Luke Skywalker goes to the Dagobah system in order to find Yoda and undergo a Jedi training, as the spirit of Obi-Wan Kenobi told him.`)

Taking a closer look at mention Йоду (*Jodu*), we notice that its computed embedding is hardly interpretable as-is. In order to describe it, we select several entities, most similar to the generated vector: Оби-Ван Кеноби (en. `Obi-Wan Kenobi`), Лея Органа (en. `Princess Leia`), Граф Дуку (en. `Count Dooku`), Падме Амидала (en. `Padmé Amidala`), Дарт Мол (en. `Darth Maul`). Even though the produced distributed representation does not exactly match embedding of Йода (en. `Yoda`) entity, it manages to capture the correct topic.

Our knowledge base provides the following potential meanings for the mention Йоду (*Jodu*), ordered by their commonness: Иод (en. `Iodine`), Йода (en.*Yoda*) and Йуд (буква еврейского алфавита) (en. `Yodh - Hebrew Yud`). Reorganizing these entities by proximity to the embedding, generated for the mention, raises the correct value—Йода (en. `Yoda`)—to the top of the list.

5   Conclusion

In this paper we introduce Deep JEDi—fully neural network approach for Entity Disambiguation. Our model is based on encoder-decoder architecture enriched with additional attention alignment vectors, making it a perfect combination of sequence translation and sequence labeling approaches. Following up on [27] findings, we implement BiLSTM encoder and improve it with convolutional gated units.

Another important contribution to contemporary Entity Disambiguation studies is the way joint token and entity embeddings are learned. Capturing relevant information from Wikipedia graph positively affects token-entity interaction and, as a consequence, final results.

To sum up, Deep JEDi is the first neural Entity Disambiguation model developed for Russian, that outperforms most existing approaches. Moreover, it is a worthy competitor to bulky GLOW [22] with smart context [27], as it does not require additional feature generation.

References

1. Bahdanau, D., Cho, K., Bengio, Y.: Neural machine translation by jointly learning to align and translate. In: Proceedings of 3rd International Conference for Learning Representations, pp. 1–15 (2015)
2. Bojanowski, P., Grave, E., Joulin, A., Mikolov, T.: Enriching word vectors with subword information. Trans. Assoc. Comput. Linguist. **5**, 135–146 (2017)
3. Bordes, A., Usunier, N., Garcia-Duran, A., Weston, J., Yakhnenko, O.: Translating embeddings for modeling multi-relational data. In: Advances in Neural Information Processing Systems, pp. 2787–2795 (2013)
4. Francis-Landau, M., Durrett, G., Klein, D.: Capturing semantic similarity for entity linking with convolutional neural networks. In: Proceedings of NAACL-HLT, pp. 1256–1261 (2016)
5. Ganea, O.E., Hofmann, T.: Deep joint entity disambiguation with local neural attention. In: Proceedings of the 2017 Conference on Empirical Methods in Natural Language Processing, pp. 2619–2629. Association for Computational Linguistics (2017)
6. Gehring, J., Auli, M., Grangier, D., Yarats, D., Dauphin, Y.N.: Convolutional sequence to sequence learning. In: International Conference on Machine Learning, pp. 1243–1252 (2017)
7. Gers, F.A., Schmidhuber, J., Cummins, F.: Learning to forget: continual prediction with LSTM. Neural Comput. **12**(10), 2451–2471 (2000)
8. Grover, A., Leskovec, J.: node2vec: scalable feature learning for networks. In: Proceedings of the 22nd ACM SIGKDD International Conference on Knowledge Discovery and Data Mining, pp. 855–864. ACM (2016)
9. Guo, X., Zhang, H., Yang, H., Xu, L., Ye, Z.: A single attention-based combination of CNN and RNN for relation classification. IEEE Access (2019)
10. Hochreiter, S., Schmidhuber, J.: Long short-term memory. Neural Comput. **9**(8), 1735–1780 (1997)
11. Huang, H., Heck, L.P., Ji, H.: Leveraging deep neural networks and knowledge graphs for entity disambiguation. CoRR abs/1504.07678 (2015)
12. Le, P., Titov, I.: Improving entity linking by modeling latent relations between mentions. In: Proceedings of the 56th Annual Meeting of the Association for Computational Linguistics (Volume 1: Long Papers), pp. 1595–1604. Association for Computational Linguistics, Melbourne, July 2018
13. LeCun, Y.: Generalization and network design strategies. In: Connectionism in Perspective, vol. 19. Citeseer (1989)
14. Li, J., Cai, Y., Cai, Z., Leung, H., Yang, K.: Wikipedia based short text classification method. In: Bao, Z., Trajcevski, G., Chang, L., Hua, W. (eds.) DASFAA 2017. LNCS, vol. 10179, pp. 275–286. Springer, Cham (2017). https://doi.org/10.1007/978-3-319-55705-2_22
15. Luong, T., Pham, H., Manning, C.D.: Effective approaches to attention-based neural machine translation. In: Proceedings of the 2015 Conference on Empirical Methods in Natural Language Processing, pp. 1412–1421 (2015)
16. Mihalcea, R., Csomai, A.: Wikify!: linking documents to encyclopedic knowledge. In: Proceedings of the Sixteenth ACM Conference on Information and Knowledge Management, pp. 233–242. ACM (2007)
17. Mikolov, T., Chen, K., Corrado, G., Dean, J.: Efficient estimation of word representations in vector space. arXiv preprint arXiv:1301.3781 (2013)

18. Mikolov, T., Sutskever, I., Chen, K., Corrado, G.S., Dean, J.: Distributed representations of words and phrases and their compositionality. In: Burges, C.J.C., Bottou, L., Welling, M., Ghahramani, Z., Weinberger, K.Q. (eds.) Advances in Neural Information Processing Systems 26, pp. 3111–3119. Curran Associates, Inc. (2013)
19. Milne, D., Witten, I.H.: Learning to link with Wikipedia. In: Proceedings of the 17th ACM Conference on Information and Knowledge Management, pp. 509–518. ACM (2008)
20. Nothman, J., Ringland, N., Radford, W., Murphy, T., Curran, J.R.: Learning multilingual named entity recognition from Wikipedia. Artif. Intell. **194**, 151–175 (2013)
21. Pennington, J., Socher, R., Manning, C.: Glove: global vectors for word representation. In: Proceedings of the 2014 Conference on Empirical Methods in Natural Language Processing (EMNLP), pp. 1532–1543 (2014)
22. Ratinov, L., Roth, D., Downey, D., Anderson, M.: Local and global algorithms for disambiguation to Wikipedia. In: Proceedings of the 49th Annual Meeting of the Association for Computational Linguistics: Human Language Technologies, vol. 1, pp. 1375–1384. Association for Computational Linguistics (2011)
23. Sergeev, A., Balso, M.D.: Horovod: fast and easy distributed deep learning in TensorFlow. arXiv preprint arXiv:1802.05799 (2018)
24. Socher, R., Chen, D., Manning, C.D., Ng, A.: Reasoning with neural tensor networks for knowledge base completion. In: Advances in Neural Information Processing Systems, pp. 926–934 (2013)
25. Sutskever, I., Vinyals, O., Le, Q.V.: Sequence to sequence learning with neural networks. In: Advances in Neural Information Processing Systems, pp. 3104–3112 (2014)
26. Sysoev, A., Andrianov, I.: Named entity recognition in Russian: the power of wiki-based approach. In: Proceedings of International Conference "Dialogue-2016", pp. 746–755 (2016)
27. Sysoev, A., Nikishina, I.: Smart context generation for disambiguation to Wikipedia. In: Ustalov, D., Filchenkov, A., Pivovarova, L., Žižka, J. (eds.) AINL 2018. CCIS, vol. 930, pp. 11–22. Springer, Cham (2018). https://doi.org/10.1007/978-3-030-01204-5_2
28. Turdakov, D., et al.: Semantic analysis of texts using Texterra system (2014). http://www.dialog-21.ru/digests/dialog2014/materials/pdf/TurdakovDY.pdf
29. Wang, X., Jiang, W., Luo, Z.: Combination of convolutional and recurrent neural network for sentiment analysis of short texts. In: Proceedings of COLING 2016, 26th International Conference on Computational Linguistics: Technical Papers, pp. 2428–2437 (2016)
30. Wang, Z., Zhang, J., Feng, J., Chen, Z.: Knowledge graph and text jointly embedding. In: Proceedings of the 2014 Conference on Empirical Methods in Natural Language Processing (EMNLP), pp. 1591–1601 (2014)
31. Yamada, I., Shindo, H., Takeda, H., Takefuji, Y.: Joint learning of the embedding of words and entities for named entity disambiguation. In: Proceedings of the 20th SIGNLL Conference on Computational Natural Language Learning, pp. 250–259. Association for Computational Linguistics, Berlin, Germany, August 2016
32. Yamada, I., Shindo, H., Takeda, H., Takefuji, Y.: Learning distributed representations of texts and entities from knowledge base. Trans. Assoc. Comput. Linguist. **5**, 397–411 (2017)

Selecting an Optimal Feature Set for Stance Detection

Sergey Vychegzhanin[1] , Elena Razova[1] ,
Evgeny Kotelnikov[1(✉)] , and Vladimir Milov[2]

[1] Vyatka State University, Kirov, Russia
`vychegzhaninsv@gmail.com`, `razova.ev@gmail.com`,
`kotelnikov.ev@gmail.com`
[2] Nizhny Novgorod State Technical University n.a. R.E. Alekseev,
Nizhny Novgorod, Russia
`vladimir.milov@gmail.com`

Abstract. Stance detection is an automatic recognition of author's view point in relation to a given object. An important stage of the solution process is determining the most appropriate way to represent texts. The paper proposes a new method of selecting an optimal feature set. The method is based on a homogenous ensemble of feature selection methods and a procedure of determining the optimal number of features. In this procedure the dependence of task performance on the number of features is approximated and the optimal number of features is determined by analyzing the growth rate of the function. There have been conducted experiments with text corpora consisting of "for" and "against" stances towards vaccinations of children, the Unified State Examination at school, and human cloning. The results demonstrate that the proposed method allows to achieve better performance in comparison with individual methods and even an overall feature set with a considerably fewer number of features.

Keywords: Stance detection · Feature selection · Ensembles · Gini index

1 Introduction

A stance detection task lies in automatically determining the position of the author in relation to a given object, for example, politicians, movements or events, socially significant or religious issues, products, organizations, entertainment [18]. The main classes of positions are "for" and "against", although some scientists distinguish a neutral position, the impossibility of determining the author's view point and an agreement with the previous view point [6, 14]. A recent review of this research area can be found in [21].

One of the key tasks in machine learning and natural language processing is feature selection – a process of obtaining a subset of relevant features from the original feature set [4]. The success of this procedure leads to increased learning accuracy, reduced learning time and improved interpretability of results. However, at present there is a lack of works on the stance detection, which explore different approaches to feature selection.

© Springer Nature Switzerland AG 2019
W. M. P. van der Aalst et al. (Eds.) AIST 2019, LNCS 11832, pp. 242–253, 2019.
https://doi.org/10.1007/978-3-030-37334-4_22

There are three approaches to feature selection [12, 15]: filter methods, wrapper methods, and embedded methods. *Filter methods* consider only the relationship between the feature and the class label and do not depend on the learning algorithm, therefore, with a high-speed operation and an ability to handle large amounts of data, the results may not be optimal for the given learning algorithm. In contrast, *wrapper methods* take into account the results obtained through the learning algorithm when evaluating the feature importance. But the work speed at the same time is very low. In *embedded methods*, the feature selection process occurs during training (for example, through regularization), and at the end of training, besides the model itself, the result is the optimal feature set. The work speed is higher than that of the wrapper methods, but there is also a rigid connection with the learning algorithm.

Often one feature selection method (as well as one classifier) is not able to select the optimal set of features in a given task. One possible way to solve this problem is to use ensembles of feature selection methods [2]. Ensembles performed well in the text categorization [8].

There are two approaches to building ensembles of feature selection methods: homogeneous and heterogeneous approaches [17]. Individual feature selection methods, that are part of an ensemble, are called base selectors. If one and the same base selector is used to select features on different subsets of training data, then such an ensemble is called *homogeneous*. If you use different base selectors on the same set of training data, the ensemble is called *heterogeneous*.

The key point to create the ensemble is the choice of the optimal number of features for the sets obtained as a result of base selectors. There are several ways to select the optimal number of features, including constant, based on some function of a number of features, based on the maximum of accuracy [3, 17]. However, as an optimality criterion, they use either the maximum of accuracy or the minimum of accuracy while maintaining the performance at some acceptable level, but not both of these criteria at the same time.

Recently, the DOFNAF procedure (Determining the Optimal Feature Number by the Approximating Function) has been proposed, which allows to obtain a minimum number of features while maintaining the quality of the classification at a high level [22]. This procedure first constructs the dependence of task performance on the number of features, then this dependence is approximated based on the Weibull distribution function, after which the optimal number of features is determined by analyzing the growth rate of this function.

In this article we propose a new method of forming an optimal feature set based on the DOFNAF and a homogenous ensemble of feature selection methods. We compare the proposed method with other feature selection methods in the stance detection task. The method allows to obtain the maximum performance with a relatively small number of features.

2 Previous Work

2.1 Feature Selection Methods

Feature selection methods can be divided into the following groups [3, p. 18; 12; 15]:

1. Filter methods:

    - similarity based methods (Laplacian Score, ReliefF, Fisher Score, Trace Ration Criterion, etc.);
    - information theoretical based methods (Information Gain, Minimum Redundancy Maximum Relevance, Fast Correlation Based Filter, etc.);
    - sparse learning based methods (Efficient and Robust Feature Selection, Multi-Cluster Feature Selection, Feature Selection Using Nonnegative Spectral Analysis, etc.);
    - statistical methods (Chi-Square Score, Gini index, T-Score, Correlation Based Feature Selection, etc.).

2. Wrapper methods (Sequential Forward Selection, Sequential Backward Elimination, Genetic Algorithms, etc.).
3. Embedded methods (Decision Trees, Weighted Naïve Bayes, Recursive Feature Elimination for Support Vector Machine, etc.).

It is impossible to make an unambiguous conclusion about the advantage of a particular feature selection method: in comparative studies, different methods are better for different tasks [1; 3, p. 18; 16, p. 233; 19; 20; 23]. To form heterogeneous ensembles, we have chosen the following methods, which have shown themselves well in comparative studies: ReliefF (RelF), Information Gain (IG), Chi-Square (χ^2), Gini Index (GI), Recursive Feature Elimination (RFE). To construct homogeneous ensembles, we have used Gini Index, which has turned out to be one of the best in the text classification [20], and in our studies it has demonstrated a high quality at low time costs.

2.2 Ensembles of Feature Selection Methods

Seijo-Pardo et al. [17] investigated homogeneous and heterogeneous ensembles with different feature selection methods for feature ranking, ways of union ranking feature sets and selecting the optimal feature number. The SVM has been used as a classification method. The results have been comparable for both approaches with much less work time for homogeneous ensembles.

Guru et al. [8] studied homogeneous, heterogeneous and hybrid ensembles. For a combination of ranking feature sets, the authors applied union and intersection operations. As a result, it has been found out that the ensembles are superior to individual methods.

Hoque et al. [11] proposed an approach to the selection of the optimal feature set, which uses ensembles of feature selection methods based on feature-class and feature-feature mutual information. First, the features are ranked by several methods. The number of features to select is set up by the user. If the rank of some feature is the same for all methods, such a feature is chosen into the result set. Otherwise, it computes feature-class and feature-feature mutual information and selections a feature that has maximum feature-class mutual information but minimum feature-feature mutual information. The results show the superiority of the proposed approach over individual methods.

Our approach has two key differences from the existing works. First, in many works the number of features in the result set must be set manually [8, 11]; our approach uses the DOFNAF procedure proposed in [22] to determine the optimal number of features. Secondly, we apply a different approach for combining feature sets obtained by different methods – not on the basis of some function that aggregates ranking lists, but by searching through all possible combinations of feature sets and by evaluating their effectiveness on the validation dataset.

3   A New Method of Creating the Optimal Feature Set

The proposed method is based on the procedure of determining the optimal number of features DOFNAF [22] and combining feature sets. We have investigated two approaches to create an optimal set – homogeneous and heterogeneous. A homogeneous version of our method is shown in Fig. 1.

Train dataset is an input of the method. On the basis of the cross-validation procedure, Gini Index, SVM classifier[1] and DOFNAF procedure, the optimal number of features N is determined. To do this, we first rank the features based on Gini Index. Then we build the dependence of classification performance on feature number using the cross-validation procedure. The obtained dependence is approximated on the basis of the Weibull distribution function. The optimal number of features is chosen at the point where the growth rate of the function stops to significantly differ from zero. In our work we take this point starting from which the derivative differs from zero by no more than $\varepsilon = 0.0001$.

Then, the train dataset is divided into 6 parts: one part is validation dataset, the other five parts are involved in forming ranking feature sets based on Gini Index (homogeneous approach). For each of these five feature sets, N features with the largest weights that form subsets S_1–S_5 are chosen. Then these subsets are combined in all possible ways – by $\{1, 2, 3, 4, 5\}$-combinations. The total number of combinations will be determined by the formula of the sum of combinations without repetitions of k from m:

$$\sum_{k=1}^{m} \binom{m}{k} = 2^m - 1. \tag{1}$$

At $m = 5$ the number of combinations will be 31.

While combining, two simple strategies of feature aggregation are used – union and intersection (thus, $2 \times 31 = 62$ feature sets are formed) and one more complex. In this

---

[1] We also tried Bagging and AdaBoost classifiers but SVM showed the best results.

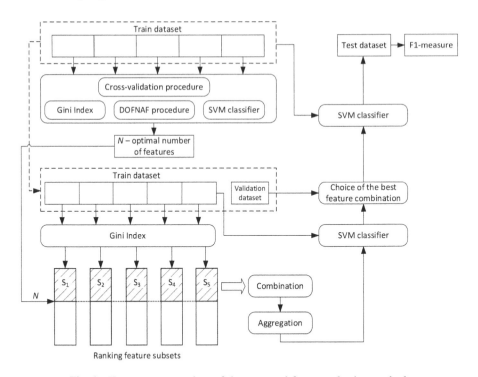

Fig. 1. Homogeneous variant of the proposed feature selection method.

strategy, you first find sets by intersecting subsets in 2-combination, 3-combination, 4-combination separately. Then, in each of these three groups, the sets are combined, resulting in 3 additional feature sets.

All 65 sets obtained as a result of combination and aggregation come to the SVM classifier, which is trained on the combination of the five parts of the train dataset, and is tested on the validation set. Then we select the set that gives the best classification performance that is used when training the final SVM classifier on the source train dataset. The method results in evaluating the classification performance (F1-measure) on the test dataset.

Figure 2 presents a heterogeneous variant of the proposed feature selection method.

The scheme of this variant has several differences from the previous one. Instead of one feature selection method, we use five methods (a heterogeneous approach) – ReliefF, Information Gain, Chi-Square, Gini Index and RFE. Based on the DOFNAF procedure, we determine the optimal numbers of features (N_{RelF}, N_{IG}, $N_{\chi2}$, N_{GI}, N_{RFE}) independently for each method. The ranked feature sets are also built independently by each method throughout the train dataset (excluding the validation set). Then optimal feature subsets S_{RelF}, S_{IG}, $S_{\chi2}$, S_{GI}, S_{RFE} are formed on the basis of the calculated optimal numbers of features. These subsets are combined and aggregated on analogy with the homogeneous variant. In the same way, the best combination is chosen, on whose basis the SVM classifier is trained, which results in evaluating the classification performance on the test dataset.

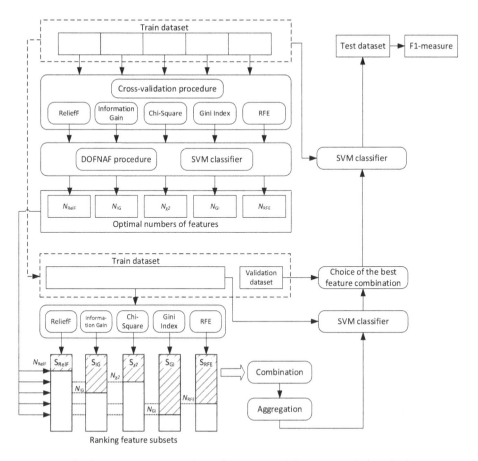

Fig. 2. Heterogeneous variant of the proposed feature selection method.

Both of the proposed variants have been investigated in comparison with separate methods in the stance detection task.

4   Text Corpora

To study the proposed method there have been formed text corpora, containing the user's view point on three issues: vaccinations for children, Unified State Examination at school (USE), and human cloning. To do this, messages have been collected from several Internet forums[2] and the social network Vkontakte[3].

---

[2] http://www.woman.ru/forum, http://www.yaplakal.com, http://www.rusforum.com, http://www.kid.ru, etc.

[3] https://vk.com.

The collected messages were annotated by three annotators (native Russian speakers) with three labels: "for", "against" and "neither in favor nor against" [14], but only messages with "for" and "against" labels have been selected. The inter-annotator agreement by means of Fleiss' kappa statistical measure [7] turned out to be high (Table 1). The formed corpora are publicly available[4].

5   Results and Discussion

We have investigated the following variants of the proposed method:

- a homogeneous ensemble of feature selection methods, the best combination of feature sets (except for the combination of all sets), the aggregation based on intersection (*homo_int*) and union (*homo_uni*);
- a homogeneous ensemble of feature selection methods, a combination of all sets of features, an aggregation based on intersection (*homo_int_all*) and union (*homo_uni_all*);
- the same variants for heterogeneous ensembles (*hetero_int, hetero_uni, hetero_-int_all, hetero_uni_all*).

The suggested method has been compared with the following feature selection methods (the SVM has been used for classification everywhere):

- Gini index. Let n be the size of the complete feature set. The following methods have been used to select the optimal number of features:
  - choice of fixed thresholds (log n, $0.1n$, $0.25n$, $0.5n$) [3, 17];
  - choice of a number of features at which the maximum performance is achieved in the cross-validation (max) procedure [5];
- Correlation-based Feature Selection (CFS) [10]. This method refers to filter methods, but is fully automatic, does not require specifying the number of features required for selection. The CFS method evaluates features subset based on the hypothesis that a good subset contains features that strongly correlate with class labels and weakly correlate with each other. According to this hypothesis, irrelevant features have low correlation with class labels and are ignored by the algorithm.
- Support Vector Machine, SVM – Recursive Feature Elimination with Cross-Validation, RFECV [9]. In this method, features are ranked on the basis of the results of training the SVM classifier. The number of features corresponding to the maximum performance of classification on the validation set is taken as optimal.

The SVM classifier has also been tested on an overall feature set. The results are given in Table 2.

Table 2 shows that the best results (F1 = 0.7667) for the corpus "Vaccinations of children" are achieved by Gini index, $0.25n$ and Gini index, max. The results of *homo_uni_all* (F1 = 0.7665) almost coincide with the maximum results. The best result (F1 = 0.7520) for the corpus "The USE at school" has been obtained by

---

[4] http://tiny.cc/cka08y.

Table 1. Characteristics of text corpora.

Topic	Stance	No. of texts	Total number of words	Average length of texts, words	Size of lexicon, words	Fleiss' kappa
Vaccinations of children	for	500	35,326	70	5,100	0.87
	against	500	34,167	68		
The USE at school	for	600	35,410	59	5,943	0.89
	against	800	40,453	51		
Human cloning	for	450	18,860	42	4,990	0.86
	against	650	27,405	42		

Table 2. Results of classification.

Method	Averaged number of selected features	F1-measure by corpora			F1-measure on average
		Vaccinations of children	The USE at school	Human cloning	
Gini, $\log n$	12	0.6654	0.6166	0.6458	0.6426
Gini, $0.1n$	535	0.7656	0.7248	0.7199	0.7368
Gini, $0.25n$	1,336	**0.7667**	0.7310	0.7429	0.7468
Gini, $0.5n$	2,673	0.7657	0.7294	0.7402	0.7451
Gini, max	3,347	**0.7667**	0.7334	0.7462	0.7488
CFS	71	0.7118	0.6996	0.6786	0.6967
RFECV	984	0.7466	0.7252	0.7351	0.7356
Overall	5,347	0.7625	0.7432	**0.7475**	0.7510
homo_int	644	0.7369	0.7334	0.7268	0.7323
homo_uni	2,006	0.7576	0.7411	0.7473	0.7486
homo_int_all	38	0.6665	0.6549	0.6155	0.6456
homo_uni_all	2,837	0.7665	**0.7520**	0.7415	**0.7533**
hetero_int	515	0.7437	0.7222	0.7066	0.7241
hetero_uni	2,428	0.7634	0.7346	0.7392	0.7457
hetero_int_all	79	0.7096	0.6992	0.6847	0.6978
hetero_uni_all	3,219	0.7625	0.7320	0.7456	0.7467

homo_uni_all. For the corpus "Human cloning" the maximum result has been obtained on an overall feature set.

On average, the *homo_uni_all* method (F1 = 0.7533) has the best result for all the three cases with the average number of features $N = 2,837$, which is 53% of the total set of features, which ranks second (F1 = 0.7510).

The analysis of strategies in the proposed method shows the following. At the intersection of all features (*homo_int_all* and *hetero_int_all*), the formed sets ($N = 38$ and $N = 79$, respectively) are too small, which do not allow to obtain a high performance classification. The *hetero_int_all* method is comparable the CFS method (F1 = 0.6978, N = 79 vs. F1 = 0.6967, N = 71).

The intersections of sets with the choice of the best combination (*homo_int* and *hetero_int*) allow to obtain compact sets with a sufficiently high classification performance (F1 = 0.7323, N = 644 and F1 = 0.7241, N = 515). These results are close to the RFECV method (F1 = 0.7356, N = 984) with fewer features. In *homo_int* method in 60% of cases, the best results are obtained by using a complex strategy of intersection union (see Sect. 3). In *hetero_int* method in 93% of cases the best combination of feature selection methods involves RFECV, and in 60% of cases it involves Gini index and Information Gain.

The strategy of combining sets (*homo_uni*, *homo_uni_all*, *hetero_uni* and *hetero_uni_all*) allows to achieve a high classification performance (from F1 = 0.7457 for *hetero_uni* to F1 = 0.7533 for *homo_uni_all*) with a relatively small number of features (from 37.5% of the total number of features for *homo_uni* to 60.2% for *hetero_uni_all*). In *hetero_uni* method in 80% of cases the best strategy involves Chi-Square, in 73% of cases – ReliefF.

We also compared the best variant (*homo_uni_all*) of proposed method with well-known TF–IDF method [13, p. 118]. We took N features with maximal TF–IDF weights where N coincides with optimal feature number for *homo_uni_all* method (N = 2,837). In Table 3 the ratio of common features for *homo_uni_all* and TF–IDF is shown. It can be seen that this ratio does not exceed 76%.

Table 3. The ratio of common features for *homo_uni_all* and TF–IDF methods.

Topic	Vaccinations of children	The USE at school	Human cloning
Ratio of common features	73%	58%	76%

Table 4 shows the classification performance of SVM classifier with features chosen by *homo_uni_all* and TF–IDF methods.

Table 4. Result of classification for feature sets chosen by *homo_uni_all* and TF–IDF methods (N = 2,837).

Method	F1-measure by corpora			F1-measure on average
	Vaccinations of children	The USE at school	Human cloning	
TF–IDF	0.7586	0.7407	0.7409	0.7467
homo_uni_all	**0.7665**	**0.7520**	**0.7415**	**0.7533**

The *homo_uni_all* method turns out to be better than TF–IDF for all the corpora and on average.

Table 5 shows Top-10 features obtained by of *homo_uni_all* method for all the topics. We took features with maximal rank averaged by five features subsets S_1–S_5 (see Fig. 1).

Table 5. Top-10 features obtained by of *homo_uni_all* method.

Vaccinations of children	The USE at school	Human cloning
– *противопоказание* (contraindication)	– *против* (against)	– *против* (against)
– *дочь* (daughter)	– *возможность* (possibility)	– *можно* (it is possible)
– *против* (against)	– *готовиться* (to prepare)	– *человек* (human)
– *отказ* (refusal)	– *позволять* (to allow)	– *деятельность* (activity)
– *индивидуальный* (individual)	– *маленький* (little)	– *вред* (harm)
– *девочка* (girl)	– *за* (for)	– *использовать* (to use)
– *дома* (at home)	– *традиционный* (traditional)	– *природа* (nature)
– *график* (schedule)	– *если* (if)	– *зачем* (why)
– *пора* (it's time)	– *документ* (document)	– *видеть* (to see)
– *больше* (more)	– *образование* (education)	– *наука* (science)

6   Conclusion

Thus, the article proposes a new method for the creating the optimal feature set for stance detection task, based on a homogeneous ensemble of feature selection methods and on the procedure of determining the optimal number of features DOFNAF. The method allows to obtain a higher classification performance in the stance detection task in comparison with individual feature selection methods, as well as in comparison with an overall feature set.

The prospects for further research may lie in testing the proposed method on other natural language processing tasks.

Acknowledgments. The reported study was funded by the Ministry of Education and Science of the Russian Federation according to the research project No. 34.2092.2017/4.6.

References

1. Adel, A., Omar, N., Al-Shabi, A.: A comparative study of combined feature selection methods for arabic text classification. J. Comput. Sci. **10**(11), 2232–2239 (2014)
2. Bolón-Canedo, V., Alonso-Betanzos, A.: Ensembles for feature selection: a review and future trends. Inf. Fusion **52**, 1–12 (2019)
3. Bolón-Canedo, V., Alonso-Betanzos, A.: Recent Advances in Ensembles for Feature Selection. Intelligent Systems Reference Library. Springer, Heidelberg (2018). https://doi.org/10.1007/978-3-319-90080-3

4. Cai, J., Luo, J., Wang, S., Yang, S.: Feature selection in machine learning: a new perspective. Neurocomputing **300**, 70–79 (2018)
5. Chen, P., Wilbik, A., van Loon, S., Boer, A.-K., Kaymak, U.: Finding the optimal number of features based on mutual information. In: Kacprzyk, J., Szmidt, E., Zadrożny, S., Atanassov, K.T., Krawczak, M. (eds.) IWIFSGN/EUSFLAT -2017. AISC, vol. 641, pp. 477–486. Springer, Cham (2018). https://doi.org/10.1007/978-3-319-66830-7_43
6. Ferreira, W., Vlachos, A.: Emergent: a novel data-set for stance classification. In: Proceedings of the 15th Annual Conference of the North American Chapter of the Association for Computational Linguistics: Human Language Technologies (NAACL-HLT 2016), San Diego, California, USA, pp. 1163–1168 (2016)
7. Fleiss, J.L.: Measuring nominal scale agreement among many raters. Psychol. Bull. **76**(5), 378–382 (1971)
8. Guru, D.S., Suhil, M., Pavithra, S.K., Priya, G.R.: Ensemble of feature selection methods for text classification: an analytical study. In: Abraham, A., Muhuri, P.K., Muda, A.K., Gandhi, N. (eds.) ISDA 2017. AISC, vol. 736, pp. 337–349. Springer, Cham (2018). https://doi.org/10.1007/978-3-319-76348-4_33
9. Guyon, I., Weston, J., Barnhill, S., Vapnik, V.: Gene selection for cancer classification using support vector machines. Mach. Learn. **46**(1–3), 389–422 (2002)
10. Hall, M.A.: Correlation-based feature selection for machine learning. Ph.D. dissertation. Department of Computer Science, Waikato University, Hamilton, NZ (1999)
11. Hoque, N., Singh, M., Bhattacharyya, D.K.: EFS-MI: an ensemble feature selection method for classification. Complex Intell. Syst. **4**(2), 105–118 (2017)
12. Li, J., Cheng, K., Wang, S., Morstatter, F., Trevino, R.P., Tang, J., Liu, H.: Feature selection: a data perspective. ACM Comput. Surv. (CSUR) **50**(6), Article 94 (2016)
13. Manning, C.D., Raghavan, P., Schütze, H.: An Introduction to Information Retrieval. Cambridge University Press, Cambridge (2009)
14. Mohammad, S.M., Kiritchenko, S., Sobhani, P., Zhu, X., Cherry, C.: SemEval-2016 task 6: detecting stance in tweets. In: Proceedings of the International Workshop on Semantic Evaluation (SemEval–2016), San Diego, California, USA, pp. 31–41 (2016)
15. Saeys, Y., Inza, I., Larrañaga, P.: A review of feature selection techniques in bioinformatics. Bioinformatics **23**(19), 2507–2517 (2007)
16. Seetha, H., Murty, M.N., Tripathy, B.K.: Modern Technologies for Big Data Classification and Clustering. IGI Global (2018)
17. Seijo-Pardo, B., Porto-Díaz, I., Bolón-Canedo, V., Alonso-Betanzos, A.: Ensemble feature selection: homogeneous and heterogeneous approaches. Knowl.-Based Syst. **118**, 124–139 (2017)
18. Sridhar, D., Foulds, J., Huang, B., Getoor, L., Walker, M.: Joint models of disagreement and stance in online debate. In: Proceedings of the 53rd Annual Meeting of the Association for Computational Linguistics and the 7th International Joint Conference on Natural Language Processing, Beijing, China, pp. 116–125 (2015)
19. Trivedi, S.K., Dey, S.: A comparative study of various supervised feature selection methods for spam classification. In: Proceedings of the 2nd International Conference on Information and Communication Technology for Competitive Strategies, Udaipur, India, Article No. 64 (2016)
20. Vora, S., Yang, H.: A comprehensive study of eleven feature selection algorithms and their impact on text classification. In: Proceedings of the Computing Conference, London, UK, pp. 440–449 (2017)

21. Wang, R., Zhou, D., Jiang, M., Si, J., Yang, Y.: A survey on opinion mining: from stance to product aspect. IEEE Access **7**, 41101–41124 (2019)

22. Vychegzhanin, S.V., Razova, E.V., Kotelnikov, E.V.: What number of features is optimal? A new method based on approximation function for stance detection task. In: Proceedings of the 9th International Conference on Information Communication and Management, Prague, Czech Republic, pp. 43–47 (2019)

23. Yang, Y., Pedersen, J.O.: A comparative study on feature selection in text categorization. In: Proceedings of the 14th International Conference on Machine Learning (ICML 1997), Nashville, Tennessee, USA, pp. 412–420 (1997)

Social Network Analysis

Analysis of Students Educational Interests Using Social Networks Data

Evgeny Komotskiy[1]([✉]), Tatiana Oreshkina[1], Liubov Zabokritskaya[1], Marina Medvedeva[1], Andrey Sozykin[1,2], and Nikolay Khlebnikov[1]

[1] Ural Federal University, Ekaterinburg, Russia
Evgeny.Komotsky@urfu.ru
[2] Krasovskii Institute of Mathematics and Mechanics, Ekaterinburg, Russia

Abstract. The paper presents an approach to analyze the structure of students educational interests based on data from social networks (subscriptions to pages and groups in the popular Russian social network Vkontakte). We collected data for 1379 students of Ural Federal University, who study at three institutes of the university. The students were clustered based on their interests in the social network and the clusters were compared with the institutes where students study. The approach allowed us to successfully separate the students who are interested in Computer Science and Humanitarian and Social Science. However, the students who study Economics and Management were not clustered successfully due to the heterogeneity of their interests. The approach could be used not only to determine the educational interests of existing students but also to recommend the most suitable educational area for prospective students based on social networks data.

Keywords: Social networks analysis · Machine learning · Students interests · Clustering · Education

1 Introduction

Social networks are becoming more and more popular source of data in Digital Humanities studies every year. This is due to many factors, but the most important of them is that social networks are a large source of conditionally open social and demographic data [1].

The potential of these data in educational studies is just beginning to be unfold. Thus, social networks as data sources are used to assess the psychological state and well-being of students [2], integration of students into social activities and daily life of the University [3] and even for the study of personality traits and psychological characteristics of students [4].

A number of researchers focuses on the fact that, despite some problems, social networks are increasingly used by companies as a source of data for prospective employees [5], and in some cases also for profiling and evaluating the professional suitability of employees [6].

© Springer Nature Switzerland AG 2019
W. M. P. van der Aalst et al. (Eds.) AIST 2019, LNCS 11832, pp. 257–264, 2019.
https://doi.org/10.1007/978-3-030-37334-4_23

As a number of studies have shown, subscriptions of users in social networks are a source of data suitable for predicting academic performance of schoolchildren [7] as well as their gender, age group and academic success of educational institution [8].

However, so far a little attention has been paid to the issue of choosing a most suitable field of education for students on the basis of open data in social networks.

In this paper we propose an approach to the selecting of the most appropriate field of education for the student based on data on his or her interests, which can be expressed in the form of subscriptions (or marks "like") in various social networks. We illustrate our approach using the case of Ural Federal University, for which three fields of education are considered: Computer Science (students of Institute of Fundamental Education), Humanitarian and Social Science (students of Institute of Public Administration and Entrepreneurship), and Economic Science (students of Graduate School of Economics and Management). We collected data about students and their interests from the Russian social network Vkontakte because Russian adolescents prefer it among all others [9]. We analyzed the users subscriptions to pages and groups in Vkontakte and created the clusters of users based on their interests in social network. After that we compare the clusters with the educational areas of the students at Ural Federal University.

The purpose of the study is to answer the question: are the interests of students of different educational directions different in one social network (Vkontakte) among themselves, and how much can this distinction (if it exists) be used for the task of classifying a student to an educational direction.

This task lies at the junction of such areas of knowledge as pedagogy, psychology and machine learning methods (applied to the field of social networking analysis).

2   Background

2.1   Social Network Vkontakte

Vkontakte is one of the most popular social networks in Russia among people aged from 14 to 30 [9]. Most of students are falling into this age category, hence, we decided to use this social network for the purpose of the research.

In addition to sending messages to users, Vkontake allows to create groups and public pages devoted to various topics. Users can subscribe to the groups and pages that are interested for them. Information about such subscriptions of the users are in open access. Hence, this information can be used to estimate the interests of the users.

Vkontakte provides an API for working with its data. In our work we used documented users methods search and users.get_Subscription. Searching and downloading of information about subscriptions of students was carried out in compliance with the "Rules of protection of information about users of the site VK.com" and "Terms of use of the Site Vkontakte".

Table 1. Metrics of cluster quality and their interpretation

N	Metric name	The interpretation of the values
1	Adjusted Rand Index (ARI)	The metric requires knowledge of a priori partitioning of objects into clusters. It takes values in the range $[-1,1]$. Values which close to zero correspond to random partitions, and positive values indicate that the two partitions are similar (coincide at ARI = 1)
2	Adjusted Mutual Information (AMI)	The metric requires knowledge of a priori partitioning of objects into clusters. It takes values in the range $[0, 1]$. Values which close to zero indicate the independence of partitions, and close to one – and about their similarity (coincidence at AMI = 1)
3	Homogenity	The metric requires knowledge of a priori partitioning of objects into clusters. It takes values in the range $[0, 1]$, and larger values correspond to more accurate clustering. This measure is not normalized and therefore depends on the number of clusters
4	Completeness	The metric requires knowledge of a priori partitioning of objects into clusters. It takes values in the range $[0, 1]$, and larger values correspond to more accurate clustering. This measure is not normalized and therefore depends on the number of clusters
5	Silhouette	The metric does not imply knowledge of the true labels of objects and allows to assess the quality of clustering, using only the (untagged) selection and the result of clustering. It takes values in the range $[-1, 1]$. Values close to -1 correspond to the bad (disjointed) clustering, values close to zero, suggests that the clusters intersect and overlap each other, values closer to 1 correspond to "tight", well-defined clusters

2.2 Metrics for Measuring the Quality of Clustering

To assess the quality of clustering algorithm we used a number of metrics presented in the sklearn library [10]. These metrics can be divided into external and internal quality metrics. External metrics use information about true clustering, while internal metrics do not use any external information and evaluate the quality of clustering based only on the dataset. The description of metrics and interpretation of their values are presented in Table 1.

3 Method

In the course of our research we used the following approach:

1. The data about subscriptions in social network for students, educated in Ural Federal University on different fields of education (institutes) was downloaded.
2. Data set for clustering was created based on the downloaded data.
3. Clustering was implemented and the results was evaluated using various supervised and non-supervised metrics for clustering (Table 1).
4. The institute/cluster adjacency matrix was formed.
5. The distribution of students in clusters based on social network subscriptions and the distribution by the institutes of the university was compared and visualized.

4   Dataset

For the purposes of the analysis, 1379 students were selected, who are enrolled in Ural Federal University at the following institutions:

1. Institute of Public Administration and Entrepreneurship (463 persons)
2. Institute of Fundamental Education (461 persons)
3. Graduate School of Economics and Management (455 persons)

For each of the students the data about his or her subscriptions were downloaded using the API methods of the social network Vkontakte.

Information about user subscriptions was presented in the form of a data structure, known as the adjacency list which is a set of key-value pairs of the form 'user'-'subscriptionID'. The resulting dataset had a dimension of 185 000 such pairs.

Further, to reduce the dimension of the task, we have excluded those communities to which less than 5 users are subscribed. This has reduced the resulting dataset to 75 268 records.

For constructing a feature vector describing each unique user (student of the Institute)to the resulting dataset we used One Hot Encoding procedure (unitary encoding). Participation in each of the communities was coded using a_j binary variables ($j = 1, \ldots, 3\ 890$), where $a_j = 0$, if the user is not subscribed to community j, and $a_j = 1$ if the user is subscribed to community j. Thus, the entire data set is a matrix of 1379×3890, (i, j)-element of which is equal to 1 if the user with number i is subscribed to the community with number j, and is equal to 0 otherwise.

5   Data Preprocessing and Model Building

For performing the clustering procedure we used the approach described in [4]. Since the resulting matrix is more sparse and dimensional, it was advisable to carry out a procedure to reduce the dimension to solve the clustering problem. To do this, we used the principal component analysis method (PCA). In order to estimate the number of required components we decided to use the rule, that is optimal is the number of components that can explain 90% of the variance. Experiments, the result of which is shown in Fig. 1, showed that this number is 604 components.

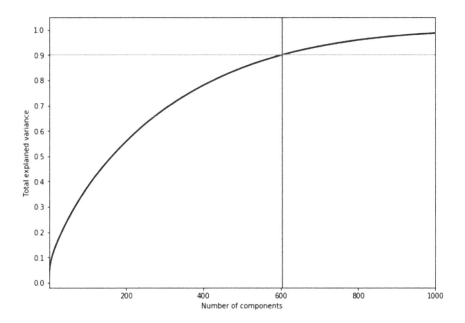

Fig. 1. Selection of the optimal number of components for PCA

The allocated 604 principal components that characterize the interests of users were used to perform the clustering procedure.

In this paper, we used well-known and described K-means clustering and Agglomerative Clustering algorithms to perform clustering. We have chosen the cluster number parameter $k = 3$ for the specified algorithms. Thus, the number of clusters planned to be created using clustering methods equal to the number of institutions initially represented in the sample.

6 Results

To compare the clusters we have build based on the interests of students with the known distribution by institutes of Ural Federal University, we used the metrics, presented at the Table 1. The results of the comparison are presented at the Table 2.

The metric values (Table 2) can be interpreted as that clusters are expressed, but weakly. In order to evaluate how the algorithm distributed the students of the institutes into the clusters based on their subscriptions, we constructed the adjacency matrix for each algorithm (Tables 3 and 4).

Table 2. A comparison of the clusters based on interest in social networks and on the educational institutions

Algorithm	ARI	AMI	Homogenity	Completeness	V-measure	Silhouette
K-means	0,25	0,23	0,23	0,42	0,30	0,23
Agglomerative	0,24	0,30	0,30	0,32	0,31	0,08

Table 3. Institute/Cluster adjacency Matrix for the k-means algorithm

Institute	Cluster 0	Cluster 1	Cluster 2
Graduate School of Economics and Management	88	0	367
Institute of Public Administration and Entrepreneurship	363	1	99
Institute of Fundamental Education	4	0	457

Table 4. Institute/Cluster adjacency Matrix for the agglomerative clustering algorithm

Institute	Cluster 0	Cluster 1	Cluster 2
Graduate School of Economics and Management	0	260	195
Institute of Public Administration and Entrepreneurship	240	39	184
Institute of Fundamental Education	0	379	82

The tables show that most of the Institute of Public Administration and Entrepreneurship students got into the Cluster 0 for the K-means and agglomerative clustering algorithms, and most of Institute of Fundamental Education students got into the Cluster 2 for the K-means algorithm and into the Cluster 1 for the agglomerative clustering algorithm. However, the students of Graduate School of Economics and Management was little affected by clustering algorithm and they were mostly distributed among other clusters.

7   Discussion and Conclusion

The results allow us to conclude that the approach made it possible to separate students based on the educational field instead of the institution. The Institute of Fundamental Education and Institute of Public Administration and Entrepreneurship have clearly defined educational areas. To be specific, that is Computer Science for Institute of Fundamental Education and Humanitarian and Social Science for Institute of Public Administration and Entrepreneurship.

The students from these two institutes got into different clusters. However, Graduate School of Economics and Management have educational programs related to Computer Science (e.g. Business Informatics) and to Economics (e.g. Marketing). As a result, the students of the institute did not form a separate cluster, but were distributed among other clusters. The students of Graduate School of Economics and Management, who are interested in Computer Science, were placed into the cluster together with the students of Institute of Fundamental Education. In contrast, the students who are interested in Economics, got into the cluster with the students of Institute of Public Administration and Entrepreneurship.

The study has both theoretical and practical significance.

Theoretically, it points to the heterogeneity of interests expressed in the form of subscriptions in the social network Vkontakte among students of different educational areas.

The practical significance of information about the characteristic interests of students of various educational areas is the possibility of pre-selection and invitation to the university of applicants and schoolchildren who look like students at the university in this area in their interests.

The results show that data on interests in social networks can be used to assess and orient students in different areas of education. In the course of further research we plan to develop and evaluate the algorithms for clustering and classifying students in the recommended areas using machine learning methods and by enriching data on interests with socio-demographic data from social networks. In addition we plan to use social network data in combination with existing university data sources [11] to predict students performance.

Since we can form individual clusters based on subscriptions that correspond to the educational interest of existing students, we can use the suggested approach to recommend the most suitable educational areas to prospective students.

References

1. Boyd, D.M., Ellison, N.B.: Social network sites: definition, history, and scholarship. J. Comput.-Mediated Commun. **13**(1), 210–230 (2008)
2. Steinfield, C., Ellison, N.B., Lampe, C.: Social capital, self-esteem, and use of online social network sites: a longitudinal analysis. J. Appl. Dev. Psychol. **29**(6), 434–445 (2008)
3. Madge, C., Meek, J., Wellens, J., Hooley, T.: Facebook, social integration and informal learning at university: 'it is more for socialising and talking to friends about work than for actually doing work'. Learn. Media Technol. **34**(2), 141–155 (2009)
4. Kosinski, M., Stillwell, D., Graepel, T.: Private traits and attributes are predictable from digital records of human behavior. Proc. National Acad. Sci. **110**(15), 5802–5805 (2013)
5. Landers, R.N., Schmidt, G.B.: Social media in employee selection and recruitment: an overview. In: Landers, R.N., Schmidt, G.B. (eds.) Social Media in Employee Selection and Recruitment, pp. 3–11. Springer, Cham (2016). https://doi.org/10.1007/978-3-319-29989-1_1

6. McDonald, P., Thompson, P., O'Connor, P.: Profiling employees online: shifting public-private boundaries in organisational life. Hum. Resour. Manag. J. **26**(4), 541–556 (2016)
7. Smirnov, I.: Predicting PISA Scores from Students' Digital Traces. In: Proceedings Of The Twelfth International Conference On Web And Social Media. American Association for Artificial Intelligence (AAAI) Press, pp. 360–364 (2018)
8. Polivanova, K., Smirnov, I.: What's in My Profile: VKontakte data as a tool for studying the interests of modern teenagers. Educ. Stud. **2**, 134–152 (2017)
9. Koroleva, D.O.: The use of social networks in education and socialization of adolescents: an analytical review of empirical studies (international experience). Psychol. Sci. Educ. **20**(1), 28–37 (2015). https://doi.org/10.17759/pse.2015200104
10. Pedregosa, F., Varoquaux, G., Gramfort, A., et al.: Scikit-learn: machine learning in python. J. Mach. Learn. Res. **12**, 2825–2830 (2011)
11. Borisov, V.I., Rabovskaya, M.Y., Syskov, A.M., Zeyde K.M.: Design an information system for student track prediction. In: 2018 Siberian Symposium on Data Science & Engineering (SSDSE), Novosibirsk, pp. 24–27 (2018)

Multilevel Exponential Random Graph Models Application to Civil Participation Studies

Valentina Kuskova, Gregory Khvatsky, Dmitry Zaytsev,
and Nikita Talovsky(✉)

National Research University Higher School of Economics, Moscow, Russia
ntalovsky@hse.ru

Abstract. Due to the development of the Internet and social networks, civil participation in political processes is taking new forms, not accounted for in the classical theoretical frameworks. In this study, we present a new methodology for researching these new, unconventional forms of civil participation. We use data collected from a social network site VK.com. Social network analysis methods, such as multilevel exponential random graph models (ERGM) were used to analyze the data. First proposed by Wang in 2013, multilevel ERGM allows us to model tie formation in a wide class of social networks. This method allows us to efficiently model group membership based on node attributes. Since the collected data is fundamentally 2-mode, this model allows us to identify the important factors that lead people to join online protest communities.

Keywords: Exponential random graph models · Multi-mode networks · Two-mode networks · Multilevel networks

1 Introduction

Major protests have occurred around the world with increasing frequency since the second half of the 2000s. The spike in global protests is becoming a major trend in international politics, but care is needed in determination of the nature and impact of the phenomenon [1]. Protests seized both the developing and developed countries. In the European countries, many protests took place in the aftermath of the global financial crisis of 2008–2009. By the year 2011, they spread over to the United States as well. In the Middle East, the "Arab Spring" protests began in 2011; in Russia, it was the year of the Bolotnaya Square protests. A lot of research has been made on this topic. However, these protests were in many aspects unique, and could not be explained with the old theoretical and methodological approaches [2].

The existing theories of civil society and participation (collective behavior, collective action, social movements) approach mass civil actions with the overall process of the *institutionalization of civil participation*. According to this theory, protests and strikes are understood as unconventional forms of civic participation, and should be seen as the first step, later transforming to conven-

© Springer Nature Switzerland AG 2019
W. M. P. van der Aalst et al. (Eds.) AIST 2019, LNCS 11832, pp. 265–275, 2019.
https://doi.org/10.1007/978-3-030-37334-4_24

tional forms, such as elections. In this trajectory, citizen participation leads to the development of civil society, making political system more democratic and stable [3].

The focus on the institutionalization is better suited for the developed Western democracies, where the development of civil society leads to the evolution of democratic institutions. Forms of citizens' activities such as interest groups, social movements, and civil associations that have arisen in the wake of mass civil actions of 2010s, later transformed into regular practices of interaction between civil society and the state, making the citizens regular participants of the political process and the actors of change in countries involved [6].

Another approach is the social movements theory that focuses on the analysis of dynamics of collective actions. C. Tilly defines social movements as "a series of contentious performances, displays and campaigns by which ordinary people made collective claims on others" [4]. The key idea is that social movements manifest their ideas though "campaigns" – a sustained, organized public effort making collective claims on public authorities. The repertoire of their actions includes many different forms of public participation – political actions, creation of coalitions, mass meetings, letter writing campaigns, rallies, demonstrations and so on. This scope of different forms of public participation requires serious organizational work before the actual campaigns. Therefore, stable organizational structures or "special purpose coalitions" should be formed beforehand to be able to ensure a sustained and organized manner of collective actions [5].

However, mass civil actions of 2010s have progressed not only in the direction of institutionalization described above, nor they affected only the developed countries. They presented a challenge to the status quo of political systems types all over the world, from a representative democracies to closed autocracies. Having no direct connection to the social and economic crises, they manifested as indicators of crises in existing political system. As such, they paved the way for newly emerging practices of relations between the state and the civil society [7].

At this point, there is a need of the development of new research optics. To analyze the mass civil actions of the 2010s as a social phenomenon of a new type, it is necessary to develop a special scientific language and new research optics, different from the existing research tools. This requires a synthesis of old theories and new approaches.

The "theory of the public" [8–12] fills the gap described above and gives a new language to describe a new phenomenon – mass civil actions of the 2010s as the action of "active publics." The concept of "active publics" shows the transformation of the participants of these actions, how their values and identity change, while subjectivity and the ability to change social reality appear.

Another important concept for modern interpretations of the new forms of civil participation is the concept of social network [13–16]. It includes the study of "network citizens" [17], mobilizing the role of the Internet for social movements [18], and the dynamics of on-line and off-line activism [19]. These are not a complete series of studies, nor are they exhaustive of the possibility of applying a network approach to the analysis of new forms of civil participation.

Most developed models in sociology and political science explain either the changes of state and society on a macro level [20], political systems and regimes, or microsocial changes like behavior, attitudes, and human identity [21,22]. However, it appears that insufficient attention is paid to institutional change on "meso-level." These changes are taking place with the direct participation of collective actors, both traditional (political parties, trade unions, churches, social movements, NGOs, large corporations) and "new" (civil communities or "active publics," "new social movements," new media, analytical networks, online communities). A study at the intersection of these topics could provide important insights into the processes of civil transformation of developing societies, above and beyond those looked into by political science studies.

This area of study related to the institutionalization of citizen participation may be relevant to the study of a specific group of countries, where one can easily observe "transitions" from authoritarianism or "hybrid regime" to democracy. But for "developing countries," which are facing relatively more advanced "nonlinear" processes of social transformation, this approach to analysis may not appropriately evaluate the changes occurring between the state and civil society. Therefore, the problem of identification of alternative areas of institutionalization of civil participation for the developing countries with a high degree of uncertainty of the transformation process of the state and society seems very important and urgent.

One way of addressing this problem is looking into developing societies such as BRICS countries since the beginning of mass civil actions in 2010s. Mass civil actions there represent complex process of transformation of the social, economic, and political systems. The changes vary depending on a country, but their complexity and the high degree of uncertainty is the outcome of the overall features of these countries, as the processes of political and economic institutional transformation in them has not yet been completed. Thus, we are facing with the problem of identifying alternative areas of institutionalization of civil participation, forms of cooperation between civil society and the state, in countries such as the BRICS countries, where there are complex "non-linear" transformation processes with a high degree of uncertainty in the outcome.

In that light, changes occurring in Russia are becoming especially interesting and have generated a lot of discussion in the international research community. The situation in Russia is different from that of "developed countries," where protests are a challenge and an attempt to change the existing embedded institutional architecture in an effort to improve the quality of democracy (citizens' participation in politics, responsive policymaking, and responsibility of government to citizens). The expected end result of such challenge is the "participatory democracy." In Russia, just as in other BRICS countries, mass citizen actions do not lead to the establishment and development of basic democratic institutions of representative democracy or "polyarchy" as it would be the case for transitional societies. Rather, as literature speculates, it may lead to the further consolidation of authoritarianism.

Based on the mixture of existing theoretical approaches to study civil participation, we propose a new quantitative methodology for the research of these new,
unconventional forms of civil participation caused in part by the development of
the Internet and social networks.

2   Methodology

2.1   Multilevel Exponential Random Graph Models

Exponential Random Graph Models (ERGM) are the probabilistic models that
allow to predict the next tie formation. Because the tie is either formed or not,
the response variable is a binary 0–1 variable, following some of the mathematics
of the logistic regression. These are the models of social selection, where the
probability of the next tie for a given node or the entire network is determined
by both the structure of the network and the attributes of the individual model.
It is a very powerful model that can explain the selection processes, shaping the
global structure of the network [23].

ERGMs can be used to also quantify the strength of the relationships, such
as the intergroup effects, understand a particular phenomenon influencing a tie
formation, or simulate new random networks that are similar to the original
network, but offer important insights. ERGMs is based on MCMC algorithms
and are probabilistic in nature.

Because it's a family of models based on probability theory (including
Bayesian probability), the set of terms that defines the model differ from other
statistical models [24]. In a traditional model, there is a set of variables, one of
which may be called an outcome variable (also a dependent variable or a criterion
variable), and a set of others, called predictors (independent variables) are used
to model the behavior of this variable. However, there is hardly ever a difference
in essence of the variables, and some of them can be used interchangeably as
predictors and criterion. In ERGM, there is only one dependent variable – a tie
that is formed between two nodes. Predictors in the model are a function of a
tie itself, and they may be network statistics, such as triangles, mutuality, etc.,
or the node attributes. Because predictors are a function of a criterion, ERGM
models are also autoregressive, adding an additional complexity to the analysis.
Below is a list of the most common predictors that can be used in an ERGM
model.

1. Basic structural terms are the terms that control the overall probability
of a tie formation. The most basic network structural characteristics we want to
account for are density (edges) and mutuality (in directed networks). A parameter estimate for the edges is also loosely equivalent to the intercept in the linear
model – it is the existing density of the network, which is then affected differently
by additional structural or attribute variables.

2. Nodal attribute main effects are the usual individual level variables, unique
for every actor in the model. When they are included, we are testing the association between a value in the attribute and the probability that this value is

important to the formation of a tie. Just as in linear regression, standard cautions should be taken when working with categorical variables, missing data, etc., because the general assumptions of the logistic regression need to hold here as well.

3. Interactions for attribute-based mixing allow to test not just the attribute main effects, but also various assortative/dissortative mixings of the attributes.

4. Relational attributes test the characteristics of dyads and edges, instead of individual nodes, and allow for a more complex examination of a network structure.

5. More complex structural terms describe structural features of the network (beyond density and dyads), and include terms for triangles, transitivity, degree distributions, k-stars, isolates, 2-paths, cycles, and shared-partner distributions. They also provide information on the effects of the existing network structure on the probability of a tie formation.

6. Actor-specific network effects evaluate network characteristics of a given node. They are looking into whether the node is a receiver (more ties are made to this node), sender (more ties are made out of this node), sociality. Because the probability of a tie can be formulated for either an entire network or a given node, and some nodes can be very influential in the network, the ability to apply predictions to an individual node is an important attribute of the ERGM.

7. Whole network operators are used to test whether the network itself is conducive to a tie formation. For example, networks with very high density do not leave much room for other structural characteristics or node attributes, because density will be the most important (and often the only significant) predictor. Networks with low density, on the contrary, do not properly account for the influence of other predictors, because the overall probability of a tie formation in a low-density network will remain low. Therefore, for such models using whole network operators could help evaluate the impact of other predictors above and beyond the effect of edges.

8. Constraint terms account for particular features of a network that are known a priori. For example, if a network is not allowed to get larger than a certain size, then network size would act as a constraint in the model, affecting the probability of the next tie.

Originally, ERGM was designed to work with single-mode networks. However, Wang et al. [25] extended the model to work with bipartite and other types of networks. The original specifications and interpretation of the model results remained unchanged.

3   Application

The data were collected from the social network website VK.com. According to sociological surveys [26], VK is the most popular social network in Russia. In addition, VK provides an easy to use API that makes it possible to download information about users and events/communities in a machine-readable format.

The purpose of the initial step of the data collection procedure was to identify events (term used by VK to describe real world gathering organized via the website) related to mass protests in Russia. During this step, we first compiled a list of keywords and keyphrases that were possibly related to protest events e.g. "For Fair Elections" and "Russia without Putin". We then used VK's search function to find events using the keyphrases as search requests.[1] In total, we have identified 261 such events, of which privacy settings of only 212 allowed us to download the list of their users. We have also identified and removed events that were not relevant to our research. The final dataset consisted of 72 events, all of which were related to various political protests in Russia.

After that, for all events in the final dataset, we have used VK API to download a list of users who said that they would definitely take part in that event. We have used this data to construct a bipartite graph of users (identified by their unique numerical identifies known as VK IDs) and events they took part in. The final network consisted of 47742 nodes, of which 72 were events and 47670 were users.

However, this network was still too large for modeling. We have further reduced it by only including nodes with degree >2. After this reduction, the final network used for further modeling consisted of 3605 users and 69 events, 3674 nodes in total.

To collect attributes for all those users, for each user we have downloaded a list of communities[2] this user was a member of. For some users, it was impossible to download their communities due to their privacy settings; those users were later removed from the analysis. After that, we have identified 1000 communities that are most popular among the users in our dataset. For each of those communities, we have coded by hand whether it is an online only community or mostly offline community. The principle of coding was based on the origin of the community: if it was originally created offline, and later developed a social platform (or some other) web page, the community was coded as an "offline" community. If it was created using social network platform to start with (with no corresponding offline outlet), this community was coded as "online." For example, "MBH Media", "Mediazone", "Echo Moscky" and "Evgenii Roizman" were classified as offline.

After that, for all users in the network, we counted the number of offline and online communities. Users whose communities were impossible to download were interpreted as having their data missing. Next, we removed all the users with missing attributes from the network. The final network consisted of 2830 users and 69 communities, 2899 nodes in total.

We have also discovered the fact that the attributes were highly correlated (Pearson's $\rho = 0.77$), which means that our model would have suffered from multicollinearity. To remedy this, we have transformed them into a pair of

---

[1] It should be noted that the VK API search function and the website search function return different results for identical requests. For this study, we have used Selenium to extract search results directly from the website.

[2] Public pages were included.

orthogonal features using Principal Component Analysis. The results of the analysis are presented in the Fig. 1.

```
                  PC1         PC2
Online    0.8641705  -0.5031991
Offline   0.5031991   0.8641705
```

Fig. 1. PCA Eigenvectors

These new features were then treated as a set of covariates for the first mode in a 2-mode (bipartite) ERG model. As we can see in the Fig. 2 the model is significant: there are less edges in this network, and more online and offline dependent ties than in random network.

```
===========================
Summary of model fit
===========================

Formula:   active.users.net ~ edges + b1cov("onoffpc1") + b1cov("onoffpc2")

Iterations:  6 out of 20

Monte Carlo MLE Results:
                 Estimate Std. Error MCMC %  z value Pr(>|z|)
edges            -3.381460   0.012722      0 -265.796  < 1e-04 ***
b1cov.onoffpc1    0.018070   0.001888      0    9.574  < 1e-04 ***
b1cov.onoffpc2    0.018112   0.006285      0    2.882  0.00396 **
---
Signif. codes:  0 '***' 0.001 '**' 0.01 '*' 0.05 '.' 0.1 ' ' 1

     Null Deviance: 270702  on 195270  degrees of freedom
 Residual Deviance:  56646  on 195267  degrees of freedom

AIC: 56652    BIC: 56683    (Smaller is better.)
```

Fig. 2. Summary of the bipartite ERG model fit

As the first component predict ties formation depend on the users attribution with online communities, and the second component - with offline organizations, we can conclude that both online and offline types of mobilization are important.

In the Appendix we present goodness-of-fit diagnostic plots for the model, so we can check how well the model fits the data. These diagnostic plots compare statistics for networks that were simulated using the model to the observed network statistics. Figure 3 shows that the model is able to capture the structure of the network in terms of node degree. Figure 4 show that the transitivity structure of the network is simulated correctly. Figure 5 shows that the distribution of minimum geodesic distances in the simulated networks is different from the observed distribution. This indicates a slight problem with model fit, which is not critical, because the model still manages to capture global network structure.

This can be remedied by inclusion of additional structural terms. Figure 6 shows that model coefficients for the simulated network are not different from what is observed, which indicates good fit. Overall we can conclude that the model captures the global network structure adequately well.

The following results is about the probabilities of ties formation if we increase of attributed groups in one:

- "Edges only: 0.0329"
- Edges + 0 online, 1 offline: 0.0335
- Edges + 1 online, 0 offline: 0.0335
- Edges + 1 online, 1 offline: 0.0341

If we increase attributed groups in maximum of its kind we will get the following results:

- Edges + 45 online, 0 offline: 0.0712
- Edges + 0 online, 28 offline: 0.0534
- Edges + 45 online, 28 offline: 0.1129

Online and offline communities and organizations have comparable potential for protest mobilization, when only one community is involved. However, in real life, people are more likely to sign up for more than one community of interest, so it becomes much more important to calculate the potential of online vs. offline mobilization for those who select multiple communities. The power of online mobilization becomes much more clear when analyzing the case of the maximum number of groups: not only the potential of online communities is substantially (over 30%) higher than offline, but the effect of joining both types is almost cumulative (which is not clear from a single community calculation). Taking into consideration that offline channels in this research is mostly mixed (we analyzed organizations that have online representation) we can estimate the power of online and mixed channels of mobilization using this model.

4   Conclusion

In this paper we demonstrated the application of multilevel ERGM for the political science problem. Using bipartite ERGM we predict tie formation in two-mode network of users in VK who went to protest events depending on the quantity of online and offline (mixed) communities that they like. We have determined that joining the online protest communities has a higher effect on the protest activity than the offline community only. This result has a number of applications in the political science research, most of which we leave for further study.

The proposed methodology and its application can be extended by adding more attributes, and not only for users, but for the events. Application of multiplex research design also can be a valuable contribution both in network methodology and social science theory testing.

Acknowledgment. The article was prepared within the framework of the Basic Research Program at the National Research University Higher School of Economics (HSE) and supported within the framework of a subsidy by the Russian Academic Excellence Project '5–100'.

A   Appendix: Goodness-of-fit diagnostics for the bipartite ERG model

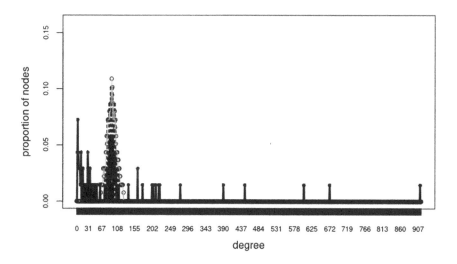

Fig. 3. ERGM goodness-of-fit (degree)

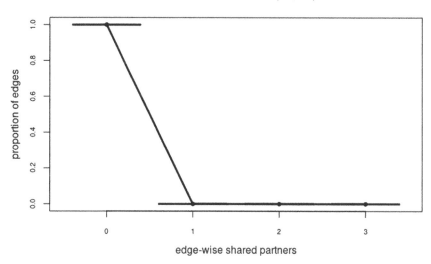

Fig. 4. ERGM goodness-of-fit (edge-wise shared partners)

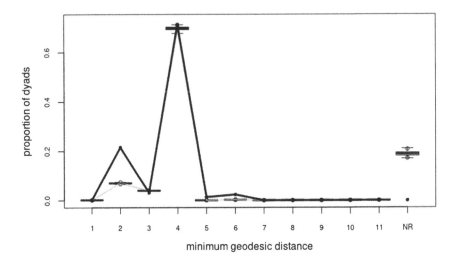

Fig. 5. ERGM goodness-of-fit (minimum geodesic distance)

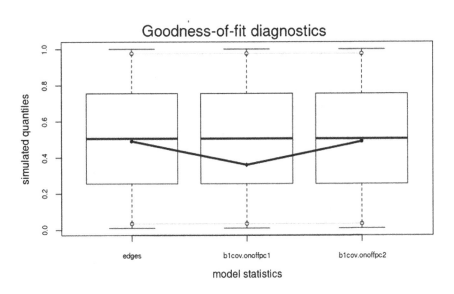

Fig. 6. ERGM goodness-of-fit (model coefficients)

References

1. Carothers, T., Youngs, R.: The Complexities of Global Protests, vol. 8. Carnegie Endowment for International Peace, Washington, DC (2015)
2. Ortiz, I., Burke, S., Berrada, M., Cortes, H.: World Protests 2006–2013. Initiative for Policy Dialogue and Friedrich-Ebert-Stiftung New York Working Paper (2013)
3. Almond, G., Verba, S.: The Civic Culture: Political Attitudes and Democracy in Five Nations. Princeton University Press, Princeton (2015)

4. Tilly, C., Tarrow, S.: Contentious Politics. Oxford University Press, Oxford (2015)
5. McAdam, D., Tarrow, S., and Tilly, C.: Comparative perspectives on contentious politics. In: Comparative politics: Rationality, culture, and structure, pp. 260–290. Cambridge University Press, Cambridge (2009)
6. Arbatli, E., Rosenberg, D. (eds.): Non-western social movements and participatory democracy. SPOT. Springer, Cham (2017). https://doi.org/10.1007/978-3-319-51454-3
7. Anufriev, A., Zaytsev, D.: "Protest Publics" in Egypt and Turkey from 2011 till present days: assessment of impact on political changes. Comp. Polit. (Russia). **7**, 34–47 (2016)
8. Warner, M.: Publics and counterpublics. Public Cult. **14**(1), 49–90 (2002)
9. Dewey, J.: Public and its Problems. Henry Holt and Company, New York (1954)
10. Wittenberg, D.: Going out in public: visibility and anonymity in Michael Warner's "publics and counterpublics". Q. J. Speech **88**(4), 426–433 (2002)
11. Mahony, N., Clarke, J.: Public crises, public futures. Cult. Stud. **27**(6), 933–954 (2013). https://doi.org/10.1080/09502386.2012.730542
12. Belyaeva, N., Dzhibladze, A.: XIV April international academic conference on economy and society development, vol. 2, pp. 377–389 (2014)
13. Castells, M., Cardoso, G.: The network society: from knowledge to policy, pp. 3–23. Johns Hopkins Center for Transatlantic Relations, Washington, DC (2006)
14. Bauman, Z.: Liquid Modernity. John Wiley and Sons, New York (2013)
15. Melucci, A.: The new social movements: a theoretical approach. Inf. (Int. Soc. Sci. Counc.) **19**(2), 199–226 (1980)
16. Gilchrist, A., Kyprianou, P.: Social networks, poverty and ethnicity. Programme paper, Joseph Rowntree Foundation, York (2011)
17. Hill, K., Hughes, J.: Cyberpolitics: Citizen Activism in the Age of the Internet. Rowman and Littlefield Publishers, Lanham (1998)
18. Earl, J., Kimport, K.: Digitally Enabled Social Change: Activism in the Internet Age. MIT Press, Cambridge (2011)
19. Kolozaridi, P., Uldanov, A.: Internet and social movements: analysis of topic development in ScienceDirect and Scopus databases. Sociol. J. **21**, 105–128 (2015)
20. Manning, P., Hopkins, T., Wallerstein, I.: World-Systems analysis: theory and methodology. African Economic History, vol. 222 (1984)
21. Mike, M.: Individualistic humans: social constructionism, identity and change. Theory Psychol. **7**(3), 311–336 (1997)
22. Stets, J., Burke, P.: Identity theory and social identity theory. Soc. Psychol. Q. **63**(3), 224–237 (2000)
23. Hunter, D., Handcock, M., Butts, C., Goodreau, S., Morris, M.: ergm: a package to fit, simulate and diagnose exponential-family models for networks. J. Stat. Softw. **24**(3), nihpa54860 (2008)
24. Morris, M., Handcock, M.S., Hunter, D.R.: Specification of exponential-family random graph models: terms and computational aspects. J. Stat. Softw. **24**(4), 1548 (2008)
25. Wang, P., Robins, G., Pattison, P., Lazega, E.: Exponential random graph models for multilevel networks. Soc. Netw. **35**, 96–115 (2013)
26. To each age - their own networks. https://wciom.ru/index.php?id=236&uid=116691

The Entity Name Identification in Classification Algorithm: Testing the Advocacy Coalition Framework by Document Analysis (The Case of Russian Civil Society Policy)

Dmitry Zaytsev[ID], Nikita Talovsky[(✉)][ID], Valentina Kuskova[ID],
and Gregory Khvatsky[ID]

National Research University Higher School of Economics, Moscow, Russia
ntalovsky@hse.ru

Abstract. We present a methodology for identification and classification of policy actors. We used network analysis and rule-based named entity recognition on a computational cluster for actor identification, and Chinese whispers algorithm with pre-specified clusters to identify probable coalitions between the identified actors. We test this methodology on the case of Russian policy towards civil society. The theory we have chosen is the Advocacy Coalition Framework, which is a public policy theory aimed at explaining the long-term policy change by understanding how and why people engage in policy-making. One of the key ideas of the theory is that people participate in policy to translate their beliefs into action, and then gather into advocacy coalitions based on the shared "beliefs system." Identification of actors is one of the most fundamental issues in political science, as it is often the first step in the research process. Another problem of interest is the classification of the actors based on latent characteristics such as shared beliefs. Most of research papers apply qualitative methodology for both of the steps. By applying our methodology, which relies heavily on the quantitative approach, we identify two coalitions – the conformist and the alternative. The leading actors in each coalition correspond with qualitative research in the field.

Keywords: Natural language processing · Unsupervised learning · Civil society

1 Introduction

Studies of public policy focus on finding regularities and laws in the processes of policy formulation and policy implementation. The field emerged in the 1980s on the idea of Howard Lasswell's policy sciences" – the inter-disciplinary, result-oriented, and value-based (normative) disciplines, designed to provide policy makers with scientific and expert knowledge [1].

© Springer Nature Switzerland AG 2019
W. M. P. van der Aalst et al. (Eds.) AIST 2019, LNCS 11832, pp. 276–288, 2019.
https://doi.org/10.1007/978-3-030-37334-4_25

The field of public policy has generated a number of theories, each offering a theoretical understanding and explanation of different policy-making issues [2]. One of the most widely used theory is the Advocacy Coalition Framework (ACF) developed by Sabatier and Smith-Jenkings in the 1980s [3]. This theory was developed to describe and explain a complex policy environment, which (1) involves many policy actors, (2) takes a lot of time to translate policy decisions into outcomes, and (3) has a different dynamics in different policy sectors [4].

The ACF is aimed at understanding and explaining the formation of coalition among policy actors, and assessing their impact on the policy-making process, especially the long-term effects on the policy change. One of the key ideas of the theory is that people engage in policy to translate their beliefs into action. In turn, people gather into an advocacy coalition based on the shared "beliefs system." Figure 1 shows a diagram of the structure of the ACF theory.

Therefore, determination of the policy actors is the key issue for the application of this theory. Who are the people who organize themselves around some belief systems and can form coalitions to try to change the policy? And how can one determine, first, the actors themselves, and then classify them into advocacy coalitions based on latent characteristic of "beliefs?"

The question of actor identification is the traditional question for the political scientists and public policy analysts. The initial question was formulated as "Who governs?" in the classical Robert Dahl's study of the power-holders in the American city [5]. He used the survey of experts, who were asked to create a list of power-holders in the city. Reliance on the experts' ability to define the main policy actors in the field of their expertise comes with its own flaws, inherent for all qualitative research studies. The understanding of words like "important," "influential," and "prominent" might be different among different experts, and between the researchers and the experts. A natural solution for this problem was to describe importance or influence in the survey, so that misunderstanding was less likely. However, making a formal definition of such concepts would be limited by the chosen criteria: formal position of the actor (affiliated/elected or not); popularity (visible or active in the media); or reputation (based on the expert's opinion). Designing a survey that would encompass every criteria is a rather challenging task, and not always possible. Also, surveys represent subjective opinions and distortions of the expert. Thus, the qualitative methodology based on expert surveys has flaws that we can't escape [7].

There is a need for quantitative methodologies in the field. According to the state-of-the-art article on the applications of the Advocacy Coalition Framework, a great majority of the articles (75%) uses some form of qualitative analysis. Most researchers also used interviews as the main method of data collection [8]. 60% of the published articles use documents (government papers, media), but mostly for the qualitative analysis.

According to Weible et al. [9, p. 126], almost half of the articles with applications of the ACF used the "methods that were unspecified and appeared to rely on unsystematic collection and analysis of existing documents and reports." Therefore, there is a need for quantitative methodologies.

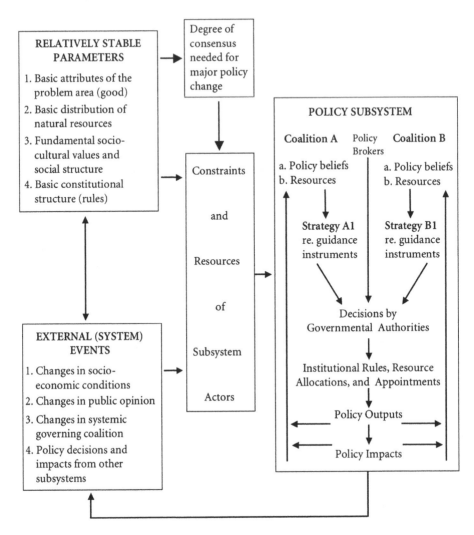

Fig. 1. The diagram of the Advocacy Coalition Framework structure [6, p. 191]

To close the existing gap in the field, we have developed a quantitative methodology based on text mining, named entities recognition, and classification analysis. In the context of the ACF theory, the task to identify actors themselves is burdened by the need to classify them into coalitions based on the belief system that they should share. Our methodology provides the solution to the problems of identification and classification of the policy actors into advocacy coalition, based on the shared belief system. We hope that the proposed methodology might be useful for the researchers who want to apply the ACF.

2   The Case of the Russian Policy Towards Civil Society

Civil society is a complex concept with many definitions, which depend on the socio-political context, political regime, and aims of the researchers. We understand civil society as "the public space that includes non-governmental organizations and other civil initiatives, but excludes political parties and business organizations aimed solely at private profit." [10, p. 5]. Then, the policy towards civil society is a policy that is aimed at regulation (via laws, political actions, legal actions) this public space.

We have decided to choose the example of the policy towards civil society as it provides a clear example of division of actors based on their beliefs. The history of development of civil society in Russia has gone from democratic development with governmental support to a polarized and politicized space, where the government sees many non-profits as agents of foreign interests that are harmful to Russia [12]. This case, then, is suitable for testing the quantitative methodology we propose in this paper.

2.1   Civil Society in Russia in the 1990s: Laying the Foundations

The 1990s in Russia was a turbulent, politically and economically unstable period of formation and transformation of many aspects of the state. The collapse of the Soviet Union led to development of democratic institutions including these of civil society, political parties, and market economy. As a result, in the first half of the 1990s, the civil society was developing rapidly, under the influence of democratic forces in Russia, which came to power in 1989–1991 [13]. For these democratic forces, civil organizations were the foundation of the civil society and of democracy, and they openly supported it by making the registration process for NGOs easier.

In the second half of the 1990s, state became less active in this field, dealing primarily with political and economic crises. The lack of support from the government was compensated by the funding from foreign sponsors. NGOs received large grants both via direct help and via regular competitive tenders. Independence from the state, combined with substantial resources, led to the transformation of NGOs into political actors which were able to influence policy-makers through the development and (sometimes) introduction of new, modern trends in the legal system and social policy. Examples of this were the initiative to develop juvenile justice, the development of anti-corruption programs ("Transparency International—Russia," the INDEM Foundation), as well as activities towards the development of the institution of the Ombudsman for Human Rights in the Federal Subjects of the Russian Federation (St. Petersburg humanitarian and political center STRATEGY) [13].

The civil society development was taking place in the general context of the lack of consolidation of political elites and Russian society. Moreover, there were problems with low institutionalization of relations between the state and civil society [15]. The good practices of civil society development were borrowed from the Western experience; the side effect was that civil associations had to compete against each other for foreign grants. In this context, and with the prevalence of foreign funding, segmentation and cleavages among the activists mirrored the splits in political elites and society [16]. Therefore, the under-institutionalized sphere of civil society had problems with policy transfer of western best practices to the Russian context. As a result, NGOs were active and relatively powerful, but restricted to provide large-scale policy change [11,15].

2.2   Civil Society in Russia in the 2000s: Biting the Bullet

In the 2000s, the policy towards civil society had gone through two stages. In the beginning, the government made an attempt to cooperate with civil associations, declaring to seek a partner for the process of modernization of Russia. Later, however, the state pulled back. The "orange revolutions" in Ukraine and Georgia were used as an excuse to stop the process of institutionalization of relations between the state and civil society. Independent civil activists were perceived as a potential threat to the Russian political regime.

The closure of the NTV TV channel in 2001, which supported Putin's opponents in the 1999–2000 elections, led to the decrease in confidence in the new president among civil society representatives. However, the state declared its interest in the civil society as the potential partner in the process of modernization of Russia. To achieve a certain level of cooperation, several formal institutions of interaction of civil society and the state were created (the Civic Forum, the Presidential Council on Civil Society and Human Rights, the Public Chamber). The government invited the civil society to become a partner, but in the end not all member of civil society consented [13].

However, the following political events destroyed the achievements of previous decades in the development of the civil society. The first event was the arrest of oligarch Mikhail Hodorkhovsky, who supported a lot of civil initiatives. The second event was the terrorist attack in Beslan, which was used as an excuse to establish non-democratic reforms (such as the constellation of governors' elections). Finally, it was the "colour revolutions" that took place on the territories of the ex-USSR countries [14], especially in Georgia (2003) and Ukraine (2004). These events led to the rise of a new policy towards civil society associations, which was reflected in the legislation. It was followed by a significant reduction in funding sources compared to the 1990s and even the beginning of the 2000s, primarily due to the gradual withdrawal of foreign funds from Russia [17].

The Russian government constructed the relations with NGOs in such a way that legislature and funding was given mainly to the GONGO (governmental non-governmental organizations), loyal to the current political regime. For the associations that proposed alternative policies, it was becoming more difficult to receive state support. In many cases they were forced to become informal associations [7].

This led to the polarization of civil society into two groups: the conformist and loyal organizations on one end, and the alternative to the official group, on the other. The policy coalitions mirror this clash. The first coalition developed a policy that supported GONGO and those NGOs that did not raise any political and policy problems as an alternative to the governmental agenda. The second coalition developed a policy that supported independent NGOs articulating alternative political and policy agenda. We assume that this dynamic became visible in 2011, when the Bolotnaya protests took place, and led to the further stiffening of the policy towards civil society. The potential clusterization based on the formal affiliated is presented in the Fig. 2.

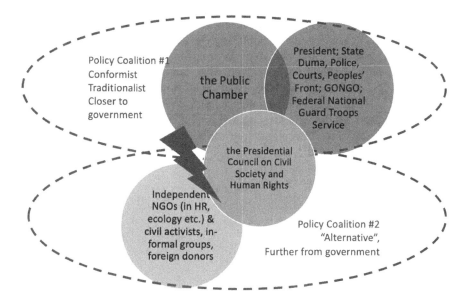

Fig. 2. Policy coalitions in policy towards civil society in Russia [7]

3   Methodology

We used media texts from TV, radio, newspapers, and the Internet as the unit of analysis. We downloaded 10 000 news articles (maximum available for download) from various mass media sources (both online and print) from year 2011. The articles were collected using the Public.ru media analytics system. The articles were found in the archive with search request "civil society" (in Russian).

After the articles were found, we have identified names of possible actors using a rules-based named entity recognition library Natasha. The identified names were then used to construct a name co-occurrence (co-mention) graph. It should be noted that only the last names were kept, as this made the analysis and interpretation of its results much simpler. The nodes in the co-mention graph were the last names, which were connected if they were mentioned it the same article, with the weight of the edge being the number of such articles.

The original co-mention graph consisted of 5186 nodes and 34070 edges. This graph, however, was not yet ready for analysis, because it contained words that were not names and were classified as such due to a false-positive in the named-entity recognition engine, and names that were not important and detrimental to the analysis.

To remedy this situation, the original graph was reduced using a two-step procedure. In the first step, links with weight less than or equal to 3 were removed. We have assumed 3 to be the least weight for a non-spurious link. In the second step, we have removed isolates, that is, nodes with degree of less than 1. The reduced co-mention graph consisted of 951 nodes and 3207 edges. This procedure was done using the large network analysis program Pajek [18].

The main part of the analysis started with clustering algorithms. However, our network was too dense for the usual clustering algorithms (including k-means, DBSCAN, Chinese whispers algorithms). We have then created a modified version of the Chinese whispers algorithm [19] that supported a feature that we call *seeded clustering*. This algorithm works as follows:

```
n := 0
colors := empty dictionary
for node in graph
    colors[node] = n
    n += 1

do n_iterations times
    newColors := empty dictionary
    for node in graph
        colorScores := empty dictionary
        for neigh in neighbors(node)
            neighColor := colors[neigh]
            colorScores[neighColor] += 1
        newColor = color with maximum score
        newColors[node] := newColor
    colors = newColors
return colors
```

This algorithm was implemented in python using the NetworkX package [20]. We have decided to use Chinese whispers algorithm because it provides consistently good results in the analysis of dense text networks, which, in our case, was an important factor (Table 1).

However, instead of starting with each node having a unique colour, we have specified the colours for some groups of more than one node. This allowed us to start with some theoretical pre-specified clusters and allow the algorithm to grow or shrink them as needed. Another feature of this algorithm is the stability of the clustering as long as the ordering of edges is preserved.

The logic behind the pre-specified dictionary follows the logic of the context of the Russian policy towards civil society. From the beginning of the 2000s, the

Table 1. Actors with clear identity, chosen for the dictionary.

Conformist coalition	Alternative coalition
Dmitry Medvedev	Mikhail Fedotov
Vladimir Putin	Mikhail Khodorkovsky
Valery Fedorov	Ludmila Alekseeva
Sergey Sobyanin	Vladimir Lukin
Vladislav Surkov	Sergei Magnitsky
Valentina Matvienko	Platon Lebedev
Ramzan Kadyrov	Svetlana Gannushkina
Andrey Isaev	Jury Dzhibladze
Boris Gryzlov	Alexey Navalny
Vladimir Zhirinovsky	Boris Nemtsov

government has been intervening the sphere of civil society trying to control this field. This led to decrease in opportunities for those members of civil society who did not want to cooperate with the government. This dynamics, by 2011, created an environment that was polarized and split between the those who agreed to the agenda proposed by the government, and those who did not. So, we define two clusters – the conformist one (closer to government), and the alternative one.

In making the dictionary, we sorted all actors by closeness centrality, selected those, who had the highest centrality measures, and whose socio-political position was public and well-known. The closeness centrality is especially useful for the identification of central nodes for the clique [21].

In the dictionary for the conformist coalition we included people affiliated with the government, including the president of Russian Federation Dmitry Medvedev, the Prime-Minister Vladimir Putin, members of the State Duma and Federation Council – Valentina Matvienko, Boris Gryzlov, Andrey Isaev, and conservative politicians – Vladimir Zhirinovsky, Ramzan Kadyrov.

In the dictionary for the alternative coalition we included well-known human rights activists, who also are often members of the Presidential Council for Civil Society and Human Rights–one of the main political institutions for cooperation between government and civil society. The Presidential Council often suggests alternative policy solutions to these of the government. There are also "prisoners of conscience" per definition of Amnesty International–people imprisoned because of their political views–Mikhail Khodorkovsky and his close associate Platon Lebedev. And the most prominent members of political opposition in Russia – Boris Nemtsov and Alexey Navalny.

For the purposes of this research, we have ignored clusters that we did not seed and nodes that did not end up in those clusters. The final network consisted of 592 nodes, 391 in one cluster and 201 in another. The rest of the analysis was conducted on this reduced network.

Figure 3 demonstrates the two coalitions, which we expected to see according to our qualitative analysis and received using quantitative methodology based on machine learning proposed in this article. Red actors represent conformist coalition, and cyan–alternative coalition. As it was mentioned above, vertices are policy actors, and edges are number of texts where these actors were mentioned together.

4   Policy Towards Civil Society in 2011: Testing for Coalitions

The public presence of the coalitions is showed in the Fig. 3, where the size of the labels is based on the closeness centrality. The network, however, is too dense–everyone is very close to the central node, Dmitry Medvedev, the President of Russia. As the Fig. 3 is not clearly visible, we provide the Table 2 with 20 members with highest closeness centrality measures of each coalition.

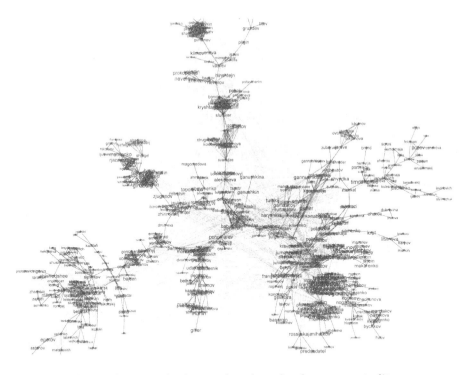

Fig. 3. The network of actors based on the closeness centrality.

In the conformist coalition, Dmitry Medvedev is the largest node, followed by Vladimir Putin, then the Prime Minister. Other members include Alexander Misharin, governor of Sverdlovsk Oblast and executive at Russian Railways

Table 2. Actors with highest closeness centrality divided by coalitions

Conformist	Centrality	Alternative	Centrality
medvedev	0,596650668	fedotov	0,486499776
putin	0,539019638	lukashenko	0,476720886
surkov	0,427973487	alekseeva	0,460894524
matvienko	0,419458171	medvedeva	0,458666187
misharin	0,410089869	hodorkovskij	0,45829689
fedorov	0,406865431	gannushkina	0,442960885
mironov	0,406284609	merkel	0,441243983
popov	0,401980747	eltsin	0,433845074
kadyrov	0,401697063	pozner	0,433184732
neverov	0,400003329	lebedev	0,43121571
zheleznjak	0,400003329	panfilova	0,425732788
bychkov	0,399441921	orlov	0,424462892
gryzlov	0,399161808	vorobev	0,4241466
javlinskij	0,399161808	lukin	0,423200549
makarov	0,398323819	timoshenko	0,423200549
kucherena	0,398045271	chirikov	0,422572188
franguljana	0,397767112	svanidze	0,416999808
bondarchuk	0,397489342	navalnyj	0,416085335
ostrovskij	0,397489342	nemtsov	0,410089869
melnikov	0,397211959	udaltsov	0,407739783

company; Sergey Mironov, Chairman of the Federation Council; Sergey Popov and Sergey Zhelevnyak, members of the State Duma.

For the alternative coalition, there is Vladimir Pozner, famous journalist; Nikolay Svanizde, member of the Presidental Council for Civil Society and Human Rights; Evgeniia Chirikova, eco-activist and politician; Sergey Udaltsov, leftist politician and oppositioner; and a number of foreign leaders including Alexander Lukashenko, the President of Belarus, Yulia Timoschenko, Ukrainian politician, Barack Obama, the President of the US, and Angela Merkel, Chancellor of Germany.

There is also an example of the noise left after the named-entities recognition algorithm–the node "Medvedeva," fourth largest in the network, is just a surname of Dmitry Medvedev in the genitive grammatical case (as in "proposed by Medvedev"). The fact that this node is in the alternative coalition might indicate that people often referred to what the President said or ordered to do. In this case, due to higher media presence of alternative coalition, it might be critique or discussion of the policy solutions he proposed (Fig. 4).

In 2011, we expected to see two rather separated coalitions of conformist and alternative as it would correspond with the development of the major trends in

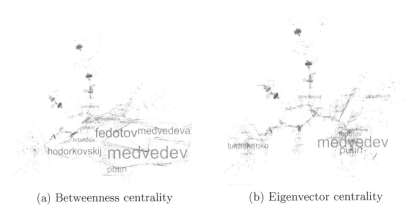

(a) Betweenness centrality   (b) Eigenvector centrality

Fig. 4. Network based on other types of centralities

the country. The results, however, show a slightly different picture that contains more information about our network. First of all, the alternative coalition is larger than the traditionalist one. We did not expect that, because the policy towards civil society have been changing in the interests of the government and often against the interests of what we call the "alternative coalition." However, it can be explained by the fact, that the government, and those, who act in the interest of the government, have more resources–both financial and administrative. They have no need to persuade public or justify their decisions. As a result, they do not need publicity as much as activists and human rights advocates. People, who are in opposition to the current policy, are further away from the governmental resources, and their main resource in this situation is publicity. The publicity is amplified by the turbulent time period of 2011–it was the year, public policy in Russia was very active and heated, culminating with Bolotnaya Square protests in December.

The data also suggests that there were more than two coalitions. It is especially visible in conformist coalition, that contains three cliques not connected with each other. The alternative coalition, however, is also split into parts that are weakly connected together. Overall, the machine-based classification of actors into coalitions and their location inside these coalitions corresponds with the theoretically justified analytical clusterization. Such results support using this approach for other attempts to classify named entities from large information sources.

The results of the analysis show that representatives of the hypothesised coalitions are closer to the persons of their coalition, and further from the representatives of opposite coalition, which correspond well with the proposed theory. However, we can also conclude that we have more than two coalitions: each big coalition is divided into small coalitions. So, we have several conformists sub-coalitions, and alternative sub-coalitions. The further analysis and interpretation, which is beyond the scope of this article can bring more lights on this smaller divisions.

5 Conclusion

In conclusion, we would like to formalize the steps for the development of the proposed methodology and outline future key research directions. The first is to increase the accuracy of names entity recognition. With names of people, the proposed methodology, consisting of two steps (exploratory and confirmatory), works decent. However, with organizations it is worth improving since the name of organization consists of several words that can be hard for the machine to recognize.

The second problem that requires human action is the fact that people with the same surnames can influence the outcomes of the algorithm's work. While it is not a major issue when we analyze the leaders and most public names in the society, it can cause interpretation issues for other research.

The final methodological problem is recognition of the sentiment and the types of ties between different actors in coalition. At this stage, our proposed methodology does not differentiate the type of connection between actors. However, they can be mentioned together in a positive, negative, or even more complicated manner, which is crucial to be defined and recognized by social scientists.

Despite some problems and issues, our proposed methodology, to the best of our knowledge, is the first attempt at using quantitative approach to classify policy actors. Given the potential that machine learning provides, it could be an important first step to advancing political science research.

Acknowledgment. The article was prepared within the framework of the Basic Research Program at the National Research University Higher School of Economics (HSE) and supported within the framework of a subsidy by the Russian Academic Excellence Project '5-100.'

References

1. Lasswell, H.: The emerging conception of the policy sciences. Policy Sci. **1**(1), 3–14 (1970)
2. Goodin, R., Moran, M., Rein, M.: The Oxford Handbook of Public Policy. The Oxford Handbooks of Political Science, vol. 6. Oxford University Press, Oxford (2006)
3. Sabatier, P.: An advocacy coalition framework of policy change and the role of policy-oriented learning therein. Policy Sci. **2–3**(21), 129–168 (1988)
4. Sabatier, P., Jenkins-Smith, H.: Evaluating the advocacy coalition framework. J. Pub. Policy **2**(14), 175–203 (1994)
5. Dahl, R.: Who Governs?: Democracy and Power in an American City. Yale University Press, New Haven (2005)
6. Sabatier, P., Weible, C.: The advocacy coalition framework: innovations and clarifications. In: Sabatier, P.A. (ed.) Theories of the Policy Process, pp. 189–223. Westview Press, Boulder (2007)
7. Zaytsev, D.: Fluctuating capacity of policy advice in Russia: testing theory in developing country context. Policy Stud., 1–21 (2019)

8. Pierce, J., Hicks, K., Giordono, L., Peterson, H.: Common approaches for studying the advocacy coalition framework: review of methods and exemplary practices. In: European Consortium for Political Research General Conference, Oslo, Norway (2017)
9. Weible, C., Sabatier, P., McQueen, K.: Themes and variations: taking stock of the advocacy coalition framework. Policy Stud. J. **37**, 121–140 (2009)
10. Henry, L., Sundstrom, L.M.: Introduction. In: Evans Jr., A.B., et al. (eds.) Russian Civil Society: A Critical Assessment, pp. 3–10. M. E. Sharpe, Armonk (2006)
11. Crotty, J., Hall, S.M., Ljubownikow, S.: Post-soviet civil society development in the Russian Federation: the impact of the NGO law. Eur. Asia Stud. **66**(8), 1253–1269 (2014)
12. Henderson, S.: Civil society in Russia: state-society relations in the post-Yeltsin era. Probl. Post-Communism **58**(3), 11–27 (2011)
13. Sungurov, A.: The models of interaction of the strcutres of civil society and authorities: Russian experience. Modernizatsia ekonomiki i globalisatzia **3**, 500–508 (2009)
14. Way, L.: The real causes of the color revolutions. J. Democracy **19**(3), 55–69 (2008)
15. Kakabadze, S., Zaytsev, D., Zvaigina, N., Karastelev, V.: Institute of civil participation: verification of the activities of subjects. POLIS. Polit. Res. **3**, 88–108 (2011)
16. Crotty, J.: Managing civil society: democratisation and the environmental movement in a Russian region. Communist Post-Communist Stud. **36**(4), 489–508 (2003)
17. Sungurov, A.: Civil society and its development in Russia (2007)
18. De Nooy, W., Mrvar, A., Batagelj, V.: Exploratory Social Network Analysis with Pajek: Revised and Expanded Edition for Updated Software, vol. 46. Cambridge University Press, Cambridge (2018)
19. Biemann, C.: Chinese whispers: an efficient graph clustering algorithm and its application to natural language processing problems. In: Proceedings of the First Workshop on Graph Based Methods for Natural Language Processing, pp. 73–80. Association for Computational Linguistics (2006)
20. Hagberg, A., Swart, P., Chult, D.S.: Exploring network structure, dynamics, and function using NetworkX. Los Alamos National Lab. (LANL), Los Alamos, NM (United States) (2008)
21. Wasserman, S., Faust, K.: Centrality and prestige. In: Social Network Analysis: Methods and Applications (Structural Analysis in the Social Sciences), pp. 169–219. Cambridge University Press, Cambridge. https://doi.org/10.1017/CBO9780511815478.006
22. The representative of sex minorities is pleased with the meeting with Lukin (2009). Interfax. http://www.interfax.ru/society/news.asp?id=90683

Analysis of Images and Video

Multi-label Image Set Recognition in Visually-Aware Recommender Systems

Kirill Demochkin$^{(\boxtimes)}$ and Andrey V. Savchenko

Laboratory of Algorithms and Technologies for Network Analysis, National Research University Higher School of Economics, Nizhny Novgorod, Russia
kvdyomochkin@edu.hse.ru

Abstract. In this paper we focus on the problem of multi-label image recognition for visually-aware recommender systems. We propose a two stage approach in which a deep convolutional neural network is firstly fine-tuned on a part of the training set. Secondly, an attention-based aggregation network is trained to compute the weighted average of visual features in an input image set. Our approach is implemented as a mobile fashion recommender system application. It is experimentally show on the Amazon Fashion dataset that our approach achieves an F1-measure of 0.58 for 15 recommendations, which is twice as good as the 0.25 F1-measure for conventional averaging of feature vectors.

Keywords: Visually-aware recommender system · Fashion recommendation · Multi-label image set recognition · Feature aggregation · Deep convolution neural networks · Mobile applications

1 Introduction

User modeling plays an essential role [1] in any recommender system. However, traditional recommender systems such as collaborative filtering [2] leverage additional information about products and users such as purchasing history, brand name, item reviews and user demographics to model the interests of a user. An example of such an approach is the matrix factorization algorithm that was used by the winners of the Netflix Prize [3] relied on representing the user item interactions as a large sparse matrix as a decomposition for smaller dense matrices. More recently recommender systems are being built using neural networks. Authors of [4] propose to train a neural network model to predict the ratings using a combination of dense and embedding layers. The most recent works deal study recommender systems that utilize visual features extracted from the pictures of items for user modeling. This approach is especially relevant for fashion recommendation systems. Authors of [5] propose an approach that achieves state of the art results in the personal recommendation task by modeling visual preferences of users and inferring specific style and visual attributes of clothing items that are important to users. Authors of [6] also used deep convolutional networks to extract aesthetic features to model user visual preferences. Authors

W. M. P. van der Aalst et al. (Eds.) AIST 2019, LNCS 11832, pp. 291–297, 2019.
https://doi.org/10.1007/978-3-030-37334-4_26

of [7] suggest using the visual features extracted from the pictures of clothes to generate unique personal recommendations by using Generative Adversarial Networks.

In this paper we investigate a setting, where the only information available is the image of an item [8], e.g., in classifying user interests based on the set of photos in the gallery of a mobile phone. Research of visually-aware recommender systems has gotten traction in the last few years [9,10]. Visual awareness is especially important in personalized clothing recommendation [10]. For example, the clothing, shoes and jewelry from Amazon product dataset are recognized in [8] by extracting of ResNet-based visual features and a special shallow net. Fashion recommendation and design with generative models using Siamese convolutional neural network (CNNs) are discussed in [9].

It is important to emphasize that during the processing of photos in a gallery we usually have access to a *set* of images, therefore multi-label image set recognition methods should be used. Modern image set classification algorithms are inspired by video classification, which usually combines the feature vectors of each image (video frame) [11] into a single descriptor of the whole image set using average or max pooling. Moreover, aggregation methods with trainable weights ("learnable pooling") are garnering a lot of attention. The winner of YouTube 8M Large-Scale Video Understanding challenge in 2017 used learnable pooling methods such as the Soft Bag-of-words, Fisher Vectors, NetVLAD and context gating (CG) to model interdependencies between different classes [12]. Sequentially connected neural attention blocks caused an increase of video face identification accuracy in a neural aggregation system [13].

Thus, in this paper we propose a novel approach to multi-label image set classification based on a modified neural aggregation module [13] previously used for video recognition. We leverage ideas from the Fire module [14] and the Transformer [15] to improve the models' space complexity to produce smaller models.

2   Proposed Approach

The task can be formulated as follows: it is required to predict the relevant products based on a collection of product images that the user has previously purchased. Every product belongs to one or more of D categories, i.e. a special case of multi-label classification is considered. Hence, it is needed to estimate the posterior probabilities that the user will order a product from each of the D categories. A training set should contain the galleries of N users. Every n-th user is associated with a collection of products' images $X_n(m), m = 1, 2, ..., M_n$ that this user has purchased or browsed. It is assumed that each image contains only a single product and is labeled with a D-dimensional binary vector $\mathbf{y} = [y_1, ..., y_D\}$, where y_d is set to 1 only if the product on the image belongs to the d-th category.

In this paper we split the N collections of images into two disjoint sets with size N_1 and N_2. The first set is used to fine-tune the CNN in order to learn the feature extractor [11,16]. The fine-tuned model is then used to obtain the K-dimensional feature vectors $\mathbf{x}_n(m)$ for each m-th image of the n-th user from the second subset. Secondly, these features are aggregated into single K-dimensional user descriptor $\mathbf{x}_n$ by computing a weighted sum of features of individual images:

$$\mathbf{x}_n = \sum_{m=1}^{M_n} w(\mathbf{x}_n(m))\mathbf{x}_n(m), \tag{1}$$

where the weights may depend on the features $x_n(m)$. If equal weights are used then conventional averaging with computation of the mean feature vector is implemented. However, in this paper we propose to learn the weights $w(\mathbf{x}_n(m))$, particularly, with the neural aggregation module with an attention mechanism originally used in video-based face recognition [13].

In the latter paper it was proposed to stack two attention blocks sequentially: the output of the first attention block (1) is fed into a fully connected (FC) layer with the $tanh$ activation and a rather large matrix of weights $W \in \mathbb{R}^{K \times K}$. However, an additional attention block causes an increase of the model size by $K(K+1)$ parameters. In order to reduce the number of trainable weights in this work we propose to add a linear dimensionality reduction operator with trainable parameters $W_s \in \mathbb{R}^{S x D}$, where $S < K$:

$$\mathbf{s}_n(m) = W_s\mathbf{x}_n(m). \tag{2}$$

The squeezed features $\mathbf{s}_n(m)$ are used in attention block instead of $\mathbf{x}_n(m)$:

$$w_s(\mathbf{x}_n(m)) = \frac{\exp(\mathbf{q}^T\mathbf{s}_n(m))}{\sum\limits_{j=1}^{M_n} \exp(\mathbf{q}^T\mathbf{s}_n(j))}, \tag{3}$$

where q is a learnable S-dimensional vector of weights. The resulting model has $S(2K+S+2)+K$ trainable parameters, which is $K(K+D)-S(S+D)$ parameters less that the original neural aggregation block.

In order to further reduce the model size, we use the squeezed vector $\mathbf{s}_n(m)$ not only for computing attention weights (3) but also for computing the final aggregated (squeezed output, SO) feature vector of size $S < K$.

$$\mathbf{x}_n = \sum_{m=1}^{M_n} w_s(\mathbf{x}_n(m))\mathbf{s}_n(m). \tag{4}$$

Finally, we consider a neural aggregation block with two sequential attention blocks that has an additional linear projection matrix for expansion $W_e \in \mathbb{R}^{S \times K}$ after the squeezing operator (3)–(4):

$$w_n^1(m) = \frac{\exp(\mathbf{q}^{1T}\mathbf{s}_n(m))}{\sum\limits_{j=1}^{M_n} \exp(\mathbf{q}^{1T}\mathbf{s}_n(j))}, \text{ where } \mathbf{q}^1 = W_e \tanh(W\mathbf{x}_n + \mathbf{b}). \tag{5}$$

The weights $w_n^1(m)$ (5) are substituted into the calculation of the final feature vector for the image set. In addition, we implemented the CG mechanism [12] that applies a scaling mask to the resulting aggregated vector for modelling the interdependencies of different features and estimating the features that often appear together. Hence, the weights for closely related features should be scaled up if they are present in a single collection. The opposite is also true: for features that are not likely to appear simultaneously the weights are reduced.

The complete model architecture is shown in Fig. 1. Here the aggregated vectors (4) are fed into a FC layer with dropout regularization. Because the resulting vector **y** of labels contains multiple non-zero elements since the product may belong to more than a single category (multi-label classification), the output layer has sigmoid activation. Therefore, the output layer predicts the posterior probabilities that the d-th category is relevant to the particular user.

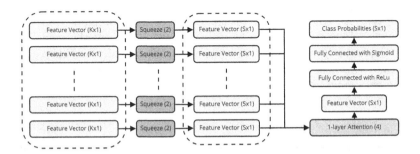

Fig. 1. Proposed architecture

The proposed pipeline was implemented in a prototype of Android mobile recommender system application (Fig. 2). We added Google maps/Google places and Amazon search functionality so that a request is made to the Google Places API for each identified category and all of the results are consolidated into a single list of store names and markers are placed on an interactive map for each identified store. Amazon search for now just opens a search page on the Amazon website with a query for all found categories. The prototype application can be used to recommend local shops based on the interests of a user inferred from the images in their gallery.

3   Experiments

In the experimental study we used the Amazon Fashion [9] dataset that contains 500000 entries of $N = 16000$ unique users interacting with 40000 products from $D = 75$ categories. The number M_n of items purchased by a user varies from 5 to 40; an average user has interacted with 8 unique items.

At first, 60% of all images selected by random split were used to fine-tune the CNN. We used the MobileNet v1 [17] due to the requirements of implementation for mobile application (Fig. 2). Since each item only belongs to a few of the categories the resulting target vectors **y** are sparse. To reduce class imbalance, we

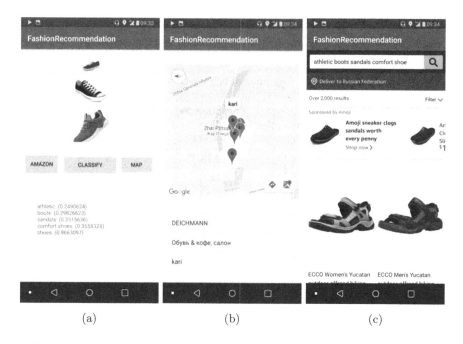

Fig. 2. Fashion recommendation: (a) recognition, (b) nearby shops; (c) search

implemented the weighted binary cross-entropy loss function. Several values were tested for the positive class weight $1, 2, 4, 8, 16$ and the best quality was achieved for the positive class weight equal to 8. The fine-tuned MobileNet was then used to extract $K = 1024$ dimensional feature vectors for all of the user images. 90% of the data in N_2 was used to learn the pooling weights while the images of products for remaining $10\%(0.1 \cdot 45184 = 4518)$ users were moved to the validation set. The algorithms were implemented using the Keras framework. All models (CNN and weights aggregation) were trained using the Adam optimizer.

The baseline for all of the experiments was chosen to be a simple averaging of individual feature vectors with equal weights $w(\mathbf{x}_n(m)) = 1/M_n$. The resulting vector was used to calculate the predictions via logistic regression (FC layer). Moreover, three types of aggregation layers were implemented: (1) original 1-Layer Attention [13]; (2) single attention block with dimensionality reduction operator (Squeezed 1-Layer Attention) to $S = 256$ (2)–(3); and (3) two sequential attention layers with dimensionality reduction and expansion (Squeeze/Expand 2-Layers Attention) (5). We tested the following configurations for each of the above-mentioned aggregation techniques: (1) adding an extra FC layer of size 256 with ReLU activation before the classifier layer; (2) passing the aggregated vector through the same dimensionality reduction operator (4); and (3) adding the CG layer [12]. The model size (excluding 13 Mb for the MobileNet-based feature extractor) and the dependence of precision, recall, F1-score and AUC (area under the ROC curve) on the number k of top categories (with the highest probabilities) for each aggregation technique and the best configuration are shown in Table 1.

Table 1. Performance of aggregation techniques

Metric	Average	1-Layer Attention + CG + FC	Squeezed Attention (2)–(3) + SO (4) + FC	Squeeze Expand 2-Layers Attention (1)–(5) + FC
Size, MB	**1.3**	4.5	2.18	3.5
Decision time, ms	**40**	43	41	47
Precision@k = 5	0.29	0.44	**0.51**	0.50
Recall@k = 5	0.56	0.61	0.66	**0.68**
F1-score@k = 5	0.38	0.51	**0.58**	**0.58**
AUC@k = 5	0.50	0.60	**0.62**	0.61
Precision@k = 10	0.18	0.43	**0.51**	0.50
Recall@k = 10	0.70	0.63	0.67	**0.68**
F1-score@k = 10	0.29	0.51	**0.58**	**0.58**
AUC@k = 10	0.52	0.51	**0.71**	**0.71**

Here one can note that the F1-score of traditional feature aggregation is 13–31% lower when compared to learnable pooling techniques. The best approach in terms of model size to accuracy among the evaluated configurations was (2)–(3) due to its small number of trainable parameters and quality metrics that are higher than almost all other methods of aggregation.

4   Conclusion

In this article we proposed a novel algorithm for classifying a set of images based on attention mechanisms which is characterized by low space (memory) and low inference latency. This algorithm was used to predict the user fashion interests based on a collection of images of clothing items found in the gallery of their mobile device. It was experimentally shown that our approach outperforms the baseline averaging method with an F1 measure of 0.58 vs 0.25. Moreover, the model together with the MobileNet feature extractor requires only 15 Mb, hence why it can be used on practically any mobile device. The main direction for future research is comparing traditional recommender systems based on collaborative filtering and matrix factorization [5] to our visual feature aggregation method. Moreover, it is important to examine more complex CNN architectures in order to maximize the quality (F1-score, AUC) of recommendations.

Acknowledgements. The paper was prepared within the framework of the Academic Fund Program at the National Research University Higher School of Economics (HSE) in 2019 (grant No. 19-04-0004) and by the Russian Academic Excellence Project 5-100.

References

1. Shankar, D., Narumanchi, S., Ananya, H., Kompalli, P., Chaudhury, K.: Deep learning based large scale visual recommendation and search for e-commerce. arXiv preprint arXiv:1703.02344 (2017)
2. Bokde, D., Girase, S., Mukhopadhyay, D.: Matrix factorization model in collaborative filtering algorithms: a survey. Procedia Comput. Sci. **49**, 136–146 (2015)
3. Zhou, Y., Wilkinson, D., Schreiber, R., Pan, R.: Large-scale parallel collaborative filtering for the Netflix Prize. In: Fleischer, R., Xu, J. (eds.) AAIM 2008. LNCS, vol. 5034, pp. 337–348. Springer, Heidelberg (2008). https://doi.org/10.1007/978-3-540-68880-8_32
4. Park, D.H., Kim, H.K., Choi, I.Y., Kim, J.K.: A literature review and classification of recommender systems research. Expert Syst. Appl. **39**(11), 10059–10072 (2012)
5. McAuley, J., Targett, C., Shi, Q., Van Den Hengel, A.: Image-based recommendations on styles and substitutes. In: Proceedings of the 38th International ACM SIGIR Conference on Research and Development in Information Retrieval, pp. 43–52. ACM (2015)
6. de Barros Costa, E., Rocha, H.J.B., Silva, E.T., Lima, N.C., Cavalcanti, J.: Understanding and personalising clothing recommendation for women. In: Rocha, Á., Correia, A.M., Adeli, H., Reis, L.P., Costanzo, S. (eds.) WorldCIST 2017. AISC, vol. 569, pp. 841–850. Springer, Cham (2017). https://doi.org/10.1007/978-3-319-56535-4_82
7. Yang, Z., Su, Z., Yang, Y., Lin, G.: From recommendation to generation: a novel fashion clothing advising framework. In: 2018 7th International Conference on Digital Home (ICDH), pp. 180–186. IEEE (2018)
8. Andreeva, E., Ignatov, D.I., Grachev, A., Savchenko, A.V.: Extraction of visual features for recommendation of products via deep learning. In: van der Aalst, W.M.P., et al. (eds.) AIST 2018. LNCS, vol. 11179, pp. 201–210. Springer, Cham (2018). https://doi.org/10.1007/978-3-030-11027-7_20
9. Kang, W.C., Fang, C., Wang, Z., McAuley, J.: Visually-aware fashion recommendation and design with generative image models. In: 2017 IEEE International Conference on Data Mining (ICDM), pp. 207–216. IEEE (2017)
10. Packer, C., McAuley, J., Ramisa, A.: Visually-aware personalized recommendation using interpretable image representations. arXiv preprint arXiv:1806.09820 (2018)
11. Goodfellow, I., Bengio, Y., Courville, A.: Deep Learning. MIT Press, Cambridge (2016)
12. Miech, A., Laptev, I., Sivic, J.: Learnable pooling with context gating for video classification. arXiv preprint arXiv:1706.06905 (2017)
13. Yang, J., et al.: Neural aggregation network for video face recognition. In: 2017 IEEE Conference on Computer Vision and Pattern Recognition (CVPR), pp. 5216–5225. IEEE (2017)
14. Iandola, F., Han, S., Moskewicz, M., Ashraf, K., Dally, W., Keutzer, K.: SqueezeNet: AlexNet-level accuracy with 50x fewer parameters and <0.5 mb model size. arXiv preprint arXiv:1602.07360 (2016)
15. Vaswani, A., et al.: Attention is all you need. In: Advances in Neural Information Processing Systems, pp. 5998–6008 (2017)
16. Savchenko, A.: Sequential three-way decisions in multi-category image recognition with deep features based on distance factor. Inf. Sci. **489**, 18–36 (2019)
17. Howard, A., et al.: MobileNets: efficient convolutional neural networks for mobile vision applications. arXiv preprint arXiv:1704.04861 (2017)

Input Simplifying as an Approach for Improving Neural Network Efficiency

Alexey Grigorev, Artem Lukoyanov, Nikita Korobov, Polina Kutsevol, and Ilya Zharikov[✉]

Moscow Institute of Physics and Technologies,
Institutsky per. 9, Dolgoprudny 141700, Russia
{grigorev.ad,lukoyanov.as,korobov.ns,
kutsevol.pn,ilya.zharikov}@phystech.edu

Abstract. With the increasing popularity of smartphones and services, symbol recognition becomes a challenging task in terms of computational capacity. To our best knowledge, existing methods have focused on effective and fast neural networks architectures, including the ones which deal with the graph symbol representation. In this paper, we propose to optimize the neural networks input rather than the architecture. We compare the performance of several existing graph architectures in terms of accuracy, learning and training time using the advanced skeleton symbol representation. It comprises the inner symbol structure and strokes width patterns. We show the usefulness of this representation demonstrating significant reduction of training time without noticeable accuracy degradation. This makes our approach the worthy replacement of conventional graph representations in symbol recognition tasks.

Keywords: Image classification · Skeleton · Graph neural network · Graph · Text recognition

1 Introduction

Text recognition in images is a fundamental computer vision problem. The great success of smartphones and different vision systems has made character recognition a critical task in the range of human and computer interaction issues. This is due to its various applications such as scene understanding, visual assistance and image retrieval. In particular, these issues can imply letters and digits classification. Current approaches to character recognition [1,2] are based on the Convolution Neural Networks (CNN) [5], bitmaps and combination of different heuristics to increase performance [6] and reduce learning and inference time.

Existing approaches allow improving solution performance and efficiency by simplifying its architecture. However, state-of-the-art solutions are still time-consuming and resource-intensive. In this work we introduce a novel symbol

---

A. Grigorev, A. Lukoyanov, N. Korobov and P. Kutsevol—Authors contributed equally and listed in alphabetical order.

classification method, which is based on the skeletonization [7] of the binarized images, the skeleton-to-graph translation and the further graph classification by means of Graph Neural Networks (GNN). This work focuses on the classification task which is more challengeable than binarization, skeletons extraction and its graph representations. The proposed technique simplifies neural network input rather than its architecture.

Our contributions are the followings. We propose a new state-of-the-art graph neural networks based method for character classification which implies the skeleton image representation. Moreover, we compare the performance of several existing graph architectures in terms of accuracy, learning and training time. To evaluate the efficiency of our method we conduct extensive experiments with MNIST datasets. As a result, we show that our method is competitive compared to existed classification methods trained on superpixels [8] in classification accuracy. Furthermore, it outperforms other GNN networks in terms of training and inference time and memory consumption.

The rest of the paper is structured as follows. The second section describes related graph neural networks for the graph classification problems. The third section presents a proposed classification method. The fourth section is devoted to the experiments. The fifth one concludes the results and contribution of this paper.

2   Related Works

In this paper, we propose to simplify neural network input image by skeletonization and its graph representation. Therefore, we restrict this discussion to related works devoted to graph neural networks and skeletonization algorithms.

2.1   Skeletonization Algorithms

The skeleton is the plain locus of points, which have at least two nearest points on the figure boundary. It should be noted that there are internal and external skeletons. In this work we consider the inner skeleton representation only. There are several different approaches to the skeleton representation and further graph obtaining. For instance, in [3] it is proposed to improve the discrete topological skeleton representation by graph simplifying and taking into account both topological and contour symbol properties. In this work we use the algorithm, similar to [7], where the skeleton representation is derived from the binary image. Thus, the problem of the contour obtaining is solved, since it is known in the binary image. The considered algorithm is also noise stable and computationally efficient compared with the previous ones.

2.2   Graph Neural Networks

It is necessary to learn state-of-the-art Graph Neural Networks to build Graph Classifier. According to the work [12], there are two types of Graph Convolution

Neural Networks (GCNN): Spectral and Spatial GCNNs. Spectral approaches require input graphs to have the same size. Since skeletons may have arbitrary size, that makes spectral GCNNs inapplicable for our task. So, in our work, we mainly focus on spatial GCNNs, what in its turn, are able to process graphs with an unfixed number of nodes and vertices.

Message Passing Neural Networks (MPNNs). Gilmer and others [9] have unified the existing approaches to the graph convolution neuron networks. The proposed method can be split into two stages: the message or hidden attributes passing through the graph several times in a row and then readout phases, where the hidden state is extracted from the graph. The authors have concluded that the variety of spatial convolution networks imply the same idea of information circulation through the graph with different message passing and data reading functions.

In [10] the authors have proposed a graph classification method with MPNN and Siamese networks. Note that the validation of this method has been conducted on the symbol graph representation database. It is seen that the proposed solution shows a 3% better quality compared with the conventional classification based on the MPNN architecture. One of the main disadvantages of MPNN is that quality is severely reduced when the graph is widened. In particular, the authors claim that this approach can not be extended to the graphs of the size more than 30 vertexes without accuracy degradation. However, this problem can be avoided with the skeleton symbol representation, since the effective graph pruning algorithms are used.

k-GNN. k-GNN [16] is a generalization of the traditional GNN architecture. The model is based on the k-dimensional WL algorithm. According to the authors, the key advantage of the proposed architecture is the ability to perform message passing directly between subgraphs instead of individual nodes. Such form of message passing makes it possible to take into account higher-order graph structures at different scales. Thus, the proposed architecture is more powerful than the classical GNN and outperforms it on a variety of graph classification and regression tasks. k-GNNs can also be stacked into a hierarchical model which is capable of combining graph representations obtained at multiple scales. The implementation of the above-mentioned idea implies that the output of the lower-dimensional model is treated as the initial messages in the higher-dimensional k-GNN.

Mixture Model Network (MoNet). MoNet [8] has extended the conventional CNN graph deep learning approach with the methods useful in non-euclidean metric spaces. For that Monti and others have introduced pseudo coordinates spaces, which are defined via the pairs of the coordinates of the vertexes and their neighbours, and the weight functions in these spaces.

Many of graphs CNN approaches can be expressed via the proposed representation with different weight functions $w(u)$ and pseudo coordinates u. The authors propose to use the Gaussian kernel as a weight function, where the mean vector and covariance matrix are learnable parameters. With MNIST database symbols as an example where every pixel is a graph node it has been shown that this architecture performs worse than the conventional CNN LeNet5. However, with 75 superpixels the accuracy of MoNet is higher than that of the spectral architecture ChebNet (91,11% against 75,62%).

SplineCNN. Having been inspired by the work of MoNet, Matthias Fey, Jan Eric Lenssen and others proposed a new class of deep neural networks, based on the so-called spline convolution for non-constant geometric structures [4]. According to the authors, SplineCNN is the first deep architecture that can be trained end-to-end on geometric data.

It is assumed that the input data are arbitrary graph structures, each of the vertices of which is associated with a point in d-dimensional space and a feature vector. The core of the proposed convolution is a continuous function defined on a certain interval in the input data space. The parameters to be taught are the coefficients with which the previously selected basic function is summed, placed in the nodes of a uniform grid over the selected interval. Thus, the convolutional layer aggregates the features of the neighboring vertices, weighing them with the learned continuous function. As a pool of layers, this article uses the Graclus method, which clusters the vertices of the graph and declares clusters for the vertices of the "squeezed" graph. At the same time, either the maximum of cluster attributes (max pooling) or their average pooling (average pooling) is taken as signs of such vertices.

This method surpasses the quality of state-of-the-art architectures in such tasks as the classification of image graphs, the classification of graph nodes and the comparison of shapes. At the same time, according to the authors, the proposed class of neural networks has a very good learning and prediction speed and is also optimized for computing on the GPU.

3   Proposed Method

In this section we present the details of the proposed method. The motivation of this approach is to make classification neural networks less resource-intensive and near state-of-the-art performance at the same time. Since the same symbols or doodles have the similar structures and the color, size of the symbol and other characteristics don't play a role it is more important to investigate this structure in a spatial way rather than visual way. It leads to the idea of using graph representation of the symbols. Processing of a graph will be more efficient especially if a graph has a lot of times fewer nodes than pixels in a standard raw image for classification.

The proposed algorithm consists of four stages: binarization, skeletonization, skeleton to graph mapping and graph classification, as shown in Fig. 1. This stages are extensively represented in the next four paragraphs.

Binarization. Standard skeleton algorithms require binary images, except neural networks skeleton generative approach proposed in [13].

A high number of binarization algorithms exists. Some of them are described in [14] responding to the ICFHR binarization competition. The standard problem statement of the binarization algorithms is the following: given image $\mathbb{I}_{m \times n \times k} \in \mathbb{R}^{m \times n \times k}$, where m and n is the image size and k is the number of channels of the image. Binarization is the function $B(\mathbb{I}) : \mathbb{R}^{m \times n \times k} \rightarrow \{0; 1\}^{m \times n \times 1}$.

Skeletonization. Skeleton algorithm is based on the approximation of connected object by a polygonal figure. The general idea of this method is to find the sequence of circles of maximum radius inscribed in the considered object. The algorithm is introduced in [7].

The algorithm comprises two stages:

1. Boundary corridor construction. In this stage the edge between white and black areas on the binary image is interpolated with the polygonal figure.
2. Construction of skeletons, which has complexity $O(n)$ for figures without holes.

It was shown that the algorithm effectively implemented on CPU as well as GPU.

Skeleton to Graph Mapping. Mapping to the graph is performed by obtained skeleton pruning. The proposed in [7] algorithm prunes skeleton representation until Hausdorff distance between the figure and its silhouette is greater than the predefined threshold.

The obtained graph is presented in the following way: $G = (E, V)$, where V is the set of vertices and E is the set of edges. The graph is matched with the feature matrix $F \in \mathbf{R}^{N \times D}$, where N is a number of vertices and D is the dimension of the feature vector F_i of the v_i vertex. Moreover, each vertex has its own coordinates u_i. Hidden representation of the feature vector is the D-dimensional vector. Note that $D = 2$ in our case. $F_i = [r_i, d_i]$, where r_i is the scale of the skeleton in this vertex and d_i is the degree of the vertex.

Graph Neural Network. The last phase is the graph classification algorithm. We are using graph classification neural networks. The short survey of the graph convolution neural networks will be given in the next section.

4 Experiments

4.1 MNIST Skeleton Dataset

We used MNIST Skeleton Dataset, which was collected by us with the help of the algorithm suggested in [7]. Since MNIST dataset is presented by gray images

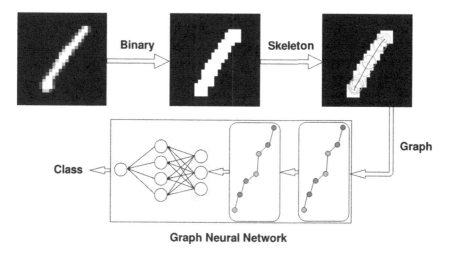

Fig. 1. Pipeline of the proposed approach

we introduced the binarization function of pixel (i, j) $B_{i,j}$ as the following:

$$B(\mathbb{I})_{i,j} = \begin{cases} 1, & \text{if } \mathbb{I}(i,j) > 0. \\ 0, & \text{otherwise.} \end{cases}, \tag{1}$$

where $\mathbb{I}(i, j)$ return the pixel value in the one-channel image.

On the Fig. 2 is shown the distribution of the number of vertices in the obtained dataset. It is clear that graphs have not big size and the sensitive to the size of the graph in terms of performance NNs can be efficiently applied to skeletons. The representative example of the dataset is shown on the Fig. 3.

5   Experimental Setup

Experiments are conducted with two datasets: MNIST Skeleton Dataset and MNIST Superpixel 75, which is described in [8]. We divide all of the datasets into three parts: Train set 55K, Test set 10K and Validation set 5K.

The time required for training till convergence and accuracy of classification requirements meeting is measured for four architectures: k-GNN, MoNet, MPNN, SplineCNN and for two datasets. The required on training memory volume and inference time are also evaluated. Inference time is measured for the whole pipeline, including binarization, skeletonization, skeleton to graph mapping and graph classification. For each architecture the following experiment is carried out. The represented architecture (see description below) is implemented. The batch size is $min(k; 64)$, where k is maximum batch size, which can be enclosed in the memory. The inference time is measured with batch size 1. The hyperparameters of the Neural networks are chosen as described in original papers. The same set of hyper-parameters is used for both models: model on

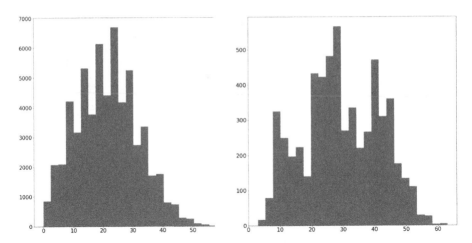

Fig. 2. Left: Distribution of graph nodes number for the MNIST Skeleton dataset. **Right:** Distribution of nodes number in "8" character in particular.

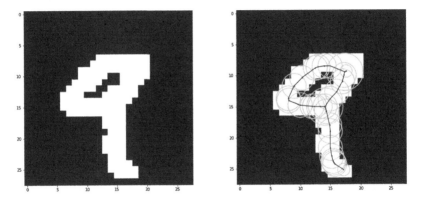

Fig. 3. Left: Binarized version of MNIST image with digit 9. **Right:** Skeleton version of the image.

skeletons and model on superpixels. The experiments are performed by means of Torch Geometric framework [11].

MPNN model is trained with a uniform random hyperparameter search. There is 4h passing steps. An aggregate function is mean-function. After message passing phase set2set [15] is used as a readout function. 8 is the parameter of proceeding steps for set2set. Two full-connected layers and log soft-max proceed after set2set. Models are trained using SGD with the ADAM optimizer with learning rate $1e-3$. We use a linear learning rate decay that starts between $1e-3$ and $1e-5$ of the way through training. Model on skeletons dataset is trained with batch size 64, and model on superpixels dataset—with batch size 25.

The k-GNN architecture is implemented as follows. The parameter k is taken equal to 1. The neural network consists of 3 convolutional layers described in the considered article with a hidden-dimension size of 64. Three fully connected layers with the ELU activation functions and dropout $p = 0.5$ after the first layer are used afterwards. Lastly, the softmax function is applied to the output of the last layer. Moreover, we employ the Adam optimizer with an initial learning rate of 10^{-2} and a rate decay based on validation results down to 10^{-5} given the patience parameter is equal to 5. The networks is trained for 100 epochs for both datasets.

The MoNet architecture is implemented in a different way comparing to the original work. It has been found that the graclus clustering and pooling layer interleaved with the convolutions as described in the original paper gives worse performance than three convolution layers with batch normalization and the ELU activation function are followed by the global mean polling layer and two full-connected layers. We use the Adam optimizer with initial learning rate 10^{-4}, dropout factor $p = 0.5$. The batch size 64 is used for both skeletons and superpixels dataset.

6   Results

The obtained training curves (dependencies between validation accuracy on each epoch and training time) are presented on the Fig. 4. It should be noted that for

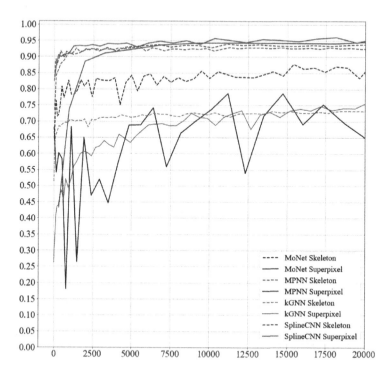

Fig. 4. Training curve for the GNNs on both skeletons and superpixels datasets

Table 1. Average inference time in milliseconds

	MPNN	k-GNN	MoNet	SplineCNN
Skeleton	**4.4**	**3.6**	**3.6**	**3.5**
Superpixel	60.5	11.2	15.8	10.9

Table 2. Maximum GPU memory usage during training process presented in gigabytes

	MPNN	k-GNN	MoNet	SplineCNN
Skeleton	**1.50**	**0.34**	**0.32**	**0.53**
Superpixel	10.76	0.85	1.54	1.04

Table 3. Classification accuracy in percents

	MPNN	k-GNN	MoNet	SplineCNN
Skeleton	94.5	74.4	**91.7**	93.4
Superpixel	**97.3**	**81.5**	89.7	**97.5**

models trained on the skeleton dataset it is always needed much less training time to achieve the same quality, but on superpixels, in a long perspective, it achieves a higher quality.

There is the average inference time represented in the Table 1. It should be noted that for skeletons we also included the prepossessing time needed for skeleton extraction but we did not include the prepossessing time for superpixel extraction. Even in such conditions skeletons show a much better performance for all the models, decreasing the inference time in 13.7, 3.1, 4.4 and 3.1 times for MPNN, k-GNN, MoNet and SplineCNN respectively. In the Table 2 one can find the maximum GPU memory used during the training for all the models. It is shown that models trained on the skeleton dataset use significantly less memory.

Finally, classification accuracy is presented in the Table 3. Mainly the number of errors is approximately 2 times higher for the skeleton dataset, except the MoNet architecture, which achieves a slightly higher accuracy on the skeletons.

7   Conclusion

In this work we have compared 4 Graph Neural Networks: MoNet, SplineCNN, MPNN and k-GNN in terms of inference time, training curves and memory usage. The comparison had been performed on the MNIST dataset with two different types of preprocessing: skeletonization and superpixel extraction. Both preprocessings represents a binarized image as graphs but skeletons have a fewer number of nodes, and thus less complex structures. The main result of our work is have shown that input simplification can significantly increase the efficiency of the neural network while sacrificing a little accuracy, what can be a desired trade-off for mobile and embedded systems. Models trained on skeleton dataset shows a least 3 times better inference time, at least 2 times less GPU memory usage and require significantly less time for training.

References

1. Palvanov, A., Im Cho, Y.: Comparisons of deep learning algorithms for MNIST in real-time environment. Int. J. Fuzzy Logic Intell. Syst. **18**(2), 126–134 (2018)
2. Zou, X., Duan, S., Wang, L., Zhang, J.: Fast convergent capsule network with applications in MNIST. In: Huang, T., Lv, J., Sun, C., Tuzikov, A.V. (eds.) ISNN 2018. LNCS, vol. 10878, pp. 3–10. Springer, Cham (2018). https://doi.org/10.1007/978-3-319-92537-0_1
3. Bai, X., Latecki, L.J., Liu, W.-Y.: Skeleton pruning by contour partitioning with discrete curve evolution. IEEE Trans. Pattern Anal. Mach. Intell. **29**(3), 449–462 (2007)
4. Fey, M., et al.: SplineCNN: fast geometric deep learning with continuous B-spline kernels. In: Proceedings of the IEEE Conference on Computer Vision and Pattern Recognition, pp. 869–877 (2018)
5. LeCun, Y., Bengio, Y.: Convolutional networks for images, speech, and time series. Handb. Brain Theory Neural Netw. **3361**(10), 255–258 (1995)
6. Ciresan, D.C., Meier, U., Gambardella, L.M., Schmidhuber, J.: Convolutional neural network committees for handwritten character classification. In: 2011 International Conference on Document Analysis and Recognition, pp. 1135–1139. IEEE (2011)
7. Mestetskiy, L., Semenov, A.: Binary image skeleton-continuous approach. VISAPP **1**, 251–258 (2008)
8. Monti, F., Boscaini, D., Masci, J., Rodola, E., Svoboda, J., Bronstein, M.M.: Geometric deep learning on graphs and manifolds using mixture model CNNS. In: Proceedings of the IEEE Conference on Computer Vision and Pattern Recognition, pp. 5115–5124 (2017)
9. Gilmer, J., Schoenholz, S.S., Riley, P.F., Vinyals, O., Dahl, G.E.: Neural message passing for quantum chemistry. In: Proceedings of the 34th International Conference on Machine Learning, vol. 70, pp. 1263–1272 (2017)
10. Riba, P., Fischer, A., Lladós, J., Fornés, A.: Learning graph distances with message passing neural networks. In: 2018 24th International Conference on Pattern Recognition (ICPR), pp. 2239–2244. IEEE (2018)
11. Fey, M., Lenssen, J.E.: Fast Graph Representation Learning with PyTorch Geometric. arXiv preprint arXiv:1903.02428 (2019)
12. Wu, Z., Pan, S., Chen, F., Long, G., Zhang, C., Yu, P.S.: A comprehensive survey on graph neural networks. arXiv preprint arXiv:1901.00596 (2019)
13. Shen, W., Zhao, K., Jiang, Y., Wang, Y., Bai, X., Yuille, A.: DeepSkeleton: learning multi-task scale-associated deep side outputs for object skeleton extraction in natural images. IEEE Trans. Image Process. **26**(11), 5298–5311 (2017)
14. Pratikakis, I., Zagoris, K., Barlas, G., Gatos, B.: ICFHR2016 handwritten document image binarization contest (H-DIBCO 2016). In: 2016 15th International Conference on Frontiers in Handwriting Recognition (ICFHR), pp. 619–623. IEEE (2016)

15. Vinyals, O., Bengio, S., Kudlur, M.: Order matters: sequence to sequence for sets. arXiv preprint arXiv:1511.06391 (2015)
16. Morris, C., et al.: Weisfeiler and leman go neural: higher-order graph neural networks. AAAI (2019)

American and Russian Sign Language Dactyl Recognition and Text2Sign Translation

Ilya Makarov$^{(\boxtimes)}$ ⓘ, Nikolay Veldyaykin, Maxim Chertkov, and Aleksei Pokoev

National Research University Higher School of Economics, Moscow, Russia
iamakarov@hse.ru, novel8mail@gmail.com, lunaticpy@gmail.com,
powerleks@yandex.ru

Abstract. Sign language is the main way to communicate for people from deaf community. However, common people mostly do not know sign language. In this paper, we overview several real-time sign language dactyl recognition systems using deep convolutional neural networks. These systems are able to recognize dactylized words gestured by signs for each letter. We evaluate our approach on American (ASL) and Russian (RSL) sign languages. This solution may help fasten the process of communication for deaf people. On the contrary, we also present the algorithm for generating sign animation from text information using text-to-sign video vocabulary, which helps to integrate sign language in dubbed TV and combining with speech recognition tool provide full translation from natural language to sign language.

Keywords: Sign language translation · Russian Sign Language · American Sign Language · Hand gesture recognition · Deep convolutional neural networks

1 Introduction

Sign language is used for communication with people having hearing or speech impairment. However, deafness as a disability rather viewed as a part of experience and certain restriction for speed and completeness of communication. The main way to overcome these limitations is to share emotions and communicate while having bi-directional translation for natural and sign languages.

Nowadays, advances in deep learning applications for computer vision and natural language processing lead to the possibility to improve automated sign language translation, which consists of typical machine learning problems.

Starting with data acquisition from the Web [15,16], gloves [10,24], phone sensors [3,34] or RGB-D sensor [6,20] researchers label the data according to the supervised tasks suitable for real-world application.

I. Makarov—The work was supported by the Russian Science Foundation under grant 17-11-01294 and performed at National Research University Higher School of Economics, Russia.

Next, machine learning models extract features from these data by training model under certain optimization criterion. Original models of Hand skin segmentation and filtering [8,23,33] require proper feature engineering. Deep convolutional neural networks, on the contrary, automatically learn proper features and are able to adapt for domain shifts on test data [22] or process video for continuous sign language recognition [4,12].

In what follows, we describe this pipeline for data acquisition and validation of trained models. Then, we proceed to text-to-sign translation and methods of animating sign language based on sign language vocabulary.

2   Sign Language Dactyl Recognition

2.1   Sign Languages Dactyl Description

Sign Languages (SL) consist of hands gestures and additional modalities, such as facial expressions and human pose. Gestures may be dynamic or static. SL dactyl represents special gestures for letters of natural language alphabet for spelling the words letter-by-letter.

ASL Datasets. American Sign Language (ASL) is used by deaf people of US and English-speaking parts of Canada. For our study, we use only static dactyl gestures and numbers. It includes 26 gestures from A to Z, excluding two of them, which are dynamic (J and Z), and 10 numbers from 0 to 9. We also take middle positions of dynamic signs for letters 'J' and 'Z' making their static analogues.

We used Massey University Researchers [1] dataset consisted of 2425 images from 5 individuals with variations shooting conditions. Additional datasets were used from Kaggle competition [27] and GitHub repositories [14,26].

Most of datasets were taken in "laboratory" conditions with low variance in shooting conditions, however they are sufficient to train light convolutional neural networks for real-time application. ASL dataset image samples are shown in Fig. 1.

RSL Dataset Creation. Russian Sign Language (RSL) is language of community in Russia and some of its neighbors. We consider RSL dactyl containing 33 gestures, among which there are 26 static gestures that we collect in our dataset [29] for RSL.

The list of observed RSL gestures represents gestures for the letters of Russian alphabet from Table 1 [15].

RSL dactyl dataset was collected from YouTube videos for sign alphabet learners and by crowdsourcing; see Fig. 2 for examples[1]. The dataset is published on the GitHub repository [29].

---

[1] https://sl-data.ddns.net/.

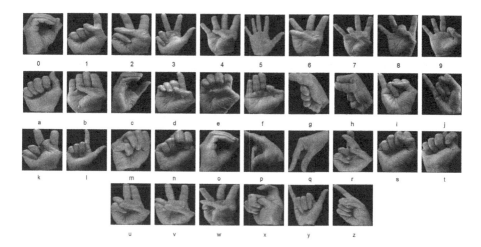

Fig. 1. Dactyl samples from ASL

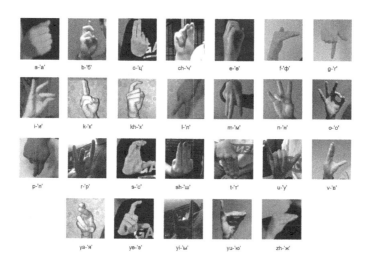

Fig. 2. Dactyl samples from RSL

Table 1. Russian alphabet and its transcriptions. "x" ("y") means that y is a Russian letter and x being its English transcription

"a" ("а")	"b" ("б")	"c" ("ц")	"ch" ("ч")	"e" ("е")	"f" ("ф")
"g" ("г")	"i" ("и")	"k" ("к")	"kh" ("х")	"l" ("л")	"m" ("м")
"n" ("н")	"o" ("о")	"p" ("п")	"r" ("р")	"s" ("ф")	"sh" ("ш")
"t" ("т")	"u" ("у")	"v" ("в")	"ya" ("я")	"ye" ("э")	"yi" ("ы")
"yu" ("ю")	"zh" ("ж")				

2.2   Convolutional Neural Networks for Sign Language Dactyl Recognition

In this paper, we study two models for dactyl recognition [13,15], evaluated for ASL [15] and RSL [16] dactyl recognition. Their architecture is described below. We compared or models with networks presented in [28] and [2]. Our models take 32×32 images as an input, while compared ones were trained on grayscale images of size 28×28. The detailed architecture of the all considered model is shown in Fig. 3.

We trained all networks on 100 epochs using Adam optimizer [11] with a learning rate of 0.003 validated during experimental setup. We focused on multiclass classification machine learning problem thus choosing cross-entropy loss function and corresponding metrics for evaluating the performance of trained models.

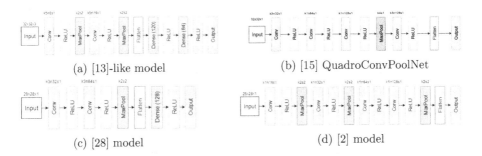

(a) [13]-like model   (b) [15] QuadroConvPoolNet

(c) [28] model   (d) [2] model

Fig. 3. Architecture of models. Each convolutional layer (as well as MaxPool layer) has detailed description of the structure: the kernel (filter) size (k), number of filters (n), and strides (s).

3   Known Phrases Recognition for Text2Sign Translation

The second significant part of the work implies the synonym phrase recognition using certain vocabularies. The recognition problem is to transform the input (a clear meaningful text without spelling and grammar mistakes) to the output (the phrases that are presented in a "written language—sign language" vocabulary and after concatenating produce the requested text or a text with synonym replacements). We define the "written language—sign language" vocabulary as the vocabulary that map a word or a phrase to the webpage or the image/video source that represents the phrase[2].

Let us describe the algorithm in a step-wise manner. For a one-word requested text, three cases may occur: (1) the word is presented in the vocabulary; (2) the word synonym is presented in the vocabulary; (3) the word is signed with dactyl. Such a system was implemented in RSL translation system 'Surdophone'[3].

---

[2] https://www.spreadthesign.com.
[3] https://www.xn--d1ascahfol.xn--p1ai/.

The algorithm also should take into account the equality between the part of speech predicted by morphological analyzer and the part of speech marked in vocabulary if it exists (it will be implied further). The final algorithm steps through every word in a query and tries to figure out in how many phrases the word itself or its synonym appears. The algorithm aims to split text into as long as possible accurately matching phrases, thus two parameters are introduced: the phrase length importance coefficient and the synonym penalty policy. The details of the algorithm that does not find synonyms are presented in Algorithm 1.

Input: request
Data: vocabulary
Result: result
result = Empty-List; words, wpos = Get-List-Of-Lemmatized-Words-From request; best-end = -1;
for *list-Index, word, pos OVER words, wpos* **do**
> **if** *list-index ¡ best-end* **then**
> > | CONTINUE;
>
> best-end = 1;
> finded, exactly-entry = Get-All-Entries-That-Starts-With word In vocabulary.words;
> **if** *exactly-entry AND (EXISTS word With pos In vocabulary)* **then**
> > | best-end = ind + 1;
>
> tind = ind;
> **while** *Length-Of finded ¿ 0 And tind ¡ Length-Of words* **do**
> > tind += 1;
> > finded, exactly-entry = Get-All-Entries-That-Starts-With word In vocabulary.words;
> > **if** *exactly-entry* **then**
> > > | best-end = tind + 1;
>
> **if** *best-end == -1* **then**
> > | Append-Dactyl-Signs-For-Word-To result;
>
> **else**
> > | Append-Word-With-Needed-POS-Or-Phrase-To result According-To ind, best-end;

Algorithm 1: Text separation on presented in vocabulary elements

---

The latter part of the algorithm is an acceptable synonym search engine.

One way to implement this part is to utilize a synonym dictionary. This approach allows to obtain accurate synonyms, but no acceptability scores will be provided. Moreover, synonyms can have the same meaning in one context and different meanings in another that may lead to semantic errors. Such a vocabulary also does not provide collocation with the word and its synonym replacement, but helps resolving the problem of phrases that are not presented in the "written language—sign language" vocabulary. A synonym dictionary can be easily found using a search engine over the Internet[4].

Another approach is to utilize the systems that can provide information about other words that used in the similar context as the requested word. They can provide a similarity score, but they can return words that have another meaning or even opposite meaning, the system guarantees only similar context. The applications that based on context-based Word2Vec [18] models.

One may combine both models: an algorithm can choose only those words with the similar context that appear in the synonym dictionary.

The other view on the problem can be implemented via the dictionary providing connections "a specific word meaning—another word meaning" and a tool that can recognize a word meaning based on a morphological analyzer and a disambiguation system. Some researches tried to develop or enhance machine-readable semantic dictionary for Russian [9,31] and American [19] languages. The application of a Russian semantic dictionary is described in [21].

In general case, the most suitable measure can be presented by the score function that encourages the usage of minimum the "written language—sign language" vocabulary entries and penalizes inaccurate meaning heavily. Every synonym approach would effect on meaning, because no guarantees about the synonym acceptability are provided. The constructing of a meaning-accuracy score function is another challenging problem, thus no simple and objective needed measures can be demonstrated nowadays.

The manual labeling of a text collection is personal and not accurate approach to compare different algorithms, because the group of the quite successful answers can be pretty big. The algorithm can find one more accurate result that differs from a label. The last way to measure algorithm quality is to grade answers by group of experts. We aim to conduct user study in order to evaluate our approaches for Text-to-Sign translation, which is the core case for Sign subtitles for TV media [7]. The animation framework[5] was used to process skeleton joints obtained from human pose estimation work [17] for sign translation, see Fig. 4.

4   Experiment Results and Discussion

We consider Sign Language dactyl recognition as multi-class classification for labelled data from ASL and RSL static dactyl gestures. We use accuracy metrics to measure quality of our models, which is a number of correct predictions made by the model over all predictions made. For the training, we used 70%/30%

---

[4] https://www.thesaurus.com/browse/.
[5] https://github.com/animate1978/MB-Lab.

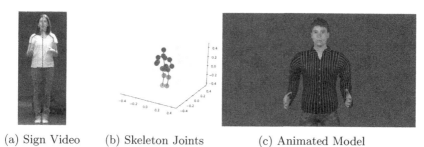

(a) Sign Video   (b) Skeleton Joints   (c) Animated Model

Fig. 4. Text2Sign Translation. Video from Sign Language vocabulary (a) is parsed by "human pose estimation" deep neural network (b) to extract skeleton joints, which are then translated into animation framework (c).

(for ASL) and 80%/20% (for RSL) train/test splits, randomizing images in each class. As a results we can cee that even light CNN architectures are able to train for the suggested task (with accuracy >90%, the details are in Table 2) on well-presented data, but on real-world dataset of RSL the quality drops due to lack of proper data for different shooting conditions, see Table 3.

Since no open RSL datasets were found, we started collection of our own RSL dataset, keeping in mind that different configurations should make the model more reluctant to different configurations of hands and exterior factors, involving lightning, clothes, shadows and background conditions.

Just the gesture recognition dataset is not enough, the hand detection is also required, for this we decided to use the YOLO: Real-Time Object Detectector [25]. To do this, we took datasets from Pascal VOC Challenges 2012 [5], from the Person Layout Taster Competition and extracted data for hand recognition and trained a model on them. The basis was taken Tiny-YOLO, so it shows the greatest performance, as well as pre-training weights for it. The Darckflow [30] framework was used to work with YOLO, and OpenCV2 for Python was used to process the video stream.

The Text2Sign problem demands on to convert various written language sentences to known elements. Every word n-gram can not be presented in the vocabulary, thus the available solution is a transforming unknown constructions to one of the known construction with same meaning. We conclude that not only the synonym approach solve the problem, but also approach that solves in the same way another well-known linguistic problem on "homonym resolution".

Table 2. Average accuracy per letters on ASL test data [15]

	[LeCun, 1998] model	QuadroConv-PoolNet	[Taskiran et al. 2018] model	[Chakraborty et al. 2018] model
0	100%	100%	87%	100%
1	100%	100%	100%	100%
2	75%	100%	87%	75%
3	100%	100%	87%	100%
4	100%	100%	100%	100%
5	100%	100%	100%	100%
6	100%	100%	100%	100%
7	100%	100%	100%	100%
8	100%	100%	100%	100%
9	100%	100%	100%	100%
a	100%	100%	100%	75%
b	100%	100%	100%	100%
c	100%	100%	100%	100%
d	100%	100%	100%	100%
e	100%	100%	100%	87%
f	100%	100%	100%	100%
g	100%	100%	100%	100%
h	100%	100%	100%	100%
i	100%	100%	100%	87%
j	100%	100%	100%	100%
k	100%	100%	100%	100%
l	100%	100%	100%	100%
m	100%	100%	100%	100%
n	87%	100%	75%	100%
o	100%	100%	100%	100%
p	100%	100%	100%	75%
q	100%	100%	100%	100%
r	100%	100%	100%	100%
s	100%	100%	100%	75%
t	87%	100%	87%	100%
u	100%	100%	100%	100%
v	87%	100%	87%	75%
w	100%	100%	100%	100%
x	87%	100%	100%	87%
y	100%	100%	100%	100%
z	100%	100%	100%	100%
Overall	97%	100%	97%	95%

Table 3. Average accuracy per letters on RSL test data [16]

	[LeCun, 1998] model	QuadroConv-PoolNet	[Taskiran et al. 2018] model	[Chakraborty et al. 2018] model
a	51%	70%	60%	20%
b	65%	75%	68%	22%
c	25%	87%	50%	25%
ch	82%	85%	85%	43%
e	75%	69%	68%	40%
f	85%	82%	85%	30%
g	68%	84%	59%	39%
i	83%	69%	78%	40%
k	83%	86%	84%	33%
kh	55%	80%	60%	45%
l	65%	78%	81%	29%
m	78%	77%	80%	27%
n	76%	82%	60%	24%
o	79%	80%	81%	21%
p	65%	61%	60%	20%
r	67%	88%	72%	45%
s	89%	93%	87%	35%
sh	62%	62%	60%	55%
t	68%	69%	65%	65%
u	82%	88%	83%	40%
v	75%	86%	70%	27%
ya	76%	81%	83%	32%
ye	58%	67%	50%	20%
yi	73%	65%	70%	26%
yu	69%	69%	68%	30%
zh	70%	83%	71%	28%
Overall	73%	78%	71%	33%

5   Conclusion

We evaluated two models for ASL and RSL static dactyl recognition. We get high quality of light real-time CNN models for "laboratory" ASL dactyl recognition, however, for real-world RSL dataset we did not achieve applicable results. We make a conclusion that we need to collect more diverse and representative data for real world scenarios for both, RSL and ASL, which we partially finalized using our developed service used for RSL dactyl dataset crowdsourcing[6]. This dataset was used to train our models and improving results from [15,16].

For the Text2Sign translation problem, the four approaches and their challenges were introduced and described. The algorithm of text splitting on known parts are demonstrated. Moreover, some intuition about measuring algorithm quality was provided. We aim to demonstrate our system of Text2Sign in practical application and compare to existing solutions [32].

References

1. Barczak, A., Reyes, N., Abastillas, M., Piccio, A., Susnjak, T.: A new 2D static hand gesture colour image dataset for ASL gestures (2011)
2. Chakraborty, D., Garg, D., Ghosh, A., Chan, J.H.: Trigger detection system for American sign language using deep convolutional neural networks. In: Proceedings of the 10th International Conference on Advances in Information Technology, p. 4. ACM, New York (2018)
3. Choe, B.W., Min, J.-K., Cho, S.-B.: Online gesture recognition for user interface on accelerometer built-in mobile phones. In: Wong, K.W., Mendis, B.S.U., Bouzerdoum, A. (eds.) ICONIP 2010. LNCS, vol. 6444, pp. 650–657. Springer, Heidelberg (2010). https://doi.org/10.1007/978-3-642-17534-3_80
4. Cui, R., Liu, H., Zhang, C.: Recurrent convolutional neural networks for continuous sign language recognition by staged optimization. In: 2017 IEEE Conference on Computer Vision and Pattern Recognition (CVPR), pp. 1610–1618. IEEE, New York (2017)
5. Everingham, M., Van Gool, L., Williams, C.K.I., Winn, J., Zisserman, A.: The pascal visual object classes challenge 2012 (voc2012) results (2012). http://www.pascal-network.org/challenges/VOC/voc2012/workshop/index.html
6. Huang, J., Zhou, W., Li, H., Li, W.: Sign language recognition using 3D convolutional neural networks. In: 2015 IEEE International Conference on Multimedia and Expo (ICME), pp. 1–6. IEEE, New York (2015)
7. Huenerfauth, M., Kacorri, H.: Best practices for conducting evaluations of sign language animation. J. Technol. Pers. Disabil. 3, 1–14 (2015)
8. Kamat, R., Danoji, A., Dhage, A., Puranik, P., Sengupta, S.: Monvoix-an android application for hearing impaired people. J. Commun. Technol. Electron. Comput. Sci. 8, 24–28 (2016)
9. Kanevskiy, E., Tuzov, V.: Some questions of subject area terms appending to a semantic dictionary. Dialogue 2, 156–160 (2002). (in Russian)
10. Kau, L.J., Su, W.L., Yu, P.J., Wei, S.J.: A real-time portable sign language translation system. In: 2015 IEEE 58th International Midwest Symposium on Circuits and Systems (MWSCAS), pp. 1–4. IEEE, New York (2015)
11. Kingma, D.P., Ba, J.: Adam: A method for stochastic optimization, pp. 1–15. arXiv preprint arXiv:1412.6980arXiv:1412.6980 (2014)
12. Koller, O., Zargaran, O., Ney, H., Bowden, R.: Deep sign: hybrid CNN-HMM for continuous sign language recognition. In: Proceedings of the British Machine Vision Conference 2016, pp. 1–12. BMVA, Durham, UK (2016)
13. LeCun, Y., Bottou, L., Bengio, Y., Haffner, P.: Gradient-based learning applied to document recognition. Proc. IEEE 86(11), 2278–2324 (1998)

[6] https://sl-data.ddns.net/.

14. Loïc Marie ASL Hand Gesture Dataset (2017). https://github.com/loicmarie/sign-language-alphabet-recognizer/tree/master/dataset
15. Makarov, I., Veldyaykin, N., Chertkov, M., Pokoev, A.: American and Russian sign language dactyl recognition. In: 12th PErvasive Technologies Related to Assistive Environments Conference (PETRA 2019), pp. 1–7. ACM, New York (2019). https://doi.org/10.1145/3316782.3316786
16. Makarov, I., Veldyaykin, N., Chertkov, M., Pokoev, A.: Russian sign language dactyl recognition. In: 42nd International Conference on Telecommunications and Signal Processing (TSP 2019), pp. 1–4. IEEE, New York (2019)
17. Mehta, D., et al.: VNect: real-time 3D human pose estimation with a single RGB camera. ACM Trans. Graph. (TOG) **36**(4), 44 (2017)
18. Mikolov, T., Sutskever, I., Chen, K., Corrado, G.S., Dean, J.: Distributed representations of words and phrases and their compositionality. In: Advances in Neural Information Processing Systems, pp. 3111–3119. Curran Associates Inc., New York (2013)
19. Mirman, D., Strauss, T.J., Dixon, J.A., Magnuson, J.S.: Effect of representational distance between meanings on recognition of ambiguous spoken words. Cogn. Sci. **34**(1), 161–173 (2010)
20. Molchanov, P., Gupta, S., Kim, K., Kautz, J.: Hand gesture recognition with 3D convolutional neural networks. In: Proceedings of the IEEE Conference on Computer Vision and Pattern Recognition Workshops, pp. 1–7. IEEE, New York (2015)
21. Mozgovoy, M.: Machine semantic analysis of Russian language and its applications, vol. 1. SPBGU, St.-Petersburg, Russia (2006). (in Russian)
22. Pigou, L., Dieleman, S., Kindermans, P.-J., Schrauwen, B.: Sign language recognition using convolutional neural networks. In: Agapito, L., Bronstein, M.M., Rother, C. (eds.) ECCV 2014. LNCS, vol. 8925, pp. 572–578. Springer, Cham (2015). https://doi.org/10.1007/978-3-319-16178-5_40
23. Prasuhn, L., Oyamada, Y., Mochizuki, Y., Ishikawa, H.: A hog-based hand gesture recognition system on a mobile device. In: 2014 IEEE International Conference on Image Processing (ICIP), pp. 3973–3977. IEEE, New York (2014)
24. Preetham, C., Ramakrishnan, G., Kumar, S., Tamse, A., Krishnapura, N.: Hand talk-implementation of a gesture recognizing glove. In: 2013 Texas Instruments India Educators, pp. 328–331. IEEE, New York (2013)
25. Redmon, J., Farhadi, A.: Yolo9000: better, faster, stronger. In: The IEEE Conference on Computer Vision and Pattern Recognition (CVPR), pp. 7263–7271. IEEE, New York, July 2017
26. rrupeshh ASL Hand Gesture Dataset (2018). https://github.com/rrupeshh/Simple-Sign-Language-Detector/tree/master/mydata
27. Sign Language MNIST (2018). https://www.kaggle.com/datamunge/sign-language-mnist
28. Taskiran, M., Killioglu, M., Kahraman, N.: A real-time system for recognition of American sign language by using deep learning. In: 2018 41st International Conference on Telecommunications and Signal Processing (TSP), pp. 1–5. IEEE, New York (2018)
29. The RSL alphabet dataset (2019). https://github.com/hse-sl/rsl-alphabet-dataset
30. thtrieu (2017). https://github.com/thtrieu/darkflow
31. Tuzov, V.: Computer semantics of Russian. Sankt-Petersburg State University, vol. 1, pp. 1–6 (2004)
32. Uchida, T., et al.: Sign language support system for viewing sports programs. In: Proceedings of the 19th International ACM SIGACCESS Conference on Computers and Accessibility, pp. 339–340. ACM (2017)

33. Viola, P., Jones, M.J.: Robust real-time face detection. Int. J. Comput. Vis. **57**(2), 137–154 (2004)
34. Wang, X., Tarrío, P., Metola, E., Bernardos, A.M., Casar, J.R.: Gesture recognition using mobile phone's inertial sensors. In: Omatu, S., De Paz Santana, J.F., González, S.R., Molina, J.M., Bernardos, A.M., Rodríguez, J.M.C. (eds.) Distributed Computing and Artificial Intelligence. AISC, vol. 151, pp. 173–184. Springer, Heidelberg (2012). https://doi.org/10.1007/978-3-642-28765-7_21

Data Augmentation with GAN: Improving Chest X-Ray Pathologies Prediction on Class-Imbalanced Cases

Tatiana Malygina$^{(\boxtimes)}$ ⓘ, Elena Ericheva ⓘ, and Ivan Drokin$^{(\boxtimes)}$ ⓘ

Intellogic Limited Liability Company (Intellogic LLC),
office 1/334/63, building 1, 42 Bolshoi blvd.,
territory of Skolkovo Innovation Center, 121205 Moscow, Russia
{tanya.malygina,ivan.drokin}@botkin.ai
https://botkin.ai

Abstract. When one applies machine learning to a real-world problem, sometimes data imbalance makes a crucial impact on the resulting model's performance. We propose to use generative adversarial network (GAN) to do data balancing through data augmentation in data preprocessing step of binary classification task. We train CycleGAN on unpaired images to be able to produce images from the opposite class for any given input image. After training we use it to produce images from the opposite class for every image in a given imbalanced dataset, thus making it fully-balanced.

The proposed augmentation technique may be used as the preprocessing step in binary classification tasks. We show that it improves performance in pneumonia presence/absence classification task on X-ray images. We inspect how binary classifier performance changes if the dataset used for GAN training differs from the dataset we measure binary classifier performance on. We also inspect its behavior on several other binary classification tasks related to medical imaging.

Keywords: X-ray medical image classification · Data augmentation · CycleGAN · Unbalanced dataset · Computer-aided diagnosis · Image analysis · Image interpretation · Computer-assisted

1 Introduction

The ability of generative adversarial networks (GAN) to generate images has made them a useful tool in many research areas and in generative art. It was shown [7] that GANs can also be used to improve model training by extending train dataset with generated previously unseen images.

Despite the constant appearance of new and expansion of old medical datasets no one of them is large enough to enable the data-hungry deep neural network to create clinically meaningful applications, including common disease pattern mining, disease correlation analysis, automated radiological report generation,

© Springer Nature Switzerland AG 2019
W. M. P. van der Aalst et al. (Eds.) AIST 2019, LNCS 11832, pp. 321–334, 2019.
https://doi.org/10.1007/978-3-030-37334-4_29

etc. Researchers need to achieve goals with limited datasets; in this case deep neural networks seem to fall short, overfitting on the training set and producing poor generalization on the test set. Another significant problem and limitation of automatic medical image processing is the fact of imbalance of classes presented within a collected datasets: the numbers of different types of target lesions are highly unbalanced. Number of samples from normal class is overwhelming [30]. For example, in CXR14 dataset of X-rays chest images [13,30] there are 84312 samples with "normal" label, only 9838 labels of "pneumonia", 10963 labels of "fibrosis" and 10963 labels of "pleural-thickening". As a result, the extraction of lesions' main characteristics is a complex process and requires many heuristic steps.

Increasing the number of training examples through the rotation, reflection, cropping, translation and scaling of existing images are common practices during the training process of algorithms. These methods allow the number of samples in a dataset to be increased by factors of thousands [15]. One of the data augmentation goals is to populate the dataset with a large amount of synthetic data in the directions of non-pertinent sources of variance. The aim of this is to reduce this variance to noise, removing any coincidental correlation with labels and preventing their use as a discriminative feature. Careful consideration of the application will inform which types of augmentation are appropriate. For instance, while random elastic deformations may be an appropriate model for microscopy images, in which the objects of interest (cells) are generally fluid and unconstrained [25], applying the same procedure to brain images could lead to disregarding certain anatomical constraints such as symmetry, rigidity, and structure [12].

Chest X-ray is the most common method of imaging examination in the world. More than 2 billion [23] procedures are performed annually. This method is crucial for screening, diagnosis and treatment of diseases of the thoracic region, many of which are among the leading causes of death worldwide [20]. Automated diagnosis using chest imaging attracts increasing attention [5] to using specialized algorithms designed to classify pulmonary tuberculosis [16] or identify pulmonary nodules [26] or using chest radiographs to detect other pathologies such as pneumonia, fibrosis etc. The release of ChestXray14 at the National Institutes of Health has led to an even greater amount of research that uses deep learning algorithms to analyze a chest X-ray [22,31].

Generative Adversarial Networks (GANs) have had a huge success since they were introduced in 2014 [9] and immediately started to represent a real conceptual progress for machine learning and, more especially, for generative models. In the case of images, this involves learning to produce images (via a generator), which are visually so similar to a set of real images, that an adversary (the discriminator) cannot distinguish them. GANs offer a potentially valuable addition to currently available augmentation techniques. A model learns a much larger invariant space than limited set of known invariants, through training a form of conditional generative adversarial network (GAN) in a different domain, typically called the source domain. This can then be applied to the low-data domain of interest, the target domain. The ideal GAN converts the discrete distribution of training samples to a continuous distribution, in this way simultaneously applying the augmentation to each source of variance in the data set. When paired training data is not available, [33] proposes an approach to translate an image from a source to a target domain with an adversarial loss in the absence of paired examples with CycleGAN.

Inspired by previous results, in this study we develop and validate a deep learning algorithm that classifies clinically important abnormalities in chest radiographs. We have designed a system for dataset balancing via oversampling with CycleGAN, that deals well with substantially unbalanced datasets.

2 Previous Work

Several articles [17,24] are devoted to the basic concepts of deep neural networks related to the analysis of medical images, and summarize more than 300 articles on such tasks as image classification, object detection, segmentation. Reviews identify problems of the successful deep learning application on medical imaging tasks as well as specific contributions that solve or circumvent these problems. Deep learning yields huge rises in performance in the medical image analysis domain for object (often human anatomical or pathological structures in radiology imaging) detection and segmentation tasks. In particular, [30] demonstrates that commonly occurring thoracic diseases can be detected, recognized and even spatially-located via a unified weakly-supervised multi-label image classification and disease localization formulation. So far, all image captioning, as well as VQA [2,29] techniques in computer vision, strongly depends on the ImageNet [11]-pretrained deep CNN models, which already performs very well in a large number of object classes and serves a good baseline for further model fine-tuning. At the same time, the results presented in [22] demonstrate that deep learning can automatically detect and localize many pathologies in chest radiographs at a level comparable to practicing radiologists. At this work a deep learning algorithm that detects concurrently 14 clinically important pathologies in chest radiographs was developed. The algorithm can also localize parts of the image most indicative for each pathology.

While working with little open source datasets to reduce training costs and, at the same time, maintaining result robustness, in [27] a novel approach is introduced. The weights of already trained pair of CNN/RNN on the domain-specific image/text dataset are applied. After this the algorithm infers joint image/text contexts for composite image labeling.

Constructing hospital-scale radiology image databases with computerized diagnostic performance benchmarks has not been addressed until ChestX-ray14 [30], which became somewhat similar to "ImageNet" in natural images. It is tremendously harder making truly large-scale medical image-based diagnosis (e.g., involving tens of thousands of patients) than making natural image dataset. In addition, the number of lesions from different categories is highly unbalanced, and many irregular regions that are visible during exam are not lesions. As a result, the extraction of lesion main characteristics is complex and requires many heuristic steps. With no addition of data source methods for detecting lesions with a wide range of phenomena are needed to improve the performance of CAD systems.

Generative Adversarial Networks (GAN) [9] offer a new way to obtain additional information from a dataset by creating synthetic samples. In [28], a method for studying a discriminating classifier from unlabeled or partially labeled data is presented. The approach is based on the objective function that exchanges mutual information between the observed examples and their predicted categorical distribution of classes.

At the same time, [4] demonstrates the feasibility of GANs to derive synthetic data to the training datasets within two brain segmentation tasks and to improve Dice Similarity Coefficient. GAN can be used to "hallucinate" additional data as a form of data augmentation in medicine image domain with Magnetic Resonance Imaging (MRI) [14], X-rays [19], Computed Tomography (CT) [7]. Data Augmentation facilitates model overfitting through more efficient use of existing data. The standard data augmentation provides only limited plausible alternative data. Given this, there is potential to generate a much wider range of additions. In [3] this type of generative model was developed. The model, based on image conditional Generative Adversarial Networks, takes data from the source domain and learns to take any data elements and generalize them to create other elements within a class.

The problem of learning mappings between domains from unpaired data has recently received increasing attention, especially in the context of image-to-image translation [33]. CycleGAN trains a generator which produces images in one domain while given images are from the other.

While the mapping that CycleGAN learns can be superficially convincing, in X-rays image analysis task we would like to learn a mapping that can capture diversity of the output. In [1] a model for learning many-to-many mappings between domains from unpaired data was proposed. Specifically, each domain is "augmented" with auxiliary latent variables and extend CycleGAN's training procedure to the augmented spaces.

3   Data and Preliminary Data Processing

A chest X-ray database, namely "ChestXray14", was presented in [30]. It comprises 112120 frontal-view X-ray images of 30805 unique patients with the text-mined 14 disease image labels (where each image can have multi-labels), from the associated radiological reports. Only 24636 images contain one or more pathologies. The remaining 84312 images are normal cases.

For experiments we have chosen 3 modalities from the database: pneumonia, fibrosis, pleural-thickening. The spatial dimensions of an chest X-ray are usually 2000×3000 pixels which impose challenges in both the capacity of computing hardware and the design of deep learning paradigm. Local pathological image regions can show hugely varying sizes or extents but often very small comparing to the full image scale. As part of the ChestXray14 database, a small number of images with pathology are provided with hand labeled bounding boxes (B-Boxes), which can be used as the ground truth to evaluate the disease localization performance.

In original dataset ChestX-ray dataset X-ray images are directly extracted from the DICOM files, resized and saved as 1024×1024 bitmap images without significantly losing the detail contents. Their intensity ranges are rescaled using the default window settings stored in the DICOM header files. For the pathology classification and localization task, the entire dataset is randomly shuffled into three subgroups: i.e. training (70%), validation (10%) and testing (20%) (Fig. 1).

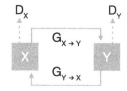

Fig. 1. CycleGAN scheme as it was defined in paper [33]

In our experiments we use the same train/validation/test split. For task of pneumonia detection for both GAN and classifier training images were resized to 512×512, for fibrosis and pleural-thickening task we resized them to 256×256 pixels to reduce required computing resources. Since the picture is a gray image with pixel values in the range $0..255$ normalization to $[-1, 1]$ interval is applied through division by 127.5 and subtraction by 1.

4 CycleGAN

Due to lack of paired or good-labeled publicly available data as well as imbalance across classes and positive samples deficiency we treat oversampling via generative adversarial network model as an augmentation technique. We have to appeal to image-to-image translation task. It is very hard, even impossible, to find paired data in medical datasets generally and in X-ray datasets particularly. In this case the unsupervised training capabilities of CycleGAN [33] are quite useful.

CycleGAN is one of the recent generative networks capable of unpaired image-to-image translation because it does not require paired training data - while an x and y set of images are still required, they do not need to directly correspond to each other. By taking image pairs from different image classes, it learns to build mapping between them. CycleGAN provides the pair of generator networks $G_{X \to Y}$ and $G_{Y \to X}$. The input image is fed directly into the encoder, which shrinks the representation size while increasing the number of channels. The encoder is composed of three convolution layers. The resulting activation is then passed to the transformer, a series of six residual blocks. It is then expanded again by the decoder, which uses two transpose convolutions to enlarge the representation size, and one output layer to produce the final image. Each layer is followed by an instance normalization and ReLU activation.

In addition to the classic adversarial loss, to further regularize the mappings different image classes, two cycle consistency losses are applied that capture the intuition that if there is translation from one domain to the other and back again one should arrive at where it started. Full objective function puts these two components: an adversarial loss and a cycle consistency loss – together, and weighting the cycle consistency loss by a hyperparameter λ.

As we consider binary classification problem on unbalanced dataset, we start with dataset X consisting of n samples $x_1, x_2, \ldots, x_n$ and corresponding labels $l_1, l_2, \ldots, l_n$, where $l_i \in \{0,1\}$ for $i = 1, \ldots, n$. In this terms we defined two networks $G_{0\to 1}$ and $G_{1\to 0}$ which are trained to minimize cycle-consistency loss, in assumption that for element x_i following rules are defined: $x_i \to G_{l_i \to 1-l_i}(x_i) \to G_{1-l_i \to l_i}(G_{l_i \to 1-l_i}(x)) \approx x_i$. Further we employ these $G_{0\to 1}$ and $G_{1\to 0}$ to produce complementary samples. E.g., for sample x_i with a corresponding label $l_i = 0$ a new sample $\tilde{x}_i = G_{0\to 1}(x_i)$ with $\tilde{l}_i = 1$ is generated, and for sample x_j with $l_j = 1$ a sample $\tilde{x}_j = G_{1\to 0}(x_j)$ with $\tilde{l}_j = 0$ is generated.

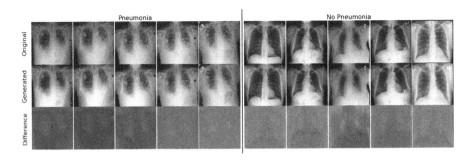

Fig. 2. Picture shows Original images of pneumonia from CheXnext and their CycleGAN-generated pairs. Difference between Original and Generated images is shown in Difference row.

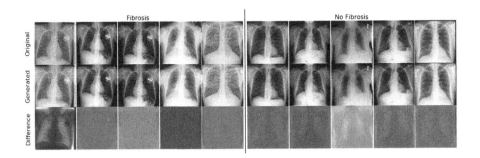

Fig. 3. Picture shows Original images of fibrosis from CheXnext and their CycleGAN-generated pairs. Difference between Original and Generated images is shown in Difference row

We use the same setup of training process for all classification tasks. Weights are initialized with Glorot initializer [8] and optimized with adaptive moment optimization algorithm (Adam). We start with learning rate value of 0.0002. We train model during 20 full epochs (the model "sees" each sample at least one time per epoch) and with batch size of 2 for input size of 512×512 (pneumonia) and 4 for input size of 256×256 (fibrosis and pleural-thickening). We use the same loss as was proposed in [33] and was described above and suggest setting $\lambda = 10$. Data goes through the preliminary data processing as was described in Sect. 3. To train discriminator we put original and generated images into network in equal proportion. For experiments marked "flip" we additionally flip images horizontally with probability of 0.5.

In our implementation of CycleGAN we use additional input with random values, which brings us uncertainty, thus giving us an opportunity to generate different images from the same input picture by providing different set of random values as a second input.

Thus, in our context for task of pneumonia dataset augmentation we have selected images with pneumonia (with label 1 in Pneumonia class or any other selected modality) or with no abnormalities (with $l_i = 0$ for all i). Then we construct both $G_{0 \to 1}$ and $G_{1 \to 0}$ generators via Cycle-GAN. We trained these generator to produce X-ray images with pneumonia from images with no pneumonia, and vice versa. Now then after finish of training process the Generators produce novel X-ray images with pneumonia from images with no pneumonia, and vice versa. Figure 2 shows several pairs of prototype images from CheXnext, corresponding complementary images generated with CycleGAN and the difference between these images. The process is the same for any other modality. Results of fibrosis generation are shown on Fig. 3.

5   DenseNet

To perform classification task we have chosen DenseNet [10] which connects each layer to every other layer in a feed-forward fashion. We start with DenseNet-121 pretrained on ImageNet from original paper [10], but change the last layer due to our binary task. As we focus on solving binary classification task, we use sigmoid at the last layer of the network and binary cross-entropy loss to optimize model during training. Classifier takes as input an X-ray grayscale image and estimates probability of pathology presence on it. As previously, we use normal Glorot distribution for weights initialization. Adam is used as an optimization algorithm. We start with learning rate value of 0.0001 and reduce it on plateau with patient 1 and factor 0.1. We train model during 15 full epochs and with batch size of 16. Input images are resized to the size of 256×256 and further horizontal and vertical flips was used as additional augmentation with probability of 0.5. Data go through the preliminary data preprocessing as was described in Sect. 3. Original validation split from CheXnext is used for testing. Only train subset is augmented with proposed framework.

6   Data Augmentation Framework

We consider binary classification problem of determining the presence of signs of a pathology on the X-ray image. We have unbalanced dataset X, where one class of labels dominates over another. To deal with imbalance we propose data oversampling using generative adversarial network (GAN).

Generator models $G_{0\to1}$ and $G_{1\to0}$ can be constructed separately with any image-to-image generative networks. We propose to use CycleGAN as was described in Sect. 4. After training, the generators produce novel complementary X-ray images with pneumonia from images with no pneumonia, and vice versa. E.g., for sample x_i with a corresponding label $l_i = 0$ a new sample $\tilde{x}_i = G_{0\to1}(x_i)$ with $\tilde{l}_i = 1$ is generated, and for sample x_j with $l_j = 1$ a sample $\tilde{x}_j = G_{1\to0}(x_j)$ with $\tilde{l}_j = 0$ is generated. We combine newly generated samples $\tilde{X}$ with samples from X to produce perfectly balanced dataset X_{aug} consisting of samples $x_1,\ldots,x_n,\tilde{x}_1,\ldots,\tilde{x}_n$, with labels $l_1,\ldots,l_n,\tilde{l}_1 = 1 - l_1,\ldots,\tilde{l}_n = 1 - l_n$. We use X_{aug} to train binary classifiers expect the rise of quality and robustness of the model due to an dataset size increase as well as due to the "smooth" balancing of the samples.

More formally and clearer framework is described on scheme (see Fig. 4) and in Algorithm 2.

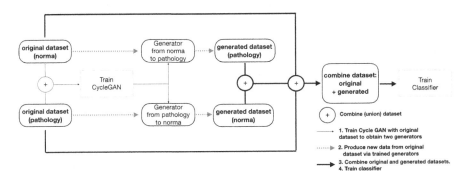

Fig. 4. Data path and main components of proposed framework for data augmentation.

7   Experiments and Results

We want to compare performances of classifier model trained on data with and without augmentation. We expect the rise of quality and robustness of the model due to dataset size increase and "smooth" balancing of the samples. To achieve this we apply our data augmentation framework to the given "pure" dataset. CXR14 [30] is utilized as our main data source with preserved original train/validation split.

For each binary classification problem data is sampled from CXR14 by selecting all images without any class labels, and combining them with samples which have associated label 1 in corresponding class. Further these data is used as train dataset. Original validation split from CheXnext is used for testing. Note that only train subset is augmented.

CycleGAN [33] method was used to construct $G_{0\to1}$ and $G_{1\to0}$ generators. We trained these generator to produce X-ray images with pneumonia from images with no pneumonia, and vice versa, as was described in Sect. 4.

Algorithm 1: Data augmentation framework for binary task

---

  input : Unbalanced dataset with N positive and M negative labels,
               Classifier to train on the dataset

1 **begin**

    1. Set dataset X with N positive and M negative labels:
- $X = \{x_i, x_j, l_i = 0, l_j = 1\}, i = 1..N, j = 1..M$;

    2. Initiate functions $G_{0 \to 1}$, $G_{1 \to 0}$ via CycleGAN

    3. Train CycleGAN with X:
- **Result:** Generators $G_{0 \to 1}$, $G_{1 \to 0}$

    4. Create perfect balanced dataset X_{aug}:
      (a) **For each** $\{x_i, l_i = 0\}$ **do:**
        generate $\{\tilde{x}_i = G_{0 \to 1}(x_i), \tilde{l}_i = 1\} \in \tilde{X}$
      (b) **For each** $\{x_j, l_j = 1\}$ **do:**
        generate $\{\tilde{x}_j = G_{1 \to 0}(x_j), \tilde{l}_j = 0\} \in \tilde{X}$
      (c) $X_{\text{aug}} = \tilde{X} \bigcup X$

    5. Train Classifier with X_{aug} instead of X:

2 **end**

  Result: Well trained binary classification neural net

As a next step we use these newly generated images both with original dataset to train DenseNet [10] classifier from Sect. 5 to detect presence of target pathology on the image.

To compare model performance with and without augmentation, we use the same validation dataset. We train two binary classifiers: on pure (without augmentation) and GAN-augmented datasets (see Fig. 4). For binary classification tasks, the receiver operating characteristic (ROC) curve and the area under this curve (AUC-ROC) are widely accepted as a general measure of classifier performance [21]. In cases with substantially unbalanced dataset, the precision-recall (PR) curve and AUC (AUC-PR) is better suited for comparing the performance of individual classifiers [6]. We evaluate proposed framework performance in terms of both.

Combined results before and after training the classifier on the augmented dataset via proposed framework are shown in Table 1.

Table 1. Metrics of model performance on test data

Pathology	Model **without augmentation**		Model **with augmentation**	
	AUC	PR AUC	AUC	PR AUC
Pneumonia	0.9745	0.9580	**0.9929**	**0.9865**
Pleural-Thickening	0.9792	0.9637	**0.9822**	**0.9680**
Fibrosis	**0.9745**	**0.9446**	0.9697	0.9294

7.1 Positive Result: Pneumonia and Pleural-Thickening Classification

We combined samples not marked with any pathology with 9838 images with "pneumonia" label, and thus collected imbalanced dataset consisting of 44972 samples, which was used to train binary classifier. In case of pleural-thickening we add 10963 images with "pleural-thickening" label to the dataset with no any pathology.

Trained generators produce X-ray images with the pathology (pneumonia or pleural-thickening) from images with no any pathology, and vice versa. Figure 2 shows several pairs of prototype images from CheXnext for pneumonia case, corresponding complementary images generated with CycleGAN and the difference between these images. We used these data to train 2 binary classifiers for each case. For the first experiment we used no data augmentation. For the second experiment we used CycleGAN to augment data, which was trained on the same subset of CXR-14.

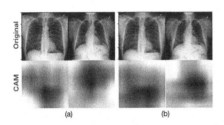

Fig. 5. Pneumonia. Class activation maps example from models trained without (a) and with (b) augmentation

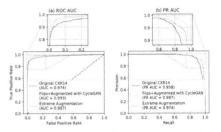

Fig. 6. Pneumonia. Test metrics for models trained without and with augmentation: (a) ROC curve; (b) Precision-Recall curve.

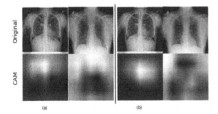

Fig. 7. Pleural-Thickening. Class activation maps example from models trained without (a) and with (b) augmentation

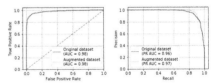

Fig. 8. Pleural-Thickening. Test metrics for models trained without and with augmentation: (a) ROC curve; (b) Precision-Recall curve.

ROC and precision-recall curves for both classifiers are shown on Fig. 6 for pneumonia and on Fig. 8 for pleural-thickening. You can see that in pneumonia pathology with our balancing method ROC AUC score increased from 0.9745 to 0.9929 PR AUC value increased from 0.9580 to 0.9865 on validation dataset. At the same time for pleural-thickening there is no significant changes in performance between 2 binary classifiers: ROC AUC score rises from 0.9792 to 0.9822 and PR AUC value changes from 0.9637 to 0.9680. To expose the implicit attention of our binary classifiers on the image, we provide examples of class activation maps [32] on Fig. 5 for pneumonia and on Fig. 7 for pleural-thickening. It highlights the most informative image regions relevant to the predicted class.

In first experiment with pneumonia we also compared our CycleGAN-based augmentation method with geometry-based augmentations (see Fig. 6). We have used imgaug library[1] for "extreme" image augmentations. For each image in train batch from 2 to 5 transformations in random order from the following list were selected randomly: Fliplr, CropAndPad, Affine, ElasticTransformation, PerspectiveTransform (each transformation had random parameters). On validation set extreme augmentation had ROC AUC score of 0.9871 and PR AUC of 0.9737. Apparently, in this case classification model overfits to extremely-augmented train dataset, which becomes too different from real X-ray images due to extreme augmentations. This way model shows results worse than our CycleGAN-based model. Our model produces images similar to the ones from the original space of X-ray images.

7.2 Negative Result: Fibrosis Classification

Since CXR14 dataset provides multiple labels, we've decided to check if data augmentation will improve classifier performance when we predict other diseases. The image generated by the chest X-ray may present shadows, which indicate scar tissue. This allows the physician to diagnose the possibility of pulmonary fibrosis. Sometimes chest X-rays may not show any scars, so further tests are required to confirm the illness. We've decided to measure fibrosis classification performance, because it is poorly distinguishable on the X-ray image [22].

For this experiment we have constructed training dataset from 35134 images with no pathologies and 10963 images with fibrosis label. Several pairs of complementary generated images are shown on Fig. 3. Further two classifiers were trained: with and without data augmentation.

Unlike pneumonia case, ROC AUC score as well as PR AUC value with our balancing method does not show any improvements on validation dataset. Contrariwise, ROC AUC score reduced from 0.9745 to 0.9697, PR AUC value increased from 0.9446 to 0.9294.

8   Conclusion and Future Work

The proposed data augmentation method has improved X-rays image classification in pathology diagnosis. Proposed framework might be useful in other

---

[1] https://github.com/aleju/imgaug.

computer vision tasks such as pulmonary nodule classification at CT scans for false positive detection tasks. In this case LUNA16 dataset [18] augmentation via framework adapted for 3D convolutions is proposed.

Despite CheXnext dataset is quite large even for binary classification task and all experiments are conducted with very limited hardware resources (single NVIDIA 1070 graphical card), developed GANs produce visually-distinguishable classes of images. Thus, this augmentation technique does not require extensive computational resources and could be applied in everyday machine learning practice.

Now when our GAN produces novel images, however, for non-radiologist it might be unclear if generated images belong to pneumonia or no-pneumonia class (see Fig. 2). Despite this, if we add these images to the train dataset, model performance improves. There are common concerns about poor interpretability of deep neural networks in biomedical tasks which might affect human lives in a bad way. That's why we plan to test both the model and augmented images against human radiologists on different datasets, including private ones. This means we'll try to check if proposal GAN produces images which cannot be distinguished from the original unadulterated X-ray image not only by its own discriminator, but also by human radiology experts.

We have shown that proposed method can significantly improve performance of the classifier for pneumonia pathology. At the same time for pleural thickening classification we couldn't achieve any considerable changes. Moreover, for more complex task of fibrosis diagnosis we observed decrease of the classifier quality. According to these results we conclude proposed GAN architecture is not complex enough for handling such complicated cases as fibrosis. In further researches we'll investigate other GAN's architectures and methods. By the way, proposed framework is still applicable to a well-diagnosable by radiologist pathology (such as pneumonia and other pathologies, which were characterized in [22]).

We plan to extend this augmentation method to solve multi-label task, i.e., to validate it on full CheXnext dataset with 14 classes as separately for each modality as for all modalities together. We plan to develop effective procedure and extend proposed augmentation method for multi-label datasets. As Cycle-GAN allows us to solve cross-domain tasks, thus we plan to research this augmentation technique further, to apply it to transfer between x-rays/fluorography domains, to improve prediction of clinically important diseases in chest radiographs. We also plan to run a more detailed analysis of $G_{0 \to 1}$ and $G_{1 \to 0}$ models and explore classification model robustness against generative model output.

References

1. Almahairi, A., Rajeswar, S., Sordoni, A., Bachman, P., Courville, A.C.: Augmented CycleGAN: learning many-to-many mappings from unpaired data. In: ICML (2018)
2. Antol, S., et al.: VQA: visual question answering. In: ICCV (2015)
3. Antoniou, A., Storkey, A., Edwards, H.: Data augmentation generative adversarial networks. arXiv:1711.04340 (2017)

4. Bowles, C., et al.: GAN augmentation: augmenting training data using generative adversarial networks. CoRR abs/1810.10863 (2018)
5. Cicero, M., et al.: Training and validating a deep convolutional neural network for computer-aided detection and classification of abnormalities on frontal chest radiographs. Investig. Radiol. **52**(5), 281–287 (2017)
6. Davis, J., et al.: View learning for statistical relational learning: with an application to mammography. In: Proceedings of the 19th International Joint Conference on Artificial Intelligence, IJCAI 2005, pp. 677–683. Morgan Kaufmann Publishers Inc., San Francisco (2005)
7. Frid-Adar, M., Diamant, I., Klang, E., Amitai, M., Goldberger, J., Greenspan, H.: GAN-based synthetic medical image augmentation for increased CNN performance in liver lesion classification. Neurocomputing **321**, 321–331 (2018)
8. Glorot, X., Bengio, Y.: Understanding the difficulty of training deep feedforward neural networks. Int. Conf. Artif. Intell. Stat. **9**, 249–256 (2010). Proceedings of Machine Learning Research
9. Goodfellow, I.J., et al.: Generative adversarial networks. arXiv:1406.2661 (2014)
10. Huang, G., Liu, Z., van der Maaten, L., Weinberger, K.Q.: Densely connected convolutional networks. In: 2017 IEEE Conference on Computer Vision and Pattern Recognition (CVPR), pp. 2261–2269, July 2017
11. ImageNet database (2018). http://www.image-net.org/
12. Kamnitsas, K., et al.: Efficient multi-scale 3D CNN with fully connected CRF for accurate brain lesion segmentation. Med. Image Anal. **36**, 61–78 (2017)
13. Kermany, D.S., et al.: Identifying medical diagnoses and treatable diseases by image-based deep learning. Cell **172**(5), 1122–1131.e9 (2018)
14. Krivov, E., Pisov, M., Belyaev, M.: MRI augmentation via elastic registration for brain lesions segmentation. In: Crimi, A., Bakas, S., Kuijf, H., Menze, B., Reyes, M. (eds.) BrainLes 2017. LNCS, vol. 10670, pp. 369–380. Springer, Cham (2018). https://doi.org/10.1007/978-3-319-75238-9_32
15. Krizhevsky, A., Sutskever, I., Hinton, G.: ImageNet classification with deep convolutional neural networks. In: Advances in Neural Information Processing Systems, pp. 1097–1105 (2012)
16. Lakhani, P., Sundaram, B.: Deep learning at chest radiography: automated classification of pulmonary tuberculosis by using convolutional neural networks. Radiology **284**(2), 574–582 (2017)
17. Litjens, G.J.S., et al.: A survey on deep learning in medical image analysis. Med. Image Anal. **42**, 60–88 (2017)
18. LUng Nodule Analysis 2016 (2016). https://luna16.grand-challenge.org/
19. Madani, A., Moradi, M., Karargyris, A., Syeda-Mahmood, T.: Chest X-ray generation and data augmentation for cardiovascular abnormality classification. In: Medical Imaging: Image Processing, vol. 10574 (2018)
20. Mathers, C.D., Loncar, D.: Projections of global mortality and burden of disease from 2002 to 2030. PLoS Med. **3**(11), e442 (2006)
21. Metz, C.E.: Evaluation of digital mammography by ROC analysis. In: Proceedings of the International Workshop on Digital Mammography, pp. 61–68 (1996)
22. Rajpurkar, P., et al.: Deep learning for chest radiograph diagnosis: a retrospective comparison of the CheXNeXt algorithm to practicing radiologists. PLOS Med. **15**(11), 1–17 (2018)
23. Raoof, S., Feigin, D., Sung, A., Raoof, S., Irugulpati, L., Rosenow, E.C.I.: Interpretation of plain chest rentgenogram. Chest **141**(2), 545–558 (2012)
24. Ravì, D., et al.: Deep learning for health informatics. IEEE J. Biomed. Health Inform. **21**(1), 4–21 (2017)

25. Ronneberger, O., Fischer, P., Brox, T.: U-Net: convolutional networks for biomedical image segmentation. In: Navab, N., Hornegger, J., Wells, W.M., Frangi, A.F. (eds.) MICCAI 2015. LNCS, vol. 9351, pp. 234–241. Springer, Cham (2015). https://doi.org/10.1007/978-3-319-24574-4_28
26. Setio, A.A.A., Ciompi, F., Litjens, G., Gerke, P., Jacobs, C., van Riel, S.J., et al.: Pulmonary nodule detection in CT images: false positive reduction using multi-view convolutional networks. IEEE Trans. Med. Imaging **35**(5), 1160–1169 (2016)
27. Shin, H.C., Roberts, K., Lu, L., Demner-Fushman, D., Yao, J., Summers, R.M.: Learning to read chest X-rays: recurrent neural cascade model for automated image annotation. In: 2016 IEEE Conference on Computer Vision and Pattern Recognition (CVPR), pp. 2497–2506, June 2016
28. Springenberg, J.T.: Unsupervised and semi-supervised learning with categorical generative adversarial networks. arXiv:1511.06390 (2015)
29. VQA Challenge (2019). https://visualqa.org/
30. Wang, X., Peng, Y., Lu, L., Lu, Z., Bagheri, M., Summers, R.M.: ChestX-ray8: hospital-scale chest X-ray database and benchmarks on weakly-supervised classification and localization of common thorax diseases. CoRR abs/1705.02315 (2017)
31. Yao, L., Poblenz, E., Dagunts, D., Covington, B., Bernard, D., Lyman, K.: Learning to diagnose from scratch by exploiting dependencies among labels. arXiv abs/1710.10501, October 2017
32. Zhou, B., Khosla, A., Lapedriza, A., Oliva, A., Torralba, A.: Learning deep features for discriminative localization. In: IEEE CVPR (2016)
33. Zhu, J.Y., Park, T., Isola, P., Efros, A.A.: Unpaired image-to-image translation using cycle-consistent adversarial networks. In: ICCV (2017)

Estimation of Non-radial Geometric Distortions for Dash Cams

Evgeny Myasnikov[(✉)] [ID]

Samara University, Moskovskoe Shosse 34A, Samara 443086, Russia
mevg@geosamara.ru

Abstract. Due to the widespread use of dash cams (car DVRs), the estimation of geometric distortions introduced by such equipment is of particular interest. However, such geometric distortions are often complex and cannot be described using classical radial models. In this paper, we propose an original method for the estimation of complex geometric distortions. At the first step of the proposed method, we build distortion models based on the polynomial approximation of the lines of a calibration pattern for each particular processed frame. At the second step, we calculate the correction field and consistently refine it using built polynomial models.

In the experimental study, we apply the proposed technique to estimate the distortion of the DVR recording using the calibration pattern. We tune the order of polynomials, select optimal aggregation technique, and show how the reconstruction error changes with the number of processed frames. The experiments show that the proposed method can be successfully applied to estimate complex geometric distortions induced by car DVRs.

Keywords: Geometric distortions · Complex distortion · Non-radial distortions · Polynomial model · Dash cam · Car DVR

1 Introduction

The estimation of geometric distortions introduced by the optical systems of a camera is an important preprocessing step in the building of a computer vision system. Due to the curvature of lens, such optical systems introduce radial distortions, which cause a distorted image to take barrel-like or pillow-like form. This type of distortions is well studied, and a number of techniques were proposed to estimate and compensate radial distortions.

All these techniques are based on the assumption that a point on the image is shifted towards or away from the optical axis with respect to its true position, and the distance from the origin to the particular point of an object in a distorted image can be described with some model function of one variable.

The reported study was funded by RFBR according to the research project no. 17-29-03190.

According to this assumption, the polynomial model [1], the quadratic model [2], the polynomial model of the fourth degree [3] and the polynomial model with nonzero coefficients of terms with even powers [4] were proposed. Another type of radial models includes rational models [5,6].

Another type of distortions is caused by the inclination of the lens optical system with respect to the photosensitive sensor. It causes the perspective distortion of an image, which can be easily estimated.

Today, a number of open and proprietary software packages exist, which allow estimating the parameters of a selected distortion model using a set of photos with a calibration pattern. Here we refer to OpenCV software [7] and Matlab toolbox [8], to name a few.

By definition, the radially symmetric models can only model a distortion radially symmetric around some distortion center [9]. Unfortunately, in some systems, due to the specific of an optical system or built-in postprocessing, the complex geometric distortions are introduced. In such cases, the distortions cannot be described using classical radial models, and common software cannot be used to estimate and compensate it.

For example, complex geometric distortions are often introduced by dash-cams (car DVRs) [10]. These systems expand the field of view in the horizontal direction. It allows to record as much related information (road condition, traffic situations) as possible.

In order to estimate and compensate the complex geometric distortions, an original method was developed in this paper. This method consists in the consistent refinement of the correction function using a set of photos with a calibration pattern. To estimate distortions using a particular image, we build a distortion model relying on the assumption that the straight lines of an ideal image can be approximated as polynomials on the corresponding distorted image.

This paper is organized as follows. The proposed approach for estimation and compensation of the complex geometrical distortions is described in Sect. 2. An experimental study of the proposed method using the DVR recording is given in Sect. 3. The paper ends up with the Conclusion and References.

2   Methods

Estimation of Complex Geometric Distortions Using One Image. In this paper, to build a distortion model, we use calibration patterns, which allow to extract regular grid of lines easily. In particular, well- known rectangular grid and chessboard can be used to extract such patterns. Let us assume that the ideal image of a calibration grid S has N_H horizontal and N_V vertical equally distant straight lines. Let us also assume that we have extracted the corresponding lines on the distorted calibration image I:

$$L_i^H = \{(x_{i,j}^H, y_{i,j}^H)\}, i = 1...N_H \qquad (1)$$

and

$$L_i^V = \{(x_{i,j}^V, y_{i,j}^V)\}, i = 1...N_V. \qquad (2)$$

Here L_i^H and L_i^V are i-th horizontal and vertical lines on the distorted image correspondingly, $(x_{i,j}^H, y_{i,j}^H)$ and $(x_{i,j}^V, y_{i,j}^V)$ are the coordinates of the j-th point (pixel) of the i-th horizontal and vertical lines, $j = 1...n_i^H$ or $j = 1...n_i^V$ correspondingly, n_i^H and n_i^H are the number of image points in the i-th horizontal and vertical lines L_i^H and L_i^V correspondingly.

In this paper we assume that the straight lines of an ideal image can be approximated as polynomials on the distorted image. Therefore, we can describe the extracted point sets L_i^H and L_i^V by polynomials $P_i^H(x)$ and $P_i^V(y)$ of some order K:

$$P_i^H(x) = p_{i,0}^H + p_{i,1}^H x + ... + p_{i,K}^H x^K \tag{3}$$

and

$$P_i^V(y) = p_{i,0}^V + p_{i,1}^V y + ... + p_{i,K}^V y^K \tag{4}$$

Here the coefficients $p_{i,k}^H$ and $p_{i,k}^V, k = 1...K$ of the polynomials can be found as a solution to the least squares problem:

$$\sum_{(x,y) \in L_i^H} (P_i^H(x) - y)^2 \underset{p_{i,k}^H}{\longrightarrow} \min, \ i = 1..N_H \tag{5}$$

$$\sum_{(x,y) \in L_i^V} (P_i^V(y) - x)^2 \underset{p_{i,k}^V}{\longrightarrow} \min, \ i = 1..N_V \tag{6}$$

The polynomials $P_i^H(x)$ and $P_i^V(y)$ allow us to estimate (reconstruct) the position (x_d, y_d) of any cross of i-th horizontal and j-th vertical lines of the pattern on the distorted image by solving the following system:

$$\begin{cases} P_i^H(x_d) = y_d; \\ P_j^V(x_d) = x_d. \end{cases} \tag{7}$$

Let us assume that some ideal (undistorted) point (x, y) falls into the cell given by lines L_i^H, L_{i+1}^H, L_j^V and L_{j+1}^V. To find the position of this point on the distorted image, we estimate the polynomials P^H and P^V corresponding to the point (x, y) using the interpolation of the coefficients:

$$p_k^H = \lambda p_{i,k}^H + (1 - \lambda)p_{i+1,k}^H, k = 1...K, \tag{8}$$

$$p_k^V = \mu p_{j,k}^V + (1 - \mu)p_{j+1,k}^V, k = 1...K. \tag{9}$$

Here coefficients $p_{i,k}^H$ and $p_{i+1,k}^H$ describe the bottom and top edges of the considered cell, $p_{j,k}^V$ and $p_{j+1,k}^V$ describe the left and right edges of the cell correspondingly. Coefficients λ and μ are linear weights:

$$\lambda = (d - d_H)/d, \ \mu = (d - d_V)/d, \tag{10}$$

where d is a linear size of the grid cell in the ideal image, and $d_H = y - y_i$ and $d_V = x - x_j$ are the distances from the considered point (x, y) to the bottom and left edges of the cell.

Thus, the correction function U, which allows us to undistort the image I can be obtained as a matrix by the following algorithm:

- Extraction of the horizontal L_i^H and vertical L_j^V lines on the distorted image of a calibration grid;
- Building the distortion model by the approximation of horizontal and vertical lines using polynomials $P_i^H(x)$ and $P_j^V(y)$;
- Construction of the correction function $U(x, y) = (x_d, y_d)$ by interpolation of the coefficients of the polynomials according to (8), (9) and solving the following system for each pixel (x, y) of the correction matrix:

$$\begin{cases} P^H(x_d) = y_d; \\ P^V(x_d) = x_d. \end{cases} \tag{11}$$

Estimation of the Distortions Using the Set of Images. The described above algorithm allows compensating distortions by one image with a calibration pattern. Applying the evaluated correction matrix to different input images, we obtain undistorted images up to the projective transformation, which is embedded in the model, and, therefore, in the correction function. In practice, this transformation can be estimated at the future steps of processing, depending on the particular applied task.

The quality of reconstructed images depends on many factors. These factors include the adequacy of a model, the size of a calibration grid, the conditions of an experiment, and so on. In this section, we describe the approach, which can be used to improve the reconstruction quality.

Let S be the ideal image of a calibration grid. Due to the misalignment in the relative position of a camera and the calibration grid, the camera observes the projectively distorted image of the calibration grid. Let T_i be the projective transform defining the observed calibration image at the i-th image. Let D be the operator of the complex distortion induced by the image registration device (camera equipment). So, we can describe the i-th registered image (frame) of the calibration grid as the result of the distortion of the projectively transformed grid:

$$I_i = DT_iS. \tag{12}$$

The undistortion function U_i (for example, the algorithm described in the previous section) performs the reconstruction of the input calibration grid by one distorted image:

$$U_iI_i = T_i^{-1}D^{-1}DT_iS = S. \tag{13}$$

Here, by construction, we assume $U_i = T_i^{-1}D^{-1}$, where T_i^{-1} and D^{-1} are inverse projective transform and undistortion operator, correspondingly.

Being applied to another distorted frame I_j, the undistortion function U_i produces projectively transformed image of the calibration grid:

$$U_i I_j = T_i^{-1} D^{-1} D T_j S = T_i^{-1} T_j S = T_{ij} S. \tag{14}$$

Here we denote this projective transform as $T_{ij} = T_i^{-1} T_j$.

Analogously, using the j-th frame I_j, we can build the undistortion function U_j, and apply it to the i-th image:

$$U_j I_i = T_{ji} S. \tag{15}$$

As the transform T_{ji} can be easily estimated (for example, using the positions of the crosses of the undistorted grid $T_{ji} S$), we can use this estimation together with the undistortion function U_j (can be obtained using another frame I_j) to make alternative estimation $U_i(j)$ of the undistortion procedure U_i:

$$T_{ji}^{-1} U_j = (T_j^{-1} T_i)^{-1} (T_j^{-1} D^{-1}) = T_i^{-1} D^{-1} = U_{i(j)}. \tag{16}$$

The latter equation allows us to progressively improve the undistortion function by aggregating a number of images (frames). The most straightforward approach can be based on averaging the correction functions obtained using the set of N images:

1. Select the reference image I_r, and build the undistortion function U_r.
2. For each other i-th image:
   (a) Build the undistortion function U_i;
   (b) Apply the procedure U_i to the reference frame I_r, and estimate the projective transform T_{ir};
   (c) Transform the correction function:

$$T_{ir} U_i = U_{r(i)}. \tag{17}$$

3. Compute the average correction function:

$$U = \frac{1}{N} \sum_i U_{r(i)}, \tag{18}$$

where $U_{r(r)} = U_r$.

3   Experiments

In this section, we provide experimental results of the developed technique. The experimental setup contained the car digital video recorder system equipped with a wide-angle camera (170° diagonal view angle, as stated by the manufacturer). The original 1920 × 1080 resolution was downgraded to 960 × 540 pixels in order to speed up computations. In our experiments, we used the set of $N = 10$ frames taken with the 50 frames interval (approx. 1.67 s) for learning the correction function. For testing, we used similar parameters, except the test sequence was shifted by the half interval relative to the learning sequence. We used the uniform grid of size 60×60 cm as a calibration pattern. We tried to have the calibration pattern cover most of the informative image area in estimation and testing.

An example image, which demonstrates the complex distortion, introduced by the considered video recording system, is shown in Fig. 1. Horizontal lines of the grid are substantially convex with relation to the center of the image, while vertical lines are slightly concave or almost straight. Besides that, local distortions of lines can be seen in a full-scale source image.

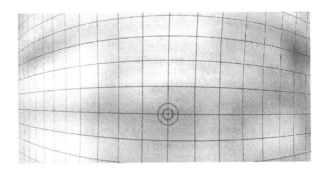

Fig. 1. An example frame with a distorted grid (the image is contrasted and truncated from the bottom to remove textual service information).

To extract lines in images of the calibration pattern we used an approach similar to that described in paper [11].

We estimated the quality of the proposed technique according to the following steps. At first, we estimated the distortion models and correction function for the learning set of frames using the method described in the previous section. At second, we undistorted each image of the test set. Then, for each resultant image, we estimated the optimal projective transform, which maps crosses of vertical and horizontal lines to their ideal positions. Finally, for each test image, we estimated the reconstruction error calculated as a mean squared error over all the projected crosses with respect to their ideal positions. Thus, we used the averaged value E of the reconstruction error as a quality measure.

In our first experiment, we estimated the dependency of the reconstruction error on the degree of the polynomials used to construct the model. The dependence of the reconstruction error E averaged over 10 test images on the degree of the polynomials is shown in Fig. 2.

As it can be seen in the figure, low orders ($K < 4$) lead to unacceptable error values. So we ended up with $K = 9$ for further experiments. In the experiments, we approximated the horizontal and vertical lines by the polynomials of the same order, but it is possible to use polynomials of different orders.

In the second experiment, we studied the dependence of the reconstruction error on the number of aggregated images (frames) used to estimate the reconstruction function. The result of this experiment is shown in Fig. 3.

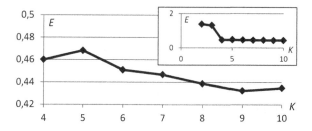

Fig. 2. The dependence of the average reconstruction error on the degree of the polynomials.

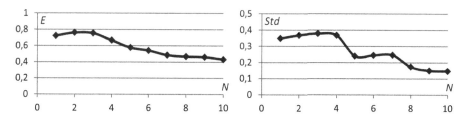

Fig. 3. The dependence of the average reconstruction error (left) and standard deviation (right) on the number of images in the learning set.

As can be seen in the figure, the error decreases as the number of learning frames grows. In this experiment, it was possible to reduce the reconstruction error by almost half. It is worth noting that the standard deviation calculated over reconstruction errors for different learning images decreases as well.

In our third experiment, we studied the effect of the parameters of aggregation procedure on the reconstruction error. In particular, we studied several aggregation methods, including the simple averaging technique (18) and two alternative techniques.

Both alternative aggregation techniques are based on the idea of giving greater or lower weights to the pixels of the aggregated correction matrices depending on our confidence in corresponding values. We introduce the weight functions $W_i(x, y)$, which show our confidence in the corresponding values of the reconstruction functions $U_i(x, y)$ obtained for images $I_i(x, y)$. Using weight functions, we assign greater values to pixels, which are closer to detected horizontal of vertical lines, and we assign lower values to pixels, which are at a distance from all the lines of a cell.

In our experiments, we considered two types of weighting functions based on linear and inverse weighting functions. These functions can be defined for one square cell of the grid with the side $d = 1$ as the following:

$$W_i^{linear}(u, v) = (2(1 - m)|u - 0.5| + m)(2(1 - m)|v - 0.5| + m), \qquad (19)$$

$$W_i^{inverse}(x, y) = \frac{1}{(1 + w(1 - |2u - 1|))(1 + w(1 - |2v - 1|))}, \quad w = \frac{1}{m} - 1. \quad (20)$$

The graphs of these functions for the one-dimensional case and different parameter values are shown in Fig. 4. The same Fig. 4 shows how the particular aggregation technique affects the reconstruction quality.

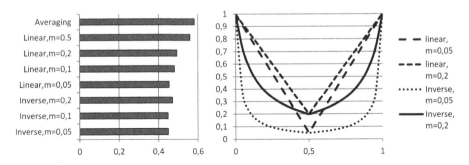

Fig. 4. The dependence of the reconstruction error on the type of the weighting function (left). Weighting functions in 1D case (right).

As can be seen, the better values of the reconstruction quality were achieved with inverse and linear functions with smaller m parameters, which correspond to functions achieving deeper minimums on the graph (see Fig. 4).

An example of the displacement matrix constructed using the refined model is shown in Fig. 5. Examples of source and undistorted images demonstrating the results of the considered technique are shown in Fig. 6.

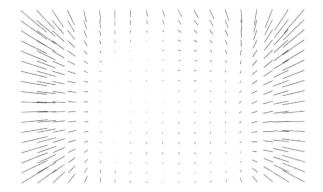

Fig. 5. An example of the displacement matrix (centered).

It is worth noting that the proposed technique is quite resistant to line breaks and noise in the images of a calibration pattern due to the approximation of lines by polynomials. Nevertheless, in practice, an operator should visually monitor line detection errors. In the case of errors, which are expressed in the deviation of the approximating line from the line in the pattern image, the operator may discard an incorrectly detected line or reject the entire frame.

Unfortunately, the described estimation technique is a very time-consuming procedure, and this is the drawback of the proposed approach. For example, it takes approximately 30 s to process one training frame on average (Intel core i7 CPU @2.5 GHz, 12 Gb RAM). Besides, this time does not include the error estimate for the test set. Fortunately, after estimating the distortions, the displacement matrix can be constructed using the refined model. The processing of images then reduces to the sequential application of this matrix to input images, which is quite fast.

Fig. 6. Examples of source and undistorted images.

4   Conclusion

The two-stage technique is proposed in this paper for the correction of complex geometric distortions induced by video recording equipment. The first stage consists in building the distortion model based on the polynomial approximation of the lines of a calibration pattern. The second step consists in calculation of a correction function by consistent aggregation of the distortion models.

It is worth noting that the second step does not depend on the nature of the model used in the first step. Therefore other distortion models can be refined using the proposed technique.

The experiments with the real DVR recording allowed to tune the parameters, and showed that the proposed technique can be successfully used to correct complex distortions induced by similar devices.

The future directions of research include the experiments with different video recording equipment and increasing the efficacy by improving the distortion models, as well as the comparison of the proposed technique to classical radial models in cases when such models can be applied.

References

1. Hartley, R., Zisserman, A.: Multiple View Geometry. Cambridge University Press, New York (2000)

2. Ma, L., Chen, Y., Moore, K.L.: Flexible camera calibration using a new analytical radial undistortion formula with application to mobile robot localization. In: IEEE International Symposium on Intelligent Control (2003)

3. Zhang, Z.: Flexible camera calibration by viewing a plane from unknown orientation. In: IEEE International Conference on Computer Vision, pp. 666–673 (1999)

4. Slama, C.C.: Manual of Photogrammetry, 4th edn. American Society of Photogrammetry, Falls Church (1980)

5. Ma, L., Chen, Y., Moore, K.L.: Rational radial distortion models of camera lenses with analytical solution for distortion correction International. J. Inf. Acquis. $\mathbf{1}$(02), 135–147 (2004)

6. Brauer-Burchardt, C., Voss, K.: A new algorithm to correct fish-eye- and strong wide-angle lens-distortion from single images. In: International Conference on Image Processing, pp. 225–228 (2001)

7. Camera calibration with OpenCV. Electronic resource. https://docs.opencv.org/2. 4/doc/tutorials/calib3d/camera_calibration/camera_calibration.html. Accessed 4 Mar 2019

8. Camera Calibrator: Electronic resource. https://www.mathworks.com/help/ vision/ref/cameracalibrator-app.html. Accessed 4 Mar 2019

9. Tang, Zh., Grompone von Gioi, R., Monasse, P., Morel, J.-M.: A precision analysis of camera distortion models. IEEE Trans. Image Process. $\mathbf{26}$(6), 2694–2704 (2017)

10. Dashcam: Electronic resource. https://en.wikipedia.org/wiki/Dashcam. Accessed 4 Mar 2019

11. Myasnikov, E.V.: Compensation of the complex geometric distortions induced by a car digital video recording equipment. In: CEUR Workshop Proceedings, vol. 2210, pp. 410–416 (2018)

On Expert-Defined Versus Learned Hierarchies for Image Classification

Ivan Panchenko$^{(\boxtimes)}$ and Adil Khan

Institute of Artificial Intelligence and Data Science, Innopolis University,
1, Universitetskaya str., Innopolis, Russia
{i.panchenko,a.khan}@innopolis.com

Abstract. For classification task involving a large number of classes, a decrease in recognition accuracy is observed for visually similar classes. We believe that forcing the model to learn appropriate features separately for each set of similar classes could improve classification performance. To justify our idea, we tried to improve classification performance by employing class hierarchy, which reflects visual similarities in data. More specifically, we used and compared two kinds of hierarchies to enhance classification performance of the model: (i) a hierarchy defined by experts (H-E), and (ii) a hierarchy created from performance results of a flat classifier and using DBScan clustering method (H-C). Moreover, we created a classification model that efficiently utilizes these hierarchies to learn appropriate features at different levels of the hierarchy. We evaluated the performance of the model on CIFAR-100 benchmark. Our results demonstrate that the hierarchical classification under H-C outperforms both H-E and the flat classifier.

Keywords: Hierarchical classification · Image classification · Confusion matrix · Class similarity

1 Introduction

In classification tasks, a trained model is used to predict the class of given data points, where classes are sometimes referred to as targets, labels, or categories. Machine learning approaches to solving a classification problem can be split into two categories. The first one is the classical machine learning approach, which uses algorithms like Decision Trees, Naive Bayes Classifiers and SVMs [9]. The second approach incorporates Deep Learning (DL) [13]. One of the main advantages of the DL models is that they require little or no manual feature engineering, meaning that the DL model is able to learn features itself. This property has given rise to machine learning models, able to solve tasks involving unstructured data, like videos, images and voice, which outperform approaches where features were created manually [12].

Functioning of a DL model can be split into three steps. The first step is features learning, which is performed by the first layers of the neural network.

© Springer Nature Switzerland AG 2019
W. M. P. van der Aalst et al. (Eds.) AIST 2019, LNCS 11832, pp. 345–356, 2019.
https://doi.org/10.1007/978-3-030-37334-4_31

These layers can be fully connected layers for simple networks [15] as well as more complex convolutional blocks [12] for image-recognition tasks and other options for different DL problems. The second step is responsible for features combination. This step is usually implemented as several fully connected layers which combine all available features and pass them into the third step. The third step involves prediction, consisting of a layer with activations having the same number of neurons as the number of classes. The output of the model is a probability distribution over all classes.

Despite their high-performance, even DL models tend to underperform in some cases. Misclassifications may occur when samples from different classes are very similar. For example, in Kinetics dataset [8] for action recognition, the top 3 misclassified class-pairs are: *riding a mule* and *riding or walking with horse*, *hockey stop* and *ice skating*, *swing dancing* and *salsa dancing*. Obviously, each of these class pairs shares many visual features.

These features allow a model to recognize mentioned classes among all other classes in a dataset. However, to distinguish among similar classes, specific features, which exists only in classes within a pair, are required. For example, features that can distinguish the kind of the animal within the first pair of actions, or the specific hand movements within the third one. Such examples lead to a statement that features that are useful for discrimination of similar samples among all categories are not necessarily discriminative for samples that fall in one compound category, and vice-versa.

To address this problem, we employ an artificial hierarchy in data in a way where classes that are hard to distinguish are united in one category. Besides, we propose a model that utilizes the learned hierarchy and performs classification in two steps. During the first step, a compound category is created. During the second step, the division of classes into categories is performed by an additional classifier. The proposed type of architecture, which exploits the similarity relationship between classes is called Hierarchical Classification (HC).

A number of works successfully utilizing HC were published in the last decade, e.g. [1,2,10,14]. This research effort contributes to the domain by addressing the question of creating an artificial hierarchy in data. More specifically, in this work we compare two hierarchies, obtained by different means on the same dataset. One hierarchy is a predefined one, or a human-labelled hierarchy. The second one is obtained by defining class separability metrics and utilizing them. We also introduce and compare a number of model architectures in order to identify the one that produces the highest accuracy on a selected dataset.

To conclude, this work aims to investigate three problems. The first problem is that of identification or creation a hierarchy in data. The second addresses building a hierarchical architecture that efficiently utilizes artificial hierarchy. The third problem involves the comparison of an expert defined hierarchy (H-E) with an artificially created hierarchy (H-C) in terms of model performance.

2   Related Works

The first question to attend to is how to create a hierarchy from the data. The simplest decision is to label classes manually or use predefined hierarchical structure [7,17,21], but we state that a hierarchy obtained in this way does not reflect separation between classes.

Another option to get a hierarchy is to create it. The process of creating an artificial hierarchy can be divided into three steps: defining a space for measurement of the distance between classes, building a hierarchical dependency tree of classes based on the defined distance and constructing a hierarchy from the dependency tree.

To define a space, where the distance between classes could be measured, two approaches are known in the literature. The first one is premature class separation, based on existing data. An example is clustering by labels' semantic [4]. The second approach is to create a hierarchy on the predictions of a trained model. This approach utilizes confusion matrix (CM) of a pretrained model as a distance matrix [6,22]. While the first approach requires additional information and has limited applicability, the second one is easy to implement and it reflects similarities of classes.

There are several widely used methods, as described in literature, to build a hierarchical dependency tree from CM. For example, Yan et al. [22] used spectral clustering method, while Silva-Palacios et al. [20] utilized an agglomerative clustering to create a dendrogram and a hierarchy from it. Both works treated CM as an inter-class distance measurement. Another option to perform clustering is DBScan algorithm [5]. An advantage of this method is its ability to detect outliers and perform grouping, based on class density.

CM values cannot be treated as distances between classes directly. There are transformations to CM that should be applied beforehand. An essential property that CM should hold is symmetry; otherwise, the distance from class A to class B is not guaranteed to be equal to the distance from class B to class A. An application of symmetric matrix is shown by Yan et al. [22]. The authors used a distance matrix, which is a transformed CM, with a symmetry property. Another valid transformation is called a similarity matrix, described in [20].

The next step after constructing a hierarchy is building a model that efficiently utilizes it. The task can be approached by: disregarding the hierarchy; connecting separated models, where each model corresponds to one node in a hierarchy; and building a single model, that has an internal hierarchical structure.

The first, and the simplest, approach ignores hierarchy and builds a flat classifier. This approach is shown to be less effective than a hierarchical classifier [22]. There are several possibilities to arrange models in the second approach. The literature distinguishes three of them: Local Classifier per Node (LCN), Local Classifier per Parent Node (LCPN) and Local Classifier per Level (LCL) [18]. A major drawback of this family of approaches is that they require separate models. To understand a problem with separate models, consider a Deep Neural Network (DNN). First layers of the DNN are responsible for learning small, low-level features like lines and edges (if we deal with pictures) [16]. These features are domain-specific, not class-specific. Therefore, learning low-level features for all classifiers is redundant. Moreover, local classifiers (every classifier that is not the root one) receives only part of training samples, which makes the model harder to train.

The third approach embeds a hierarchy directly inside the model. This idea aims to improve the shortcomings of other approaches. It was utilized by Yan et al. [22]. The authors build an architecture, called HD-CNN. HD-CNN has an intrinsic model that predicts coarse labels and a model for each set of fine labels. Coarse label prediction is responsible for choosing the most relevant subclass models and perform weighting on their output. Overall, the architecture is similar to an ensemble of models, but low-level features are shared.

Later work by Bilal et al. [3] utilizes hierarchy internally. The authors went further with the process of training a hierarchical model. They made a preliminary analysis of mistakes made by the model in hidden layers during training. Based on gained insights, the authors added intermediate outputs in the model architecture to enforce hierarchical structure. The disadvantage of this work is that building such hierarchy-aware model requires complicated preliminary analysis.

3   Methodology

3.1   Creation of a Hierarchy

We took a dataset called CIFAR-100 [11]. It consists of tiny 32×32 color images. The dataset has 100 classes with 500 train and 100 test images in each. Each class is also assigned to a superclass. A superclass is a disjoint group of classes. For CIFAR-100, superclasses were split by a human expert, relying on common knowledge. A label that identifies an image among 100 classes is called a fine label, while a label that maps an image to a superclass is called a coarse label. The list of classes and superclasses is presented in Table 1. Predefined mapping into fine and coarse categories implies a hierarchical structure of the data. This hierarchy is used as a baseline.

A better approach is to construct a hierarchy that reflects class similarity from the viewpoint of a classification model. The first step in the creation of such a hierarchy is getting a measurement of inter-class distance. The distance is obtained from CM, which in turn is obtained from the classification model. A well-known architecture for image classification is AlexNet [12]. Its architecture is shown in Fig. 1. A trained model with this architecture is used to create a CM.

Bare CM cannot be used to perform clustering on classes in order to build a hierarchy. A suitable matrix that can be obtained through transformation is symmetrical CM: F. To get an F, the CM (represented as C) is summed up with its transpose to make matrix symmetric and multiplied by a factor of 0.5 to normalize coefficients (Eq. 1a). Another option is distance matrix D which is

Table 1. Given hierarchy of classes

Coarse class	Fine classes
Aquatic mammals	Beaver, dolphin, otter, seal, whale
Fish	Aquarium fish, flatfish, ray, shark, trout
Flowers	Orchids, poppies, roses, sunflowers, tulips
Food containers	Bottles, bowls, cans, cups, plates
Fruit and vegetables	Apples, mushrooms, oranges, pears, sweet peppers
Household electrical devices	Clock, computer keyboard, lamp, telephone, television
Household furniture	Bed, chair, couch, table, wardrobe
Insects	Bee, beetle, butterfly, caterpillar, cockroach
Large carnivores	Bear, leopard, lion, tiger, wolf
Large man-made outdoor things	Bridge, castle, house, road, skyscraper
Large natural outdoor scenes	Cloud, forest, mountain, plain, sea
Large omnivores and herbivores	Camel, cattle, chimpanzee, elephant, kangaroo
Medium-sized mammals	Fox, porcupine, possum, raccoon, skunk
Non-insect invertebrates	Crab, lobster, snail, spider, worm
People	Baby, boy, girl, man, woman
Reptiles	Crocodile, dinosaur, lizard, snake, turtle
Small mammals	Hamster, mouse, rabbit, shrew, squirrel
Trees	Maple, oak, palm, pine, willow
Vehicles 1	Bicycle, bus, motorcycle, pickup truck, train
Vehicles 2	Lawn-mower, rocket, streetcar, tank, tractor

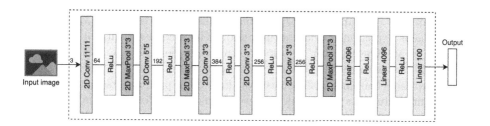

Fig. 1. Simplified AlexNet architecture

a modification of F (Eq. 1b).

$$F = 0.5 * (C + C^T) \tag{1a}$$

$$D = 0.5 * (D_{tmp} + D_{tmp}^T), \text{ where } D_{tmp} = 1 - C \tag{1b}$$

The last approach is called a similarity matrix, which is described by Silva et al. [20]. Similarity matrix S is calculated as in Eq. 2.

$$\text{Normalization} : C_{ij} = \frac{C_{ij}}{\sum_{j=1}^{n} C_{ij}}$$

$$\text{Overlap} : O_{ij} = \begin{cases} \frac{C_{ij}+C_{ij}}{2} & \text{if } i \neq j \\ 1 & \text{if } i = j \end{cases} \qquad (2)$$

$$\text{Similarity matrix} : S_{ij} = 1 - O_{ij}$$

The next step is building a hierarchy, based on inter-class distances. To produce a hierarchy, we performed clustering on classes, using transformed CM as a measure. For the classes that occur in the same cluster, it is true that they are often confused with each other or they are visually similar. Classes in one cluster are said to belong to the same superclass. Moreover, for each outlier class, produced by the clustering method, its own superclass is created. Outliers are treated this way because not belonging to any cluster means the class is not often confused and is simple to predict.

3.2   Building a Model

We built two models to investigate and compare performance on classification task involving class hierarchy. The first model is unaware of hierarchy and is referred to as the flat model. Architecture is the same as the one in Fig. 1.

The second model that we created is a modification of HD-CNN architecture [22]. The structure of its architecture is built upon basic blocks, depicted in a Fig. 2, whereas the complete model architecture is shown in Fig. 3.

There are three out of five possible types of blocks in the Fig. 2. Block Fig. 2a is responsible for learning low-level features. Block Fig. 2c is responsible for internal feature combination, which we assume is superclass specific. Finally, block Fig. 2c outputs probability vector for the sample over target distribution. Moreover, there are two kinds of blocks that are not in the picture. The first one is a fine category or coarse category prediction (shown in red in Fig. 3). It is just a convenient depiction of the output vector. The second non-drawn block also referred to as "Ones vector" block (shown in purple in Fig. 3) just returns a vector of ones, with the size equal to the input vector.

The architecture in Fig. 3, consists of three main parts: coarse category prediction, fine categories prediction, and weighting block. Convolutional layers are shared between the coarse category model and fine category models. Shared features block was used to address the problem of retraining the same low-level features. Coarse category classifier produces a probability distribution over coarse classes for an input sample. Fine category classifier outputs a distribution over fine classes in one superclass if a superclass consists of more than one class. If a superclass consists of one class, then fine classifier outputs vector with ones. Finally, inside a weighting block, each vector from fine category classifier is multiplied by the corresponding probability of a coarse category, obtained from coarse category classifier.

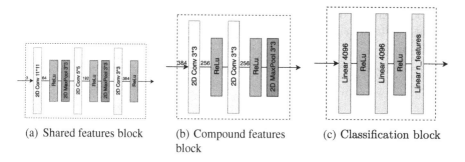

(a) Shared features block   (b) Compound features block   (c) Classification block

Fig. 2. Architecture building blocks: Numbers before and after 2D Convolutions are the number of input and output channels correspondingly. If no number specified, then the number of channels is propagated from previous output.

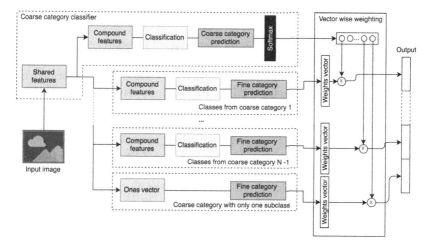

Fig. 3. Hierarchical model architecture (Color figure online)

There are differences between the proposed model and the existing work. Firstly, we didn't select fine classifiers after a prediction of the coarse category; instead, we passed an input sample to all available fine category classifiers. Secondly, we allowed the hierarchy to contain categories with one class, which leads to the presence of fine category classifiers with only one element and constant output.

4 Implementation

4.1 Hierarchy Creation

To create a hierarchy from data, first, we trained AlexNet (Fig. 1) using implementation from PyTorch [19] framework. This model also acts as a flat classifier.

The model achieved 61.04% accuracy on the test set. For this and further models training results are shown in Table 3.

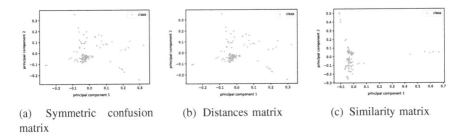

(a) Symmetric confusion   (b) Distances matrix   (c) Similarity matrix
matrix

Fig. 4. 2D scatter plots after PCA decomposition of the three matrices.

Next, the data from the training set were passed through the model again to obtain a CM of the classes. We applied three different methods to transform the matrix into distances spaces. Then PCA decomposition was performed on all transformed matrices: symmetric CM F (Fig. (4a), distances matrix D (Fig. 4b) and similarity matrix S (Fig. 4c). While decomposed F is similar to decomposed D and both of plots are diffused, similarity matrix has perceptible clusters. Therefore, similarity matrix is chosen for further processing.

The next step after choosing distances space is clustering. To perform clustering DBScan algorithm was utilized. We search for optimal epsilon value and a minimum number of neighbors for DBScan by finding the combination that returns a minimum number of clusters, while not allowing very small and very big ones. Epsilon value is a radius of a neighborhood around a point. A minimum number of neighbors is a restriction of a size for a valid cluster. So, the search space was bounded by minimum and maximum distances between points and a minimum number of neighbors. Distribution of the number of clusters over epsilon with a fixed number of neighbors is shown in Fig. 5a. The highest number of clusters is 9, according to the plot, but this marginal value is not stable, therefore, we decided to use 8 clusters. An epsilon, in this case, is equal to 1.374 and distance metric is Euclidean distance. A value for the minimum number of neighbors is 5.

The resulting clusters are plotted in Fig. 5b. The title of a cluster is the name of one of the classes, that belongs to it. Moreover, one cluster is marked as "other". These are outliers produced by DBScan. In our setup, it means that classes that occurred in this category are easily distinguishable from each other. A hierarchy that clustering produces is shown in Table 2.

4.2 Models Training and Testing

We trained two hierarchical models. The first model uses experts hierarchy (H-E), and the second one uses the created hierarchy (H-C). The training process

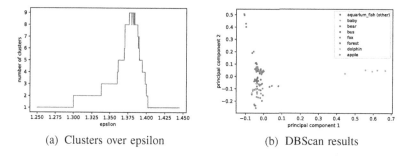

(a) Clusters over epsilon   (b) DBScan results

Fig. 5. The distribution of number of clusters over epsilon obtained using DBScan, and the resulting clusters.

of each hierarchical model was split into three steps: training a coarse classifier, training full classifier, and fine-tuning fine-label models.

The first step is training a coarse classifier. Coarse category classifier part of the model is isolated to perform training. Labels for training are coarse labels from the hierarchy. Therefore, the number of classes for H-C is equal to the number of superclasses added to the number of outliers, or $56 + 7 = 63$ and 20 for H-E. H-E achieved 76.1%, while H-C achieved 75.79% accuracy. The second step is model training on fine classification. Shared features block is frozen to prevent low-level features from training.

Investigation of trained H-C model showed that it performs better in grouping classes into supergroups, compared to the flat model. Considering samples that belong to groups in an artificial hierarchy, 965 samples are confused with a sample from its group, when passed through the flat model. In the hierarchical model, there are 1380 confused samples. While a total number of samples, belonging to groups, is 4400.

Another insight is a prediction of non-grouped classes in the hierarchical model. Samples that don't belong to any group have the same labels for both coarse and fine classification. But the result of fine classification is weighted and affected by weighting layer. Still, the difference in the number of correctly classified samples equals 153, which is 0.015 from the original dataset size. It means that overall accuracy does not suffer from weighting layer.

Therefore, the next step in training models is improving performance of each group classifier. For this purpose, shared features are directly passed to a group classifier and backpropagation starts from the output of the group classifier. Fine tuning for the model improved the performance of both models. H-E scored 59.72% and H-C scored 63.67% accuracy.

We assume that the reason why H-C performed better than H-E is differences in the hierarchy. Expert defined hierarchy is built upon common knowledge of grouping. Created hierarchy is based on a model perception of the classes, which implies classes within a group share more class-specific features. Therefore, fine labels classifiers are able to concentrate on distinguishing features, thus improving overall accuracy.

Table 2. Created hierarchy of classes

Coarse class	Fine classes
Outliers	Aquarium_fish, bed, bee, beetle, bicycle, bottle, bowl, bridge, butterfly, camel, can, castle, caterpillar, cattle, chair, clock, cloud, cockroach, couch, crab, crocodile, cup, dinosaur, flatfish, hamster, house, keyboard, lamp, lawn_mower, leopard, lizard, lobster, motorcycle, mountain, mouse, mushroom, plain, plate, possum, raccoon, road, rocket, sea, skunk, skyscraper, snail, snake, spider, sunflower, table, tank, telephone, television, turtle, wardrobe, worm
0	Baby, boy, girl, man, woman
1	Bear, beaver, chimpanzee, elephant, otter, porcupine, seal, shrew
2	Bus, pickup_truck, streetcar, tractor, train
3	Fox, kangaroo, lion, rabbit, squirrel, tiger, wolf
4	Forest, maple_tree, oak_tree, palm_tree, pine_tree, willow_tree
5	Dolphin, ray, shark, trout, whale
6	Apple, orange, orchid, pear, poppy, rose, sweet_pepper, tulip

Table 3. Results

Model	Validation accuracy for coarse classes	Validation accuracy for fine classes
Flat model	–	0.6259
H-E	0.76101	0.5972
H-C	0.75792	0.6367

5   Conclusion

In this work we performed an experiment, proving that forcing a model to learn appropriate features separately for each set of similar classes could improve classification performance. There are several works in the field, that utilize the same idea [3,20,22]. In comparison with other works, we applied a novel approach to building a hierarchy using DBScan clustering and allowing the hierarchy to have single-element groups. Moreover, we changed an existing model architecture (HD-CNN) to simplify internal fine classifier selection and proposed an approach for fine-tuning hierarchical models.

The experiment results demonstrate that hierarchical classification with created hierarchy outperforms hierarchical classification with expert-created hierarchy, as well as flat classification. It means that introducing an artificial hierarchy into a classification task could improve the overall accuracy, regardless of the presence of a given hierarchy.

References

1. Babbar, R., Partalas, I., Gaussier, E., Amini, M.R.: On flat versus hierarchical classification in large-scale taxonomies. In: Advances in Neural Information Processing Systems, pp. 1824–1832 (2013)
2. Bennett, P.N., Nguyen, N.: Refined experts: improving classification in large taxonomies. In: Proceedings of the 32nd international ACM SIGIR Conference on Research and Development in Information Retrieval, pp. 11–18. ACM (2009)
3. Bilal, A., Jourabloo, A., Ye, M., Liu, X., Ren, L.: Do convolutional neural networks learn class hierarchy? IEEE Trans. Vis. Comput. Graph. **24**(1), 152–162 (2018)
4. Deng, J., Dong, W., Socher, R., Li, L.J., Li, K., Fei-Fei, L.: ImageNet: a large-scale hierarchical image database. In: 2009 IEEE Conference on Computer Vision and Pattern Recognition, pp. 248–255. IEEE (2009)
5. Ester, M., Kriegel, H.P., Sander, J., Xu, X., et al.: A density-based algorithm for discovering clusters in large spatial databases with noise. KDD **96**, 226–231 (1996)
6. Griffin, G., Perona, P.: Learning and using taxonomies for fast visual categorization. In: 2008 IEEE Conference on Computer Vision and Pattern Recognition, pp. 1–8. IEEE (2008)
7. Jia, Y., Abbott, J.T., Austerweil, J.L., Griffiths, T., Darrell, T.: Visual concept learning: combining machine vision and Bayesian generalization on concept hierarchies. In: Advances in Neural Information Processing Systems, pp. 1842–1850 (2013)
8. Kay, W., et al.: The kinetics human action video dataset. arXiv preprint arXiv:1705.06950 (2017)
9. Kotsiantis, S.B., Zaharakis, I., Pintelas, P.: Supervised machine learning: a review of classification techniques. Emerg. Artif. Intell. Appl. Comput. Eng. **160**, 3–24 (2007)
10. Kowsari, K., Brown, D.E., Heidarysafa, M., Meimandi, K.J., Gerber, M.S., Barnes, L.E.: HDLTex: hierarchical deep learning for text classification. In: 2017 16th IEEE International Conference on Machine Learning and Applications (ICMLA), pp. 364–371. IEEE (2017)
11. Krizhevsky, A., Hinton, G.: Learning multiple layers of features from tiny images. Technical report, Citeseer (2009)
12. Krizhevsky, A., Sutskever, I., Hinton, G.E.: ImageNet classification with deep convolutional neural networks. In: Advances in Neural Information Processing Systems, pp. 1097–1105 (2012)
13. LeCun, Y., Bengio, Y., Hinton, G.: Deep learning. Nature **521**(7553), 436 (2015)
14. Li, T., Zhu, S., Ogihara, M.: Hierarchical document classification using automatically generated hierarchy. J. Intell. Inf. Syst. **29**(2), 211–230 (2007)
15. Lin, M., Chen, Q., Yan, S.: Network in network. arXiv preprint arXiv:1312.4400 (2013)
16. Ma, C., Huang, J.B., Yang, X., Yang, M.H.: Hierarchical convolutional features for visual tracking. In: Proceedings of the IEEE International Conference on Computer Vision, pp. 3074–3082 (2015)

17. Marszalek, M., Schmid, C.: Semantic hierarchies for visual object recognition. In: 2007 IEEE Conference on Computer Vision and Pattern Recognition, pp. 1–7. IEEE (2007)
18. Naik, A., Rangwala, H.: Large Scale Hierarchical Classification: State of the Art. SCS. Springer, Cham (2018). https://doi.org/10.1007/978-3-030-01620-3
19. Paszke, A., et al.: Automatic differentiation in PYTorch (2017)
20. Silva-Palacios, D., Ferri, C., Ramírez-Quintana, M.J.: Improving performance of multiclass classification by inducing class hierarchies. Procedia Comput. Sci. **108**, 1692–1701 (2017)
21. Verma, N., Mahajan, D., Sellamanickam, S., Nair, V.: Learning hierarchical similarity metrics. In: 2012 IEEE Conference on Computer Vision and Pattern Recognition, pp. 2280–2287. IEEE (2012)
22. Yan, Z., et al.: HD-CNN: hierarchical deep convolutional neural networks for large scale visual recognition. In: Proceedings of the IEEE International Conference on Computer Vision, pp. 2740–2748 (2015)

A Switching Morphological Algorithm for Depth Map Recovery

Alexey N. Ruchay[1,2(✉)] ⓘ, Konstantin A. Dorofeev[2], and Vsevolod V. Kalschikov[2]

[1] Federal Research Centre of Biological Systems and Agro-Technologies of the Russian Academy of Sciences, Orenburg, Russian Federation
[2] Department of Mathematics, Chelyabinsk State University, Chelyabinsk, Russian Federation
ran@csu.ru, kostuan1989@mail.ru, vkalschikov@gmail.com

Abstract. In this paper, we propose a switching morphological filter for RGB-D depth map recovery. The switching algorithm consists of the following steps: detection of noisy pixels and hollow areas (holes) using morphological filtering; correction of the detected noisy and hole pixels. With the help of computer simulation, we show that the proposed algorithm is able to fast recover depth maps. So, the accuracy of 3D surface reconstruction with the proposed filtering noticeably increases. The performance of the proposed algorithm is compared in terms of the accuracy of 3D surface reconstruction and processing time with that of common depth filtering algorithms.

Keywords: Depth map · Morphological filter · 3D surface reconstruction

1 Introduction

The 3D object reconstruction is a generally scientific problem of a wide variety of fields, such as computer vision, computer graphics, computer animation, computer aided geometric design, medical imaging, computational science, virtual reality, digital media, etc. [1–7].

In this paper, we are interested to improve the depth map quality by morphological filtering. A typical depth map is noisy because of infrared light reflections, and missing pixels without any depth value appear as black holes in depth maps. Noise and holes affect the accuracy of 3D object reconstruction, therefore, denoising and hole-filling algorithms should be used for 3D reconstruction [8–10].

Traditional 3D depth denoising methods are focused on fusing multiple consecutive noisy depths to get a higher quality [11]: a deep-learning based approach utilizes aligned gray images to denoise depth data [12]; a method based on the correlation between aligned color and depth frames provided by sensors [13]; spatial-temporal denoising approaches [14,15]. Enhancing the quality of the depth map using a depth frame is a popular research task: total variation

© Springer Nature Switzerland AG 2019
W. M. P. van der Aalst et al. (Eds.) AIST 2019, LNCS 11832, pp. 357–366, 2019.
https://doi.org/10.1007/978-3-030-37334-4_32

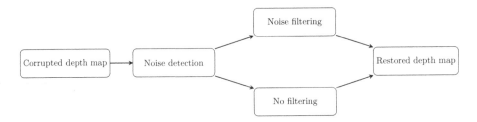

Fig. 1. Switching filtering scheme.

regularization [16]; wavelet denoising [17]; median filtering based on adaptive weighted Gaussian [18]; bilateral filter [19]; non-Local-Mean method [20].

In the last years, the following algorithms were proposed: a depth filtering scheme based on exploiting the temporal information and color information [21]; an effective divide-and-conquer method for handling disocclusion of the synthesized image [22]; a nonlinear down/upsampling filtering and a depth reconstruction multilateral filtering using a spatial resolution, boundary similarity, and coding artifacts features [23]; a 3D collaborative filtering in graph Fourier transform domain [24]; a weighted mode filter and joint bilateral filter where the joint bilateral kernel provides an optimal solution with the help of the joint histogram [25]; an adaptive method to denoise depth using Differential Histogram of Normal Vectors features along with a linear SVM [11]; a three-phase depth map correction, including eliminating anomalies, segmentation, amendment and finally inter-frame and intra-frame filtering [26]; a method based on utilizing a combination of Gaussian kernel filtering and anisotropic filtering [27].

Bilateral filtering is a technique to smooth images while preserving edges [28]. The base idea of the bilateral filter is that for a pixel to influence another pixel, it should not only occupy a nearby location but also have a similar value. The bilateral filter might not be the most advanced denoising technique but its strength lies in its simplicity and flexibility. The following modifications of the bilateral filter were proposed: Adaptive Bilateral Filter (ABF) [11], Fast Bilateral Filter (FBF) [29], Joint Bilateral Filter (JBF) [30] and Joint Bilateral Upsampling (JBU) [19].

Mathematical morphology can get fairly good effect by applying the grayscale morphology in the image denoising aspect, having the characteristics of nonlinearity and parallelism [31,32].

In this article, we propose a novel switching morphological filter for denoising depth map. We apply the denoising filter only part of pixels of the depth map, but only in those where noise and holes are possible, that is, we detect noisy pixels and holes using morphological filtering (see Fig. 1).

We consider denoising depth algorithms for 3D object reconstruction, therefore, we use the raw depth map as noisy data and we evaluate the performance of the denoising methods based on the enhancement achieved in the accuracy of 3D object reconstruction. In contrast to this approach, a common approach of noise reduction is that the raw depth map represented the ground truth, added an artificial noise such as additive or impulse, and then proposed a method to remove the noise [11]. Although this common approach can be used for quantitative comparison, wherein proposed methods reduce only the artificial noise but not the original noise contained in the raw depth. Therefore, our main goal is to evaluate the denoising methods to enhance reconstruction accuracy which depends on the quality of the captured raw depth map. We use the metric of evaluation as the root mean square error (RMSE) of measurements in the iterative closest point (ICP) algorithm and Hausdorf distance [33] between input cloud and filtered cloud.

The performance of the proposed algorithm is compared in terms of the accuracy of 3D object reconstruction and speed with the following depth denoising algorithms: ABF [11], FBF [29], JBF [30], JBU [19], Noise-aware Filter (NF) [34], Weight Mode Filter (WMF) [35], Anisotropic Diffusion (AD) [36], Markov Random Field (MRF) [37], Markov Random Field(Second Order Smoothness) (MRFS) [38], Markov Random Field(Kernel Data Term) (MRFK) [38], Markov Random Field(Tensor) (MRFT) [38], Layered Bilateral Filter (LBF) [39], Kinect depth normalization (KDN) [40], Roifill filter (RF) [41], Median filter (MF), Bilateral Filter (BF), Okada filter (OF) [42].

The paper is organized as follows. In Sect. 2, we describe the proposed depth denoising algorithm based on switching morphological filter. Computer simulation results are provided in Sect. 3. Finally, Sect. 4 summarizes our conclusions.

2   Proposed Algorithm

In this section, we describe the proposed depth denoising algorithm based on switching morphological filter (DSMF).

First, we describe two fast and reliable algorithms of noise detection based on morphological operations: switching median and morphological filter [43] and switching filter of impulse noise with morphological filtering [44].

Mathematical morphology describes the shape and structure of certain objects, and is used to extract the useful components in the image. It is utilized for image filtering, image segmentation, image measurement, area filling [45,46].

Let the the structuring element be b, then denote two discrete-value functions defined on a two-dimensional discrete space. The erosion and the dilation can be defined as

$$(D \ominus b)(i,j) = min\{D(i+s, j+t) - b(s,t) \, \| \, (i+s, j+t) \in D_d, (s,t) \in D_b\},$$

$$(D \oplus b)(i,j) = max\{D(i-s, j-t) + b(s,t) \, \| \, (i-s, j-t) \in D_d, (s,t) \in D_b\},$$

where $\ominus$ and $\oplus$ denote the erosion operator and the dilation operator, respectively. D_d and D_b denote the domain of the gray-scale image D and the domain of the structuring element b, respectively.

With the above erosion and dilation operators, two morphological gradients are determined as follows:

$$g_e(i,j) = D(i,j) - (D \ominus b)(i,j),$$

$$g_d(i,j) = (D \oplus b)(i,j) - D(i,j).$$

The hybrid gradient $g_h(i,j)$ is defined as

$$g_h(i,j) = g_d(i,j) \cdot g_e(i,j)$$

based on the two morphological gradients $g_e(i,j)$ and $g_d(i,j)$.

This detection noise method SMMF(D) for the depth map D in SMMF can be defined as follows:

$$\text{SMMF} = \begin{cases} 1, & \text{if } g_h(i,j) = 0, \\ 0, & \text{otherwise.} \end{cases}$$

We propose the following SMF1(D) method of the noise detection for depth map D with a morphological filter [44]:

$$\text{SMF1}(D) = (\text{set1}(D, mset) \star b) \bigcup (\text{set2}(D, pset) \star b),$$

where B is the square structuring element, $\star$ is morphological "bottom hat" operation defined as an image minus the morphological closing of an image, set1(A,mset) is subtraction of all the pixels A value mset, set2(A,pset) is subtraction of the values pset of all the pixels A. Extensive experiments revealed that good noise detection results can't be achieved using the original switching filter SMF SMF(D) [44], and the original switching filter requires serious time-consuming costs, therefore we do not present the results for the original switching filter [44].

The proposed switching morphological filter for the depth map (SMF1MF, SMF1AMF, SMF1RF and SMMFMF, SMMFAMF, SMMFRF) is defined by

$$y = \begin{cases} x_{MF\,(or\,AMF\,or\,RF)}, & \text{if } \text{SMMF}(x) = 1\,(or\,\text{SMF1}(x) = 1), \\ x, & \text{otherwise,} \end{cases}$$

where y is the switching filter output, x is the central pixel of the filtering window W, x_{MF} is the Median Filter (MF) output computed over the pixels declared by the detector as uncorrupted, x_{AMF} is the Arithmetic Mean Filter (AMF) output computed over the pixels declared by the detector as uncorrupted, and x_{RF} is the Random Filter (RF) output computed randomly over the pixels declared by the detector as uncorrupted.

Extensive experiments revealed that very good denoising results can't be achieved using the following filters: ABF, FBF, JBU, NF, WMF, AD, MRFS, MRFT, LBF, KDN, RF, and OF. The main reason of this is highly distorted point cloud after filtering, therefore, we don't use these filters for our next experiments and comparisons.

A common algorithm for counting RMSE by using the ICP algorithm consists of the following steps:

1. Registration a RGB and depth data.
2. Use a depth denoising algorithm: JBF, BF, SMMFAMF, SMMFMF, SMM-FRF, SMF1AMF, SMF1MF, SMF1RF, MRF, MRFK, MF.

3. Make point clouds using denoising depth data.
4. Detection and matching of keypoints in point clouds with the keypoint detection algorithm SIFT [47].
5. Remove outliers with correspondence rejectors RANSAC [47].
6. Count transformation matrix and RMSE with ICP using the associate 3D points of the inliers.

3   Computer Simulation

In this section, computer simulation results of the accuracy of 3D object reconstruction based on the proposed depth denoising algorithm using real data are presented and discussed.

As previously stated, we evaluate the performance of our proposed denoising filter against other state-of-the-art filters based on the enhancement of reconstruction accuracy achieved by each filter. We have experimental results for evaluation of the performance of the ICP algorithm for object 3D reconstruction. The metric of evaluation is the root mean square error (RMSE) of measurements and Hausdorf distance between input cloud and filtered cloud. We choose the special RGB-D dataset [3], where different statue models (a lion, an anatomy, an apollo) captured by the Kinect and the HD RGB camera with rotating counterclockwise on a turntable. The dataset has an accurate laser models for comparing and calculating of Hausdorf distance.

In our experiments, we select 11 different depth denoising algorithms which are widely cited and used in comparison: JBF, BF, SMMFAMF, SMMFMF, SMMFRF, SMF1AMF, SMF1MF, SMF1RF, MRF, MRFK, MF. The experiments are carried out on a PC with Intel(R) Core(TM) i7-4790CPU @ 3.60 GHz and 8 GB memory.

To evaluate the performance of 3D object reconstruction based on the proposed depth denoising algorithm in our experiments, we carried out 3D reconstruction of an anatomy from dataset [3]. Figure 2 shows RGB images and depth maps of an anatomy.

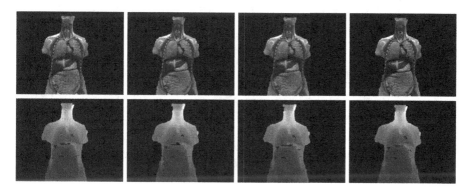

Fig. 2. The RGB images and depth maps of an anatomy are taken by a Kinect sensor.

Table 1. Results of measurements using a common ICP algorithm with JBF, BF, SMMFAMF, SMMFMF, SMMFRF, SMF1AMF, SMF1MF, SMF1RF, MRF, MRFK, MF depth denoising algorithms (DDA). This table presents RMSE, Hausdorf distance (HD), and an average time of processing in sec. (Time).

DDA	RMSE	HD	Time
Without	0.008203	0.555646	0.2778
JBF	0.008291	0.555258	12.602
BF	0.010598	0.461604	5.2472
SMMFAMF	0.008416	0.555165	36.922
SMMFMF	0.008420	0.555116	36.626
SMMFRF	0.008423	0.555120	35.824
SMF1AMF	0.008103	0.514745	0.4193
SMF1MF	0.008206	0.555605	0.4396
SMF1RF	0.008206	0.555602	0.4790
MRF	0.010459	0.485138	32.340
MRFK	0.010865	0.482792	112.30
MF	0.008197	0.555761	5.2985

Corresponding RMSE values calculated for each pair with a step of 1 in the ICP algorithm with JBF, BF, SMMFAMF, SMMFMF, SMMFRF, SMF1AMF, SMF1MF, SMF1RF, MRF, MRFK, MF depth denoising algorithms are shown in Table 1.

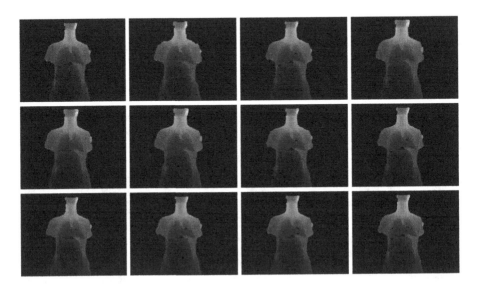

Fig. 3. The restored depth maps of an anatomy without filtering and after denoising JBF, BF, SMMFAMF, SMMFMF, SMMFRF, SMF1AMF, SMF1MF, SMF1RF, MRF, MRFK, MF filters (from left to right from top to bottom).

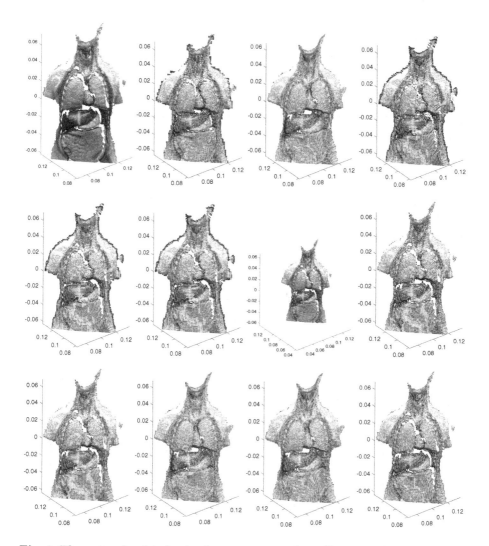

Fig. 4. The restored point clouds of an anatomy without filtering and after denoising JBF, BF, SMMFAMF, SMMFMF, SMMFRF, SMF1AMF, SMF1MF, SMF1RF, MRF, MRFK, MF filters (from left to right from top to bottom).

The quality of depth denoising we can also evaluate visually looking at the restored point cloud. Figures 3 and 4 shows the depth maps and the 3D point clouds of an anatomy after denoising JBF, BF, SMMFAMF, SMMFMF, SMM-FRF, SMF1AMF, SMF1MF, SMF1RF, MRF, MRFK, MF filters. The proposed SMF1RF yield the best result in terms of RMSE, Hausdorf distance, speed and visual evaluation among all depth denoising algorithms.

4   Conclusion

In this paper, we presented a novel switching morphological algorithm for depth map recovery based on the switching morphological filter. In other words, we apply the filter not at all pixels of the depth map, but only in those where noise and holes are possible, that is, we detect noisy pixels and holes using morphological filtering. We evaluated the performance of the ICP algorithm with the proposed depth denoising algorithm for object 3D reconstruction using real data. Also, the performance of the proposed algorithm is compared in terms of the accuracy of 3D object reconstruction, Hausdorf distance between input cloud and filtered cloud, and speed with that of common successful depth filtering algorithms. The experiment has shown that the proposed switching morphological algorithm filter yields the best result in terms of RMSE, Hausdorf distance, speed and visual evaluation among all depth denoising algorithms.

Acknowledgments. This work was supported by the Russian Science Foundation, grant no. 17-76-20045.

References

1. Zheng, L., Li, G., Sha, J.: The survey of medical image 3D reconstruction. In: Proceedings of SPIE, vol. 6534, pp. 6534–6536 (2007)
2. Echeagaray-Patron, B.A., Kober, V.: 3D face recognition based on matching of facial surfaces. In: Proceedings of SPIE, vol. 9598, p. 95980V-8 (2015)
3. Lee, K., Nguyen, T.Q.: Realistic surface geometry reconstruction using a hand-held RGB-D camera. Mach. Vis. Appl. **27**(3), 377–385 (2016)
4. Echeagaray-Patron, B.A., Kober, V.: Face recognition based on matching of local features on 3D dynamic range sequences. In: Proceedings of SPIE, vol. 9971, pp. 9971–9976 (2016)
5. Echeagaray-Patrón, B.A., Kober, V.I., Karnaukhov, V.N., Kuznetsov, V.V.: A method of face recognition using 3D facial surfaces. J. Commun. Technol. Electron. **62**(6), 648–652 (2017)
6. Ruchay, A., Dorofeev, K., Kober, A.: 3D object reconstruction using multiple Kinect sensors and initial estimation of sensor parameters. In: Proceedings of SPIE, vol. 10752, pp. 1075222–1075228 (2018)
7. Ruchay, A., Dorofeev, K., Kolpakov, V.: Fusion of information from multiple Kinect sensors for 3D object reconstruction. Comput. Opt. **42**(5), 898–903 (2018)
8. Tihonkih, D., Makovetskii, A., Voronin, A.: A modified iterative closest point algorithm for noisy data. In: Proceedings of SPIE, vol. 10396, pp. 10396–10397 (2017)
9. Makovetskii, A., Voronin, S., Kober, V.: An efficient algorithm of 3D total variation regularization. In: Proceedings of SPIE - The International Society for Optical Engineering, vol. 10752, p. 107522V (2018)
10. Voronin, S., Makovetskii, A., Voronin, A., Diaz-Escobar, J.: A regularization algorithm for registration of deformable surfaces. In: Proceedings of SPIE - The International Society for Optical Engineering, vol. 10752, p. 107522S (2018)
11. Boubou, S., Narikiyo, T., Kawanishi, M.: Adaptive filter for denoising 3D data captured by depth sensors. In: 2017 3DTV Conference: The True Vision - Capture, Transmission and Display of 3D Video (3DTV-CON), pp. 1–4 (2017)

12. Zhang, X., Wu, R.: Fast depth image denoising and enhancement using a deep convolutional network. In: 2016 IEEE International Conference on Acoustics, Speech and Signal Processing (ICASSP), pp. 2499–2503 (2016)
13. Milani, S., Calvagno, G.: Correction and interpolation of depth maps from structured light infrared sensors. Sig. Process. Image Commun. **41**, 28–39 (2016)
14. Fu, J., Wang, S., Lu, Y., Li, S., Zeng, W.: Kinect-like depth denoising. In: 2012 IEEE International Symposium on Circuits and Systems, pp. 512–515 (2012)
15. Lin, B.S., Chou, W.R., Yu, C., Cheng, P.H., Tseng, P.J., Chen, S.J.: An effective spatial-temporal denoising approach for depth images. In: 2015 IEEE International Conference on Digital Signal Processing (DSP), pp. 647–651 (2015)
16. Makovetskii, A., Voronin, S., Kober, V.: An efficient algorithm for total variation denoising. In: Ignatov, D.I., et al. (eds.) AIST 2016. CCIS, vol. 661, pp. 326–337. Springer, Cham (2017). https://doi.org/10.1007/978-3-319-52920-2_30
17. Moser, B., Bauer, F., Elbau, P., Heise, B., Schoner, H.: Denoising techniques for raw 3D data of ToF cameras based on clustering and wavelets. In: Proceedings of SPIE, vol. 6805, pp. 6805–6812 (2008)
18. Frank, M., Plaue, M., Hamprecht, F.A.: Denoising of continuous-wave time-of-flight depth images using confidence measures. Opt. Eng. **48**(7), 077003 (2009)
19. Kopf, J., Cohen, M.F., Lischinski, D., Uyttendaele, M.: Joint bilateral upsampling. ACM Trans. Graph. **26**, 3 (2007)
20. Georgiev, M., Gotchev, A., Hannuksela, M.: Real-time denoising of ToF measurements by spatio-temporal non-local mean filtering. In: 2013 IEEE International Conference on Multimedia and Expo Workshops (ICMEW), pp. 1–6 (2013)
21. Bhattacharya, S., Venkatesh, K.S., Gupta, S.: Depth filtering using total variation based video decomposition. In: 2015 Third International Conference on Image Information Processing (ICIIP), pp. 23–26 (2015)
22. Lei, J., Zhang, C., Wu, M., You, L., Fan, K., Hou, C.: A divide-and-conquer hole-filling method for handling disocclusion in single-view rendering. Multimed. Tools Appl. **76**(6), 7661–7676 (2017)
23. Zhang, Q., Chen, M., Zhu, H., Wang, X., Gan, Y.: An efficient depth map filtering based on spatial and texture features for 3D video coding. Neurocomputing **188**, 82–89 (2016)
24. Chen, R., Liu, X., Zhai, D., Zhao, D.: Depth image denoising via collaborative graph Fourier transform. In: Zhai, G., Zhou, J., Yang, X. (eds.) IFTC 2017. CCIS, vol. 815, pp. 128–137. Springer, Singapore (2018). https://doi.org/10.1007/978-981-10-8108-8_12
25. Fu, M., Zhou, W.: Depth map super-resolution via extended weighted mode filtering. In: 2016 Visual Communications and Image Processing (VCIP), pp. 1–4 (2016)
26. Pourazad, M.T., Zhou, D., Lee, K., Karimifard, S., Ganelin, I., Nasiopoulos, P.: Improving depth map compression using a 3-phase depth map correction approach. In: 2015 IEEE International Conference on Multimedia Expo Workshops (ICMEW), pp. 1–6 (2015)
27. Liu, S., Chen, C., Kehtarnavaz, N.: A computationally efficient denoising and hole-filling method for depth image enhancement. In: Proceedings of SPIE, vol. 9897, pp. 9897–9899 (2016)
28. Paris, S., Kornprobst, P., Tumblin, J.: Bilateral Filtering. Now Publishers Inc., Hanover (2009)
29. Durand, F., Dorsey, J.: Fast bilateral filtering for the display of high-dynamic-range images. ACM Trans. Graph. **21**(3), 257–266 (2002)

30. Petschnigg, G., Agrawala, M., Hoppe, H., Szeliski, R., Cohen, M., Toyama, K.: Digital photography with flash and no-flash image pairs. ACM Trans. Graph. **23**(3), 664–672 (2004)

31. Jakhar, A., Sharma, S.: A novel approach for image enhancement using morphological operators. Int. J. Adv. Res. Comput. Sci. Technol. (IJARCST) **2**, 300–302 (2014)

32. Yoshitaka, K.: Mathematical morphology-based approach to the enhancement of morphological features in medical images. J. Clin. Bioinform. **1**, 33 (2011)

33. Alexiou, E., Ebrahimi, T.: On subjective and objective quality evaluation of point cloud geometry. In: 2017 Ninth International Conference on Quality of Multimedia Experience (QoMEX), pp. 1–3 (2017)

34. Chan, D., Buisman, H., Theobalt, C., Thrun, S.: A noise-aware filter for real-time depth upsampling. In: Workshop on Multi-camera and Multi-modal Sensor Fusion Algorithms and Applications (2008)

35. Min, D., Lu, J., Do, M.N.: Depth video enhancement based on weighted mode filtering. IEEE Trans. Image Process. **21**(3), 1176–1190 (2012)

36. Liu, J., Gong, X.: Guided depth enhancement via anisotropic diffusion. In: Huet, B., Ngo, C.-W., Tang, J., Zhou, Z.-H., Hauptmann, A.G., Yan, S. (eds.) PCM 2013. LNCS, vol. 8294, pp. 408–417. Springer, Cham (2013). https://doi.org/10.1007/978-3-319-03731-8_38

37. Diebel, J., Thrun, S.: An application of Markov random fields to range sensing. In: Proceedings of the 18th International Conference on Neural Information Processing Systems, NIPS 2005, pp. 291–298 (2005)

38. Harrison, A., Newman, P.: Image and sparse laser fusion for dense scene reconstruction. In: Howard, A., Iagnemma, K., Kelly, A. (eds.) Field and Service Robotics. Springer Tracts in Advanced Robotics, vol. 62, pp. 219–228. Springer, Heidelberg (2010). https://doi.org/10.1007/978-3-642-13408-1_20

39. Yang, Q., Yang, R., Davis, J., Nister, D.: Spatial-depth super resolution for range images. In: 2007 IEEE Conference on Computer Vision and Pattern Recognition, pp. 1–8 (2007)

40. Newcombe, R.A., et al.: KinectFusion: real-time dense surface mapping and tracking. In: IEEE ISMAR (2011)

41. Fuhrmann, S., Goesele, M.: Fusion of depth maps with multiple scales. ACM Trans. Graph. **30**(6), 148:1–148:8 (2011)

42. Okada, M., Ishikawa, T., Ikegaya, Y.: A computationally efficient filter for reducing shot noise in low S/N data. PLoS ONE **11**(6), e0157595 (2016)

43. Yuan, C., Li, Y.: Switching median and morphological filter for impulse noise removal from digital images. Optik Int. J. Light Electron Opt. **126**(18), 1598–1601 (2015)

44. Ruchay, A., Kober, V.: Impulsive noise removal from color images with morphological filtering. In: van der Aalst, W.M.P., et al. (eds.) AIST 2017. LNCS, vol. 10716, pp. 280–291. Springer, Cham (2018). https://doi.org/10.1007/978-3-319-73013-4_26

45. Soille, P.: Morphological Image Analysis: Principles and Applications, 2nd edn. Springer, Heidelberg (2004). https://doi.org/10.1007/978-3-662-05088-0

46. Najman, L., Talbot, H.: Mathematical Morphology: From Theory to Applications. ISTE-Wiley, Hoboken (2010)

47. Rusu, R.B., Cousins, S.: 3D is here: point cloud library (PCL). In: 2011 IEEE International Conference on Robotics and Automation, pp. 1–4 (2011)

Learning to Approximate Directional Fields Defined Over 2D Planes

Maria Taktasheva, Albert Matveev$^{(\boxtimes)}$, Alexey Artemov, and Evgeny Burnaev

Skolkovo Institute of Science and Technology, Moscow, Russia
albert.matveev@skoltech.ru

Abstract. Reconstruction of directional fields is a need in many geometry processing tasks, such as image tracing, extraction of 3D geometric features, and finding principal surface directions. A common approach to the construction of directional fields from data relies on complex optimization procedures, which are usually poorly formalizable, require a considerable computational effort, and do not transfer across applications. In this work, we propose a deep learning-based approach and study the expressive power and generalization ability.

Keywords: Neural networks · Image vectorization · Directional fields

1 Introduction

Many spatially varying geometric and physical properties of objects, such as surface curvature, stress tensors, or gradients of scalar fields, are commonly described by directional *(vector)* fields: certain quantities (generally vector-valued) assigned to respective points in some spatial domain. While real-world measurements correspond to our most intuitive notion of directional fields, they can just as easily be *synthesized* via optimization by computational models accounting for constraints such as physical realizations or alignment conditions. Corresponding to such requirements, different representations for directional fields have been proposed, differing, *e.g.*, by several directional entities per point of the domain, or symmetries between them.

In the area of 2D and 3D computer graphics and geometry processing, directional fields have been utilized for a vast number of applications, such as mesh generation using distinct 3D data modalities [4,6], texture mapping [16], and image-based tracing of line drawings [2], to name a few. Appearing datasets [10] open more possibilities for this setup with various problems to solve. With the guidance of an appropriately designed directional field, both topological (*e.g.*, placement of singularity points) and geometric (*e.g.*, smoothness) properties of the underlying geometric structure may be efficiently derived. Other applications which could benefit from learnable directional fields include remote sensing [3,9,11,14], RGBD data processing [19] and related applications [1,5], shape retrieval [13].

M. Taktasheva and A. Matveev—These two authors contribute equally to the work.

© Springer Nature Switzerland AG 2019
W. M. P. van der Aalst et al. (Eds.) AIST 2019, LNCS 11832, pp. 367–374, 2019.
https://doi.org/10.1007/978-3-030-37334-4_33

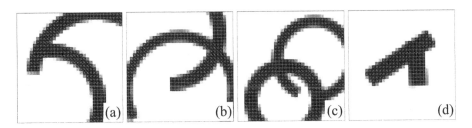

Fig. 1. (a)–(d): examples of rasterized vector primitives with accompanying discretized ground-truth 2-PolyVector field derived according to our scheme (see text).

However, obtaining a robust approximation of a directional field from raw input data is a challenging problem in many instances. Current approaches to computing directional fields require optimization of non-trivial targets with custom optimizers (*e.g.*, ADMM, L-BFGS, and their versions), which may be computationally demanding and may yield unstable solutions, where significant noise, occlusions, or gaps exist in the data [2].

On the other hand, modern deep convolutional neural networks (CNNs) have proven themselves effective in learning arbitrarily complex functions from both theoretical and practical standpoints [8,12]. However, little work has been done to approximate directional fields using learning-based approaches; thus, questions exist of whether training a conventional CNN to approximate a directional field would be easy or even feasible, and whether it can learn to produce highly robust directional fields.

In this work, we conduct an initial feasibility study, aiming to establish a principled learning setup for an approximation of directional fields. For our study, we take a setting simplified in several aspects. First, we restrict ourselves to directional fields defined over subsets of 2D planes, *i.e.* over rectangular regions $\Omega \subset \mathbb{R}^2$, enabling us to use conventional CNN architectures. Moreover, in $\mathbb{R}^2$, an explicit field representation using Euclidean coordinates is straightforward, which is not the case on, *e.g.*, curved surfaces [18]. Next, without loss of generality, we focus on 2-PolyVector fields [6] as a structure of our probe directional field. Lastly, to obtain a particular ground-truth for our experiments, we opt to derive the field from geometric primitives encountered in vector computer graphics, as they allow for constructing an explicit analytical description of the result. The overall setup represents a conceptually simple but important instance, as it corresponds to several practical applications, such as image tracing [2]. We carry out two experiments, aiming to (1) establish the effectiveness of a deep CNN in an approximation setting, and (2) evaluate its generalization ability.

This short paper is organized as follows. In Sect. 2, we detail the design of our probe directional field, our choice of the learning algorithm, and the formation of our training data. Section 3 presents two experiments evaluating our architecture. We conclude in Sect. 4 with a brief discussion.

2   Directional Field Design and the Approximation Setup

2.1   Our Probe Directional Field

For the sake of simplicity, we define our directional fields over gray-scale images. Specifically, we assign two complex-valued vectors $\{u, v\}$ to each pixel $x \in \Omega$ which is above the given threshold. Following the established nomenclature [7, 18], we denote our directional fields as *2-PolyVector fields*. The field is obtained through an optimization procedure on a complex polynomial [18]

$$f(z) = (z^2 - u^2)(z^2 - v^2) = z^4 + c_2 z^2 + c_0,$$

where the coefficients are indexed by the corresponding power of the variable.

While directional fields such as ours can be defined over arbitrary images, we define them on a particular class of images: line drawings, *i.e.*, collections of 1D geometric primitives such as straight and curved lines. For points on curves, the first field component is aligned with the curve tangent; the second field component is an interpolation between the curve normal and the tangent of a nearest intersecting curve. For points not on curves, we leave the field undefined. Thus, we obtain an intuitive interpretation of field components $\{u, v\}$; the PolyVector structure allows us to express the directionality of curves. A similar parameterization is adopted in [2].

Leveraging the analytic expressions for the primitives constituting the line drawing image (*e.g.*, lines, arcs, and Bézier curves), we can define the first field vector as a unit tangent vector of a curve. The unit tangent is uniquely determined by its slope angle α, with its complex coordinates $u = (\cos \alpha, \sin \alpha)$. The second vector $v = (\cos \beta, \sin \beta)$ is also determined by its slope angle β; v can be obtained based on the following simple rules:

– if two curves intersect in one pixel, as a tangent vector of a second curve;
– if the intersection is relatively far, as a normal vector of a curve;
– if the intersection is not far—as an interpolation between these two cases.

We assume that junctions of three and more are rare and only consider junctions of two curves in our model. The definition of a PolyVector field is straightforward to extend to capture other types of junctions.

To become independent to the order of primitives and the chosen sign of a tangent, we follow a well-known workaround and perform variable substitution $c_0 = u^2 v^2, c_2 = -(u^2 + v^2)$, which determines two vectors u and v up to relabeling and sign [2]. The final vector field representation has four parameters: two complex coordinates for each of the variables c_0 and c_2. We interpolate between the field vectors by smoothing a field within channels related to the same complex vector with Gaussian kernel. We display typical raster line-drawing images along with the corresponding ground-truth directional fields in Fig. 1.

2.2 Our Neural Network-Based Direction Field Approximator

Architecture. We formulate our field approximation as a multivariate regression task, where one needs to obtain an output of the same shape as the input, except for the number of output channels. When designing our directional field approximation architecture, we take inspiration from recent progress in semantic segmentation and derive our network from the vanilla U-Net model [17].

U-Net architecture consists of the encoder and decoder parts with skip connections between them. Each encoder layer has two convolutional layers with the increasing number of channels, followed by batch normalization and ReLU activation function. Each encoder layer output is fed through max pooling. The decoder has a symmetric structure, where the numbers of channels of convolutional layers decrease, and the upsampling layer is put instead of max pooling. The output of each of the encoder layers is concatenated with the input of the corresponding decoder layer so that the information is passed through skip connections. We modify the architecture by replacing the softmax layer at the end of the network with a linear layer having four output channels, two for real parts of c_0, c_2, two for imaginary parts.

We implemented the network in pytorch [15] and trained it on one NVIDIA Tesla P100 with 16 Gb GPU memory. The training process took 15 min for single image experiments and 40 min for the generalization study.

Loss Function. We base our loss function on the results from [2], where the PolyVector field is found as a solution of a complex optimization problem and c_0, c_2 are treated as complex functions defined at each pixel. The optimizer searches for a solution which is aligned with a tangent field approximation obtained by applying a Sobel filter. As an alignment term, we use mean squared error loss for real and imaginary parts of complex coefficients c_0 and c_2. The smoothness of the resulting field is guaranteed by a smoothness regularization term, formulated as the difference in c_k, $k = 0, 2$ between the neighbouring pixels (if these are available), where $\nabla c_k(x_{i,j})$ is a complex number:

$$\int \|\nabla c_k(x)\|^2 dx = \sum_i \sum_j \|\nabla c_k(x_{i,j})\|^2$$

$$= \sum_i \sum_j Re(\nabla c_k(x_{i,j}))^2 + Im(\nabla c_k(x_{i,j}))^2,$$

$$\nabla c_k(x_{i,j}) = \text{vertical change} + \text{horizontal change}, \ k = 0, 2$$

where the real and imaginary part are represented as separate channels.

The overall loss we optimize is the following:

$$L(c_0, c_2, c_0^*, c_2^*) = \sum_{k=0,2} \sum_i \sum_j \|c_k(x_{i,j}) - c_k^*(x_{i,j})\|^2 + \gamma \sum_{k=0,2} \sum_i \sum_j \|\nabla c_k(x_{i,j})\|^2$$

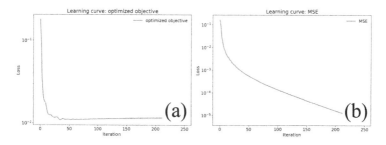

Fig. 2. Approximation progress, single image: (a) optimized objective, (b) MSE loss.

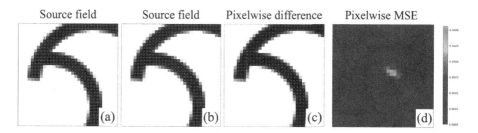

Fig. 3. Approximation results, single image: (a) input drawing and ground truth field, (b) approximated directional field, (c) difference between ground truth and approximated fields, (d) error heatmap. Best viewed in zoom.

3   Experimental Results

3.1   Expressive Power of a Conventional CNN

Here we show that a common CNN architecture optimized with stochastic gradient descent is capable of representing a directional field. We select a random synthetic 64×64 image with two primitives and optimize network parameters for 50 iterations. We display learning curves in Fig. 2 and the resulting approximations in Fig. 3. We conclude that our network learns to represent the input directional field with an alignment error of 10^{-5} order of magnitude and regularized loss of 10^{-2} order of magnitude.

3.2   Generalization Study

The second experiment is a proof of concept to the learnability of a directional field on a dataset of sufficient size. We synthesize and rasterize 5500 64×64-pixel line drawings, splitting them into 5000 training and 500 validation images. We train the network using our data for 100 epochs. In Fig. 4, we plot learning curves, showing the resulting predictions on validation samples in Fig. 5. We conclude that our model can learn the underlying directional field for the generated dataset: we report validation MSE loss 0.00152 and regularized loss 0.01321.

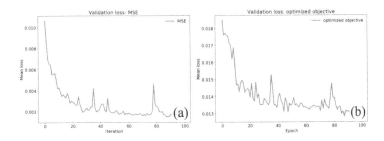

Fig. 4. Learning progress, 5000 images: (a) MSE loss, (b) optimized objective.

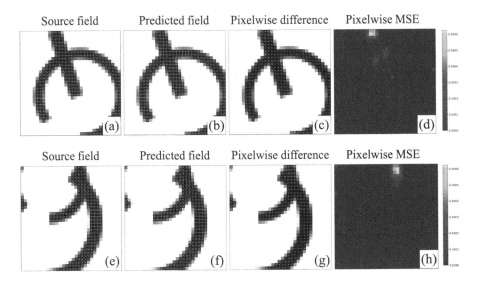

Fig. 5. Learning progress, 5000 images: (a), (e) input drawing and ground truth field, (b), (f) approximated directional field, (c), (g) difference between ground truth and approximated fields, (d), (h) error heatmap. Best viewed in zoom.

4 Conclusion

In this work, we have established two simple but important facts that (1) a general PolyVector field can be efficiently represented using an off-the-shelf CNN, and (2) the same CNN can generalize to unseen instances by training on a synthetic dataset of line drawings. These findings strongly motivate the need for further research on learnable methods for directional fields approximation and processing.

Acknowlegement. The work was supported by the Russian Science Foundation under Grant 19-41-04109.

References

1. Barabanau, I., Artemov, A., Murashkin, V., Burnaev, E.: Monocular 3D object detection via geometric reasoning on keypoints (2019). arXiv preprint arXiv:1905.05618
2. Bessmeltsev, M., Solomon, J.: Vectorization of line drawings via polyvector fields. ACM Trans. Graph. (TOG) **38**(1), 9 (2019)
3. Bokhovkin, A., Burnaev, E.: Boundary loss for remote sensing imagery semantic segmentation. In: Lu, H., Tang, H., Wang, Z. (eds.) ISNN 2019. LNCS, vol. 11555, pp. 388–401. Springer, Cham (2019). https://doi.org/10.1007/978-3-030-22808-8_38
4. Bommes, D., et al.: Quad-mesh generation and processing: a survey. Comput. Graph. Forum **32**, 51–76 (2013)
5. Burnaev, E., Cichocki, A., Osin, V.: Fast multispectral deep fusion networks. Bull. Pol. Acad. Tech. **66**(4), 875–880 (2018)
6. Diamanti, O., Vaxman, A., Panozzo, D., Sorkine-Hornung, O.: Designing N-PolyVector fields with complex polynomials. Comput. Graph. Forum **33**, 1–11 (2014)
7. de Goes, F., Desbrun, M., Tong, Y.: Vector field processing on triangle meshes. In: ACM SIGGRAPH 2016 Courses, p. 27. ACM (2016)
8. Hornik, K., Stinchcombe, M., White, H.: Multilayer feedforward networks are universal approximators. Neural Netw. **2**(5), 359–366 (1989)
9. Ignatiev, V., Trekin, A., Lobachev, V., Potapov, G., Burnaev, E.: Targeted change detection in remote sensing images. In: Proceedings of SPIE (2019)
10. Koch, S., et al.: ABC: a big CAD model dataset for geometric deep learning. In: Proceedings of the IEEE Conference on Computer Vision and Pattern Recognition, pp. 9601–9611 (2019)
11. Kolos, M., Marin, A., Artemov, A., Burnaev, E.: Procedural synthesis of remote sensing images for robust change detection with neural networks. In: Lu, H., Tang, H., Wang, Z. (eds.) ISNN 2019. LNCS, vol. 11555, pp. 371–387. Springer, Cham (2019). https://doi.org/10.1007/978-3-030-22808-8_37
12. Mhaskar, H.N., Poggio, T.: Deep vs. shallow networks: an approximation theory perspective. Anal. Appl. **14**(06), 829–848 (2016)
13. Notchenko, A., Kapushev, Y., Burnaev, E.: Large-scale shape retrieval with sparse 3D convolutional neural networks. In: van der Aalst, W.M.P., et al. (eds.) AIST 2017. LNCS, vol. 10716, pp. 245–254. Springer, Cham (2018). https://doi.org/10.1007/978-3-319-73013-4_23
14. Novikov, G., Trekin, A., Potapov, G., Ignatiev, V., Burnaev, E.: Satellite imagery analysis for operational damage assessment in emergency situations. In: Abramowicz, W., Paschke, A. (eds.) BIS 2018. LNBIP, vol. 320, pp. 347–358. Springer, Cham (2018). https://doi.org/10.1007/978-3-319-93931-5_25
15. Paszke, A., et al.: Automatic differentiation in PyTorch (2017)
16. Ray, N., Nivoliers, V., Lefebvre, S., Lévy, B.: Invisible seams. Comput. Graph. Forum **29**, 1489–1496 (2010)

17. Ronneberger, O., Fischer, P., Brox, T.: U-Net: convolutional networks for biomedical image segmentation. In: Navab, N., Hornegger, J., Wells, W.M., Frangi, A.F. (eds.) MICCAI 2015. LNCS, vol. 9351, pp. 234–241. Springer, Cham (2015). https://doi.org/10.1007/978-3-319-24574-4_28
18. Vaxman, A., et al.: Directional field synthesis, design, and processing. Comput. Graph. Forum **35**, 545–572 (2016)
19. Voinov, O., et al.: Perceptual deep depth super-resolution (2018). arXiv preprint arXiv:1812.09874

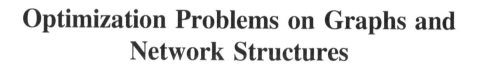

Optimization Problems on Graphs and Network Structures

Fast and Exact Algorithms for Some NP-Hard 2-Clustering Problems in the One-Dimensional Case

Alexander Kel'manov[1,2](✉) [ID] and Vladimir Khandeev[1,2](✉) [ID]

[1] Sobolev Institute of Mathematics, 4 Koptyug Avenue, 630090 Novosibirsk, Russia
[2] Novosibirsk State University, 2 Pirogova Street, 630090 Novosibirsk, Russia
{kelm,khandeev}@math.nsc.ru

Abstract. We consider several well-known optimization 2-clustering (2-partitioning) problems of a finite set of points in Euclidean space. All clustering problems considered are induced by applied problems in Data analysis, Data cleaning, and Data mining. In the general case, all these optimization problems are strongly NP-hard. In the paper, we present a brief overview of the results on the problems computational complexity and on their solvability in the one-dimensional case. We present and propose well-known and new, simple and fast exact algorithms with $\mathcal{O}(N \log N)$ and $\mathcal{O}(N)$ running times for the one-dimensional case of these problems.

Keywords: Euclidean space · 2-clustering · 2-partitioning · NP-hardness · Polynomial-time solvability in the 1D case · Fast exact algorithms

1 Introduction

The subject of this research is some hard-to-solve discrete optimization problems that model some simplest applied problems of cluster analysis and data interpretation. Our goal is to analyze and systematize the issues of constructing efficient algorithms that ensure fast and exact problems-solving in the one-dimensional case.

It is known that in terms of applied problem-solving the computer processing of large-scaling data [1] the existing exact and approximate polynomial-time algorithms having theoretical accuracy guarantee but quadratic and higher running time are often unclaimed in applications [2,3]. In other words, these strongly justified polynomial-time algorithms are not used or are rarely used in practice due to the "large" (quadratic and higher) running-time. On the other hand, many hard-to-solve computer geometric problems arising in the data analysis and interpretation [4,5] are solvable in polynomial time when the space dimension is fixed. At the same time, fast polynomial algorithms for solving problems are efficient tools for finding out the structure (i.e. Data mining) [4] of large

© Springer Nature Switzerland AG 2019
W. M. P. van der Aalst et al. (Eds.) AIST 2019, LNCS 11832, pp. 377–387, 2019.
https://doi.org/10.1007/978-3-030-37334-4_34

data by projecting it into spaces of lower dimension (for example, into three-dimensional space, or a plane, or a number line). These projective mathematical tools are popular among data analytics [6] since these tools allow one to interpret the data by the visual representation. In this connection, the construction of fast algorithms having almost linear or linear running-time to solve special cases (in which the space dimension is fixed) of the problems are important mathematical research directions. This paper belongs to these directions.

The paper has the following structure. Section 2 presents the mathematical formulations of the problems under consideration. Interpretations of problems are presented in the next section for demonstrating their origins and their connection with the problems of data analysis. Section 4 provides a brief overview of the results of the computational complexity of problems. In the next section, we present existing results on the problems polynomial solvability in the case of fixed space dimension. Finally, in Sect. 6, the existing and new algorithms are presented, which find the exact solution of the problems in linear or almost linear time in the one-dimensional case.

2   Problems Formulations

Everywhere below $\mathbb{R}$ denotes the set of real numbers, $\|\cdot\|$ denotes the Euclidean norm, and $\langle\cdot,\cdot\rangle$ denotes the scalar product.

All the problems considered below are the problems of 2-partitioning of input points set. In the problems of searching for one subset, the second cluster is understood as the subset that complements this cluster to the input set. A point in the d-dimensional space is interpreted as the measuring result of a set of d characteristics (features) of an object or as the vector (force), i.e. as the segment directed from the origin to this point in the space.

The problems under consideration have the following formulations.

Problem 1 (Longest Normalized Vector Sum). Given: an N-element set $\mathcal{Y}$ of points in d-dimensional Euclidean space. *Find:* a nonempty subset $\mathcal{C} \subseteq \mathcal{Y}$ such that

$$F(\mathcal{C}) = \frac{1}{|\mathcal{C}|}\left\|\sum_{y \in \mathcal{C}} y\right\|^2 \to \max.$$

Problem 2 (1-Mean and Given 1-Center Clustering). Given: an N-element set $\mathcal{Y}$ of points in d-dimensional Euclidean space. *Find:* a 2-partition of $\mathcal{Y}$ into clusters $\mathcal{C}$ and $\mathcal{Y} \setminus \mathcal{C}$ such that

$$S(\mathcal{C}) = \sum_{y \in \mathcal{C}} \|y - \overline{y}(\mathcal{C})\|^2 + \sum_{y \in \mathcal{Y}\setminus\mathcal{C}} \|y\|^2 \to \min, \tag{1}$$

where $\overline{y}(\mathcal{C}) = \frac{1}{|\mathcal{C}|}\sum_{y \in \mathcal{C}} y$ is the centroid of $\mathcal{C}$.

Problem 3 (Longest M-Vector Sum). Given: an N-element set $\mathcal{Y}$ of points in d-dimensional Euclidean space and some positive integer M. *Find:* a nonempty subset $\mathcal{C} \subseteq \mathcal{Y}$ of size M such that

$$H(\mathcal{C}) = \left\| \sum_{y \in \mathcal{C}} y \right\| \to \max. \tag{2}$$

Problem 4 (Constrained 1-Mean and Given 1-Center Clustering). Given: an N-element set $\mathcal{Y}$ of points in d-dimensional Euclidean space and some positive integer M. *Find:* a 2-partition of $\mathcal{Y}$ into clusters $\mathcal{C}$ and $\mathcal{Y} \setminus \mathcal{C}$ minimizing the value of (1) under constraint $|\mathcal{C}| = M$.

Problem 5 (Longest Vector Sum). Given: an N-element set $\mathcal{Y}$ of points in d-dimensional Euclidean space. *Find:* a subset $\mathcal{C} \subseteq \mathcal{Y}$ maximizing the value of (2).

Problem 6 (M-Variance). Given: an N-element set $\mathcal{Y}$ of points in d-dimensional Euclidean space and some positive integer M. *Find:* a subset $\mathcal{C} \subseteq \mathcal{Y}$ of size M such that

$$Q(\mathcal{C}) = \sum_{y \in \mathcal{C}} \|y - \overline{y}(\mathcal{C})\|^2 \to \min.$$

Problem 7 (Maximum Size Subset of Points with Constrained Variance). Given: an N-element set $\mathcal{Y}$ of points in d-dimensional Euclidean space and some real number $\alpha \in (0, 1)$. *Find:* a subset $\mathcal{C} \subset \mathcal{Y}$ of the largest size such that

$$Q(\mathcal{C}) \leq \alpha \sum_{y \in \mathcal{Y}} \|y - \overline{y}(\mathcal{Y})\|^2,$$

where $\overline{y}(\mathcal{Y}) = \frac{1}{|\mathcal{Y}|} \sum_{y \in \mathcal{Y}} y$ is the centroid of $\mathcal{Y}$.

Problem 8 (Smallest M-Enclosing Ball). Given: an N-element set $\mathcal{Y}$ of points in d-dimensional Euclidean space and some positive integer number M. *Find:* a minimum radius ball covering M points.

3   Interpretations and Origins of the Problems

All of the formulated optimization problems have simple interpretations in the geometric, statistical, physical, biomedical, geophysical, industrial, economic, anti-terrorism and social terms. One can find some interpretations in the papers cited below. An interested reader can easily give his own interpretation. In this paper, we limit ourselves to a few simple interpretations.

Problems 1–4 arose in connection with the solution of an applied signal processing problem, namely, the problem of joint detecting a quasiperiodically repeating pulse of unknown shape and evaluating this shape under Gaussian noise with zero mean [7–9]. Apparently, the first note on these problems was made in [7]. In these problems, the cluster center specified at the origin corresponds to the mean equal to zero.

It should be noted that simpler optimization problems, which are induced by the applied problems of detecting and discriminating of pulses with given forms in the noise conditions, are characteristic, in particular, for radar, electronic reconnaissance, hydroacoustics, geophysics, technical and medical diagnostics, and Space monitoring (see, for example, [10–12]).

The Problem 1 objective function can be rewritten as

$$F(\mathcal{C}) = \frac{1}{|\mathcal{C}|} \sum_{y \in \mathcal{C}} \sum_{x \in \mathcal{C}} \langle y, x \rangle = \frac{1}{|\mathcal{C}|} \sum_{y \in \mathcal{C}} \left\langle y, \sum_{x \in \mathcal{C}} x \right\rangle = \sum_{y \in \mathcal{C}} \langle y, \overline{y}(\mathcal{C}) \rangle.$$

Therefore, maximization Problem 1 can be interpreted as a search for a subset $\mathcal{C}$ of objects (forces) that are most similar to each other in the terms of an average value of the sum of all scalar products. Another interpretation is the search for a subset of forces that are most co-aimed with the vector from the origin to the point $\overline{y}(\mathcal{C})$, i.e. from the origin to an unknown centroid. Maximization Problems 1 and 5 have similar interpretations.

Apparently, in [13], Problem 6 was first formulated. This problem models a simplest data analysis problem, namely finding a subset of M similar objects in the set of N objects. In this problem, the similarity of objects is interpreted as the minimum total quadratic scatter of points in a set with respect to some unknown "average" object (centroid), which may not belong to the input set. Equivalent treatment of similarity is minimum of the sum of squares of all possible pairwise distances between objects since for the objective function of the Problem 6, the following equality holds

$$Q(\mathcal{C}) = \frac{1}{2|\mathcal{C}|} \sum_{x \in \mathcal{C}} \sum_{y \in \mathcal{C}} \|x - y\|^2.$$

Problem 7 models [14] the search for the largest subset of similar objects under the upper bound (restriction) on the similarity criterion of Problem 6, i.e. on the total quadratic scatter of points in the desired cluster. In accordance with this restriction, Problem 7 can be interpreted as clearing data from so-called outliers that violate intracluster homogeneity (see, for example, [15–17]). As a result of solving the problem, all data that have significant quadratic scatter will belong to the complementary cluster. In this problem, the degree of the desired cluster homogeneity is governed by the given number α.

Finally, Problem 8, formulated in [18], as a generalization of the well-known problem of the Chebyshev center, has a simple geometric formulation that does not require any explanation. On the other hand, in applications, this problem arises whenever it is necessary to cover (for example, surround or locate in the territory) a given number of objects in the conditions of limited resources (for example, financial or energy).

4 The Computational Complexity of Problems: Existing Results

In [19] the authors proved the strong NP-hardness of Problem 1 by polynomial reducibility of the known NP-hard problem 3-Satisfiability [20] to Problem 1. This result implies the strong NP-hardness of Problem 3 since the objective functions of these problems are related by equality

$$F(\mathcal{C}) = \frac{1}{|\mathcal{C}|} H^2(\mathcal{C}).$$

Indeed, it follows from this equality that polynomial solvability of Problem 3 would imply polynomial solvability of Problem 1 (it would be sufficient to iterate over the finite number of admissible M). Note that chronologically, NP-hardness of Problem 3 was first proved (however, NP-hardness of Problem 3 does not imply NP-hardness of Problem 1). Recall that for proving the intractability of Problem 3, the authors of [8,9,21] constructed polynomial-time reduction of the known NP-hard Clique problem [20] to Problem 3.

By virtue of equality

$$S(\mathcal{C}) = \sum_{y \in \mathcal{Y}} \|y\|^2 - F(\mathcal{C}), \tag{3}$$

the strong NP-hardness of Problem 2 follows from its polynomial equivalence to Problem 1. From the strong NP-hardness of Problem 2 follows the strong NP-hardness of Problem 4 for the same reason as the strong NP-hardness of Problem 3 follows from the strong NP-hardness of Problem 1. Indeed, polynomial solvability of Problem 4 would imply polynomial solvability of Problem 2.

In [22] the authors proved the strong NP-hardness of Problem 5 by polynomial reducibility of the known NP-hard problem 3-Satisfiability [20] to Problem 5.

Further, the paper [23] presents the proof of the strong NP-hardness of Problem 6. In this paper there is a simple proof of polynomial reducibility to Problem 6 of the well-known [24] NP-hard Clique problem on a homogeneous graph with non-fixed degree.

The authors of [14] proved the strong NP-hardness of Problem 7 by showing that decision forms of Problems 6 and 7 are equivalent.

Finally, the paper [18] presents the proof of the strong NP-hardness of Problem 8. To do this, the authors of the cited paper have shown polynomial reducibility to Problem 8 of Clique problem.

5 Exact Algorithms for the Problems in the Multidimensional Case: Existing Results

Exact algorithms of exponential time complexity were constructed for multidimensional case of Problems 1–6 in a number of papers. These algorithms are polynomial for the case of fixed space dimension (or for the case of space dimension bounded from above by some constant).

For Problem 1, an algorithm given in [25] finds the exact solution of the problem in $\mathcal{O}(d^2 N^{2d})$ time. The authors of [29] proposed an accelerated algorithm with $\mathcal{O}(dN^{d+1})$ running time.

Since Problem 2 is polynomially equivalent to Problem 1, the exact solution of Problem 2 can be found in the same time as the exact solution of Problem 1.

Polynomial solvability of Problem 3 in the case of fixed space dimension follows from [28]. The authors of [25] and [29] presented exact algorithms for Problem 3 with running time $\mathcal{O}(d^2 N^{2d})$ and $\mathcal{O}(dN^{d+1})$, respectively.

Problem 4 is polynomially equivalent to Problem 3. Therefore, the solution of Problem 4 can be found in the same time as the solution of Problem 3.

Further, polynomial solvability of Problem 5 in the case of fixed space dimension follows from [27]. The authors of [26] presented an algorithm with $\mathcal{O}(d^2 N^d)$ running time. An improved algorithm with $\mathcal{O}(dN^{d+1})$ running time is proposed in [29]. In addition, the author of [30] presented a faster algorithm with $\mathcal{O}(N^{d-1}(d + \log N))$ running time for the case $d \geq 2$.

Algorithms proposed in [13,29] find exact solution of Problem 6 in $\mathcal{O}(dN^{d+1})$ time. The feature of the algorithm proposed in [29] is that it allows one to find solutions for all admissible values of M at once.

An exact algorithm for Problem 7, obviously, can be obtained from the exact algorithm proposed in [29] for Problem 6. Running time of this algorithm is $\mathcal{O}(dN^{d+1})$.

Finally, the issue of polynomial time solvability of Problem 8 for an arbitrary but fixed dimension d of space is open till now.

It follows from the above results that for $d = 1$ these algorithms find solutions in time which quadratically depends on the power N of the input set.

Below we present simple and fast exact algorithms that find solutions of one dimensional case of the problems in $\mathcal{O}(N)$ (for Problems 3, 4, 5) or $\mathcal{O}(N \log N)$ (for Problems 2, 7, 8) time. Here, for completeness, we present known algorithms for Problems 1 and 6, the time complexity of which is $\mathcal{O}(N \log N)$.

6   Fast and Exact Algorithms for the Problems in the One-Dimensional Case

Hereafter one-dimensional ($d = 1$) cases of Problems 1–8 we will denote as Problem $X - 1D$, where X is the number of the problem.

Let us formulate algorithms for solving the problems.

Algorithm $\mathcal{A}_1$ for Problem 1-1D.
Input: the set $\mathcal{Y}$.
Step 1. Split $\mathcal{Y}$ into the two subsets $\mathcal{Y}^+ = \{y \in \mathcal{Y} \mid y > 0\}$ and $\mathcal{Y}^- = \{y \in \mathcal{Y} \mid y < 0\}$. Sort their elements so that $\mathcal{Y}^+ = \{y_1^+ \geq y_2^+ \geq \ldots \geq y_{|\mathcal{Y}^+|}^+ > 0\}$ and $\mathcal{Y}^- = \{y_1^- \leq y_2^- \leq \ldots \leq y_{|\mathcal{Y}^-|}^- < 0\}$.

Step 2. Calculate $F_i^+ = F(\{y_1^+, y_2^+, \ldots, y_i^+\})$, $i = 1, \ldots, |\mathcal{Y}^+|$; put $F^+ = \max\{F_1^+, \ldots, F_{|\mathcal{Y}^+|}^+\}$ and $U^+ = \{\{y_1^+, y_2^+, \ldots, y_i^+\}| F_i^+ = F^+\}$.

Step 3. Calculate $F_i^- = F(\{y_1^-, y_2^-, \ldots, y_i^-\})$, $i = 1, \ldots, |\mathcal{Y}^-|$; put $F^- = \max\{F_1^-, \ldots, F_{|\mathcal{Y}^-|}^-\}$ and $U^- = \{\{y_1^-, y_2^-, \ldots, y_i^-\}| F_i^- = F^-\}$.

Step 4. Put $F_A = \max\{F^+, F^-\}$, and also $\mathcal{C}_A = U^+$ if $F^+ \geq F^-$, and $\mathcal{C}_A = U^-$ if $F^+ < F^-$.

Output: the subset $\mathcal{C}_A$, the value F_A.

Proposition 1. *Algorithm $\mathcal{A}_1$ finds an optimal solution of Problem 1-1D in $\mathcal{O}(N \log N)$ time.*

This algorithm was proposed in [19] for construction of the approximation scheme which is polynomial in the case of fixed space dimension. The same paper established the accuracy and running time (determined by the sorting time) of the algorithm.

Algorithm $\mathcal{A}_2$ for Problem 2-1D.
Input: the set $\mathcal{Y}$.
Step 1. Find the solution $\mathcal{C}_A$ of Problem 1 using Algorithm $\mathcal{A}_1$.
Step 2. Calculate $S_A = \sum_{y \in \mathcal{Y}} y^2 - F_A(\mathcal{C}_A)$.
Output: the subset $\mathcal{C}_A$, the value S_A.

Proposition 2. *Algorithm $\mathcal{A}_2$ finds an optimal solution of Problem 2-1D in $\mathcal{O}(N \log N)$ time.*

The validity of the statement follows from the fact that, in accordance with (3), in the one-dimensional case the following holds

$$S(\mathcal{C}) = \sum_{y \in \mathcal{Y}} y^2 - F(\mathcal{C}).$$

Algorithm $\mathcal{A}_3$ for Problem 3-1D.
Input: the set $\mathcal{Y}$, positive integer M.
Step 1. Form a subset $\mathcal{C}_1$ of the M largest elements of $\mathcal{Y}$. Calculate $H(\mathcal{C}_1)$.
Step 2. Form a subset $\mathcal{C}_2$ of the M smallest elements of $\mathcal{Y}$. Calculate $H(\mathcal{C}_2)$.
Step 3. Put $\mathcal{C}_A = \mathcal{C}_1$ and $H_A = H(\mathcal{C}_1)$ if $H(\mathcal{C}_1) \leq H(\mathcal{C}_2)$. Otherwise put $\mathcal{C}_A = \mathcal{C}_2$ and $H_A = H(\mathcal{C}_2)$.
Output: the subset $\mathcal{C}_A$, the value H_A.

Proposition 3. *Algorithm $\mathcal{A}_3$ finds an optimal solution of Problem 3-1D in $\mathcal{O}(N)$ time.*

The accuracy of the algorithm follows from the fact that in the one-dimensional case for the function (2) the following holds

$$H(\mathcal{C}) = \left| \sum_{y \in \mathcal{C}} y \right|. \tag{4}$$

The time complexity of selecting M largest (or smallest) elements determines the running time of the algorithm. This selecting can be made in $\mathcal{O}(N)$ operations without sorting (see, for example, [31]).

Algorithm $\mathcal{A}_4$ for Problem 4-1D.
Input: the set $\mathcal{Y}$, positive integer M.
Step 1. Find the solution $\mathcal{C}_A$ of Problem 3 using Algorithm $\mathcal{A}_3$.
Step 2. Calculate $S_A = \sum_{y \in \mathcal{Y}} y^2 - F_A(\mathcal{C}_A)$.
Output: the subset $\mathcal{C}_A$, the value S_A.

Proposition 4. *Algorithm $\mathcal{A}_4$ finds an optimal solution of Problem 4-1D in $\mathcal{O}(N)$ time.*

The validity of the statement follows from the polynomial equivalence of Problems 3 and 4.

Algorithm $\mathcal{A}_5$ for Problem 5-1D.
Input: the set $\mathcal{Y}$.
Step 1. Form the subset $\mathcal{C}_1 = \{y \in \mathcal{Y} \mid y \geq 0\}$. Calculate $H(\mathcal{C}_1)$.
Step 2. Form the subset $\mathcal{C}_2 = \{y \in \mathcal{Y} \mid y \leq 0\}$. Calculate $H(\mathcal{C}_2)$.
Step 3. Put $\mathcal{C}_A = \mathcal{C}_1$ and $H_A = H(\mathcal{C}_1)$ if $H(\mathcal{C}_1) \leq H(\mathcal{C}_2)$. Otherwise put $\mathcal{C}_A = \mathcal{C}_2$ and $H_A = H(\mathcal{C}_2)$.
Output: the subset $\mathcal{C}_A$, the value H_A.

Proposition 5. *Algorithm $\mathcal{A}_5$ finds an optimal solution of Problem 5-1D in $\mathcal{O}(N)$ time.*

The accuracy of the algorithm follows from (4). The running time of the algorithm follows from the fact that constructing subsets $\mathcal{C}_1$ and $\mathcal{C}_2$ can be done in time $\mathcal{O}(N)$.

Algorithm $\mathcal{A}_6$ for Problem 6-1D.
Input: the subset $\mathcal{Y}$, positive integer M.
Step 0. Put $m = 1$; $Q_A = +\infty$; $\mathcal{C}_A = \emptyset$.
Step 1. Using sorting form the tuple $\mathcal{Y}_{1,N} = (y_1, \ldots, y_N)$, where $y_1 < \ldots < y_N$.
Step 2. Calculate $f_{m,\,m+M-1}$ using formula

$$f_{i,j} = \sum_{k=i}^{j} (y_k - \overline{y}(\mathcal{Y}_{i,j}))^2 \equiv \sum_{k=i}^{j} y_k^2 - \frac{1}{j-i+1} \left(\sum_{k=i}^{j} y_k \right)^2, \tag{5}$$

where

$$\mathcal{Y}_{i,j} = (y_i, \ldots, y_j),$$

and

$$\overline{y}(\mathcal{Y}_{i,j}) = \frac{1}{j-i+1} \sum_{k=i}^{j} y_k$$

is the centroid of $\mathcal{Y}_{i,j}$, at $i = m$ and $j = m + M - 1$.

Step 3. If $f_{m,m+M-1} \le Q_A$ then put $Q_A = f_{m,m+M-1}$, $\mathcal{C}_A = \mathcal{Y}_{m,m+M-1}$.
Step 4. If $m < N - M + 1$ then put $m = m + 1$ and go to Step 1; otherwise go to output.
Output: the subset $\mathcal{C}_A$, the value Q_A.

Proposition 6. *Algorithm $\mathcal{A}_6$ finds an optimal solution of Problem 6-1D in $\mathcal{O}(N \log N)$ time.*

This statement is based on the fact that each value of $\sum_{k=i}^{j} y_k$ and $\sum_{k=i}^{j} y_k^2$ can be found in $\mathcal{O}(1)$ time using sliding window sums. This algorithm was recently justified in [32].

Algorithm $\mathcal{A}_7$ for Problem 7-1D.
Input: the set $\mathcal{Y}$, real number α.
Step 0. Put $m = 1$, $M = 0$, $M_A = 0$. Calculate $B = \alpha \sum_{y \in \mathcal{Y}} \|y - \bar{y}(\mathcal{Y})\|^2$.
Step 1. Using sorting form the tuple $\mathcal{Y}_{1,N} = (y_1, \ldots, y_N)$, where $y_1 < \ldots < y_N$.
Step 2. Calculate $f_{m,m+M}$ using formula (5) at $i = m$ and $j = m + M$. If $f_{m,m+M} \le B$ then go to Step 3. Otherwise go to Step 5.
Step 3. Put $M = M + 1$. If $M > M_A$ then put $M_A = M$, $\mathcal{C}_A = \mathcal{Y}_{m,m+M-1}$.
Step 4. If $m + M \le N$ then go to Step 2. Otherwise go to output.
Step 5. If $m < N$ then put $m = m + 1$, $M = M - 1$ and go to Step 2. Otherwise go to output.
Output: the subset $\mathcal{C}_A$, the value M_A.

Proposition 7. *Algorithm $\mathcal{A}_7$ finds an optimal solution of Problem 7-1D in $\mathcal{O}(N \log N)$ time.*

The algorithm accuracy follows from the monotonicity property [14] of the function $Q(\mathcal{C})$. The sorting determines the algorithm running time since the calculations in Step 2 can be performed in $\mathcal{O}(1)$ time using prefix summation and this step is performed no more than $\mathcal{O}(N)$ times.

Algorithm $\mathcal{A}_8$ for Problem 8-1D.
Input: the set $\mathcal{Y}$, positive integer M.
Step 0. Put $m = 1$; $r_A = 0$; $\mathcal{C}_A = \emptyset$.
Step 1. Using sorting form the tuple $\mathcal{Y}_{1,N} = (y_1, \ldots, y_N)$, where $y_1 < \ldots < y_N$.
Step 2. Calculate $r_{m,m+M-1} = (y_{m+M-1} - y_m)/2$.
Step 3. If $r_{m,m+M-1} > r_A$ then put $r_A = r_{m,m+M-1}$, $\mathcal{C}_A = \mathcal{Y}_{m,m+M-1}$.
Step 4. If $m < N - M + 1$ then put $m = m + 1$ and go to Step 1; Otherwise go to output.
Output: the subset $\mathcal{C}_A$, the value r_A.

Proposition 8. *Algorithm $\mathcal{A}_8$ finds an optimal solution of Problem 8-1D in $\mathcal{O}(N \log N)$ time.*

The algorithm accuracy follows from the fact that in the one-dimensional case a minimum radius ball enclosing points $y_i, y_{i+1}, \ldots, y_j$, where $y_i < y_{i+1} < \ldots < y_j$, has a radius of $(y_j - y_i)/2$. The algorithm running time is determined by sorting.

7   Conclusion

The paper provides a brief overview of the complexity of some recently identified optimization problems of 2-clustering a finite set of points in Euclidean space. We present fast and exact algorithms for the one-dimensional case of these problems. In our opinion, these algorithms will serve as a good tool for solving the problems of projective analysis and interpretation of big data.

Acknowledgments. This work was supported by the Russian Foundation for Basic Research, project nos. 18-31-00398, 19-01-00308, and 19-07-00397, by the Russian Academy of Science (the Program of basic research), project 0314-2019-0015, and by the Russian Ministry of Science and Education under the 5-100 Excellence Programme.

References

1. Chen, M., Mao, S., Zhang, Y., Leung, V.C.: Big Data Related Technologies, Challenges and Future Prospects. Springer, Cham (2014). https://doi.org/10.1007/978-3-319-06245-7
2. Inmon, W.H.: Oracle: Building High Perfomance Online Systems. QED Information Sciences, Wellesley (1989)
3. Inmon, W.H.: Building the Data Warehouse, 4th edn. Wiley, Hoboken (2005)
4. Aggarwal, C.C.: Data Mining: The Textbook. Springer, Cham (2015). https://doi.org/10.1007/978-3-319-14142-8
5. Bishop, C.M.: Pattern Recognition and Machine Learning. Springer, New York (2006)
6. Inmon, W.H., Nesavich, A.: Tapping into Unstructured Data: Integrating Unstructured Data and Textual Analytics into Business Intelligence. Prentice-Hall, Upper Saddle River (2008)
7. Kel'manov, A.V., Khamidullin, S.A., Kel'manova, M.A.: Joint finding and evaluation of a repeating fragment in noised number sequence with given number of quasiperiodic repetitions. In: Book of Abstract of the Russian Conference "Discrete Analysis and Operations Research" (DAOR 2004), p. 185. Sobolev Institute of Mathematics SB RAN, Novosibirsk (2004). (in Russian)
8. Gimadi, E.Kh., Kel'manov, A.V., Kel'manova, M.A., Khamidullin, S.A.: A posteriori detection of a quasi periodic fragment in numerical sequences with given number of recurrences. Sib. J. Ind. Math. **9**(1(25)), 55–74 (2006) (in Russian)
9. Gimadi, E.Kh., Kel'manov, A.V., Kel'manova, M.A., Khamidullin, S.A.: A posteriori detecting a quasiperiodic fragment in a numerical sequence. Pattern Recogn. Image Anal. **18**(1), 30–42 (2008)
10. Kel'manov, A.V., Khamidullin, S.A.: Posterior detection of a given number of identical subsequences in a quasi-periodic sequence. Comput. Math. Math. Phys. **41**(5), 762–774 (2001)
11. Kel'manov, A.V., Jeon, B.: A posteriori joint detection and discrimination of pulses in a quasiperiodic pulse train. IEEE Trans. Sig. Proc. **52**(3), 645–656 (2004)
12. Carter, J.A., Agol, E., et al.: Kepler-36: a pair of planets with neighboring orbits and dissimilar densities. Science **337**(6094), 556–559 (2012)
13. Aggarwal, H., Imai, N., Katoh, N., Suri, S.: Finding k points with minimum diameter and related problems. J. Algorithms. **12**(1), 38–56 (1991)

14. Ageev, A.A., Kel'manov, A.V., Pyatkin, A.V., Khamidullin, S.A., Shenmaier, V.V.: Approximation polynomial algorithm for the data editing and data cleaning problem. Pattern Recogn. Image Anal. **27**(3), 365–370 (2017)
15. Osborne, J.W.: Best Practices in Data Cleaning: A Complete Guide to Everything You Need to Do Before and After Collecting Your Data, 1st edn. SAGE Publication, Inc., Los Angeles (2013)
16. de Waal, T., Pannekoek, J., Scholtus, S.: Handbook of Statistical Data Editing and Imputation. Wiley, Hoboken (2011)
17. Greco, L.: Robust Methods for Data Reduction Alessio Farcomeni. Chapman and Hall/CRC, Boca Raton (2015)
18. Shenmaier, V.V.: Complexity and approximation of the smallest k-enclosing ball problem. J. Appl. Ind. Math. **7**(3), 444–448 (2013)
19. Kel'manov, A.V., Pyatkin, A.V.: On a version of the problem of choosing a vector subset. J. Appl. Ind. Math. **3**(4), 447–455 (2009)
20. Garey, M.R., Johnson, D.S.: Computers and Intractability: A Guide to the Theory of NP-Completeness. Freeman, San Francisco (1979)
21. Baburin, A.E., Gimadi, E.Kh., Glebov, N.I., Pyatkin, A.V.: The problem of finding a subset of vectors with the maximum total weight. J. Appl. Ind. Math. **2**(1), 32–38 (2008)
22. Pyatkin, A.V.: On complexity of a choice problem of the vector subset with the maximum sum length. J. Appl. Ind. Math. **4**(4), 549–552 (2010)
23. Kel'manov, A.V., Pyatkin, A.V.: NP-completeness of some problems of choosing a vector subset. J. Appl. Ind. Math. **5**(3), 352–357 (2011)
24. Papadimitriou, C.H.: Computational Complexity. Addison-Wesley, New-York (1994)
25. Gimadi, E.K., Pyatkin, A.V., Rykov, I.A.: On polynomial solvability of some problems of a vector subset choice in a Euclidean space of fixed dimension. J. Appl. Ind. Math. **4**(1), 48–53 (2010)
26. Baburin, A.E., Pyatkin, A.V.: Polynomial algorithms for solving the vector sum problem. J. Appl. Ind. Math. **1**(3), 268–272 (2007)
27. Hwang, F.K., Onn, S., Rothblum, U.G.: Polynomial time algorithm for shaped partition problems. SIAM J. Optim. **10**(1), 70–81 (1999)
28. Onn, S., Schulman, L.J.: The vector partition problem for convex objective functions. Math. Oper. Res. **26**(3), 583–590 (2001)
29. Shenmaier, V.V.: Solving some vector subset problems by Voronoi diagrams. J. Appl. Ind. Math. **10**(4), 560–566 (2016)
30. Shenmaier, V.V.: An exact algorithm for finding a vector subset with the longest sum. J. Appl. Ind. Math. **11**(4), 584–593 (2017)
31. Wirth, I.: Algorithms + Data Structures = Programs. Prentice Hall, Upper Saddle River (1976)
32. Kel'manov, A.V., Ruzankin, P.S.: Improved exact algorithm for M-variance problem in the one-dimensional case. Pattern Recogn. Image Anal. (2019, accepted)

Efficient PTAS for the Euclidean Capacitated Vehicle Routing Problem with Non-uniform Non-splittable Demand

Michael Khachay[1,2,3]([⊠]) and Yuri Ogorodnikov[1,2]([⊠])

[1] Krasovsky Institute of Mathematics and Mechanics, Ekaterinburg, Russia
{mkhachay,yogorodnikov}@imm.uran.ru
[2] Ural Federal University, Ekaterinburg, Russia
[3] Omsk State Technical University, Omsk, Russia

Abstract. The Capacitated Vehicle Routing Problem (CVRP) is the well-known combinatorial optimization problem having numerous relevant applications in operations research. As known, CVRP is strongly NP-hard even in the Euclidean plane, APX-hard for an arbitrary metric, and can be approximated in polynomial time with any accuracy in the Euclidean spaces of any fixed dimension. In particular, for the several special cases of the planar Euclidean CVRP there are known Polynomial Time Approximation Schemes (PTAS) stemming from the seminal papers by M. Haimovich, A. Rinnooy Kan and S. Arora. Although, these results appear to be promising and make a solid contribution to the field of algorithmic analysis of routing problems, all of them are restricted to the special case of the CVRP, where all customers have splittable or even unit demand, which seems to be far from the practice.

In this paper, to the best of our knowledge, we propose the first Efficient Polynomial Time Approximation Scheme (EPTAS) for this problem in the case of non-uniform non-splittable demand.

Keywords: Capacitated Vehicle Routing Problem · Non-splittable demand · Polynomial Time Approximation Scheme

1 Introduction

We consider the Capacitated Vehicle Routing Problem with non-splittable demand (CVRP-NSD), whose instance is given by a set of *customers* $X = \{x_1, x_2, \ldots, x_n\}$, each of them has an integer demand and an unbound fleet of identical *vehicles* of same capacity q located initially in a single depot y. For any customer x, the demand $d(x)$ is assumed to be non-splittable, i.e. it should be serviced once, by a single vehicle *route*. Similarly to the classic CVRP, each route departs from and arrive to the depot y and fulfills the *capacity constraint*, i.e. the accumulated demand serviced by this route should not exceed the bound q. For the sake of simplicity, we assume that q is divisible by any customer demand $d(x)$. The goal is to construct a number of routes of the minimum total transportation cost that service all the demand and fulfill the capacity constraint.

© Springer Nature Switzerland AG 2019
W. M. P. van der Aalst et al. (Eds.) AIST 2019, LNCS 11832, pp. 388–398, 2019.
https://doi.org/10.1007/978-3-030-37334-4_35

1.1 Problem Statement

Mathematically, an instance of the CVRP-NSD is specified by a complete node- and edge-weighted graph $G(X \cup \{y\}, E, d, w)$, a capacity bound q, and a finite subset $D = \{d^1, \ldots, d^J\} \subset \mathbb{N}$ of feasible demands. The node-weighting function $d: X \to D$ assigns to any customer x its demand $d(x)$. In turn, the symmetric edge-weighting function $w: E \to \mathbb{R}_+$ defines transportation cost $w(v_1, v_2)$ for any pair $\{u, v\}$ of nodes. Although, some results presented can be easily extended to the general case, in this paper, we assume that $w(u, v) = \|u - v\|_2$ and each $d^j \in D$ is a divisor of the capacity bound q.

A *feasible route* is a simple cycle $R = y, x_{i_1}, \ldots, x_{i_s}, y$ in the graph G satisfying the capacity constraint, i.e.

$$d(x_{i_1}) + \ldots + d(x_{i_s}) \leq q.$$

Assuming that, for any feasible route R, its transportation cost $w(R)$ is defined by the equation

$$w(R) = w(y, x_{i_1}) + w(x_{i_1}, x_{i_2}) + \ldots + w(x_{i_s}, y),$$

it is required to construct a set of feasible routes $S = \{R_1, \ldots, R_l\}$ of the minimum total transportation cost that fulfill the entire customer demand.

In the sequel, we use the following notation: CVRP-NSD(X') and TSP(X') for subinstances induced by some customer subset $X' \subset X$ of the problem in question and the auxiliary Traveling Salesman Problem (TSP), respectively; CVRP-NSD* and TSP* for optimum values of the appropriate problems.

1.2 Related Work

The CVRP-NSD is an extension of the famous Capacitated Vehicle Routing Problem (CVRP), introduced by Dantzig and Ramser [7]. Significance of CVRP for the theory of combinatorial optimization and operations research (see, e.g., [29]) adopts high research interest to this problem both among algorithm designers and practitioners.

There are known three main approaches to design the algorithms for the CVRP. The first approach is based on the reduction of a given instance to an appropriate Integer or Mixed Integer Program (IP or MIP) [5,6,15], which can be solved to optimality by some kind of universal solver. Since the CVRP is strongly NP-hard, this approach is restricted to the instances of rather small size.

The second approach deals with design and adaptation of a numerous heuristics and meta-heuristics, among them are local-search [13], genetic [30], memetic [4,23], ant [24] and bee [25] colony algorithms and their combinations (see, e.g., [11]). It is known that heuristic methods can solve successfully some instances of really big size that come from the practice. Unfortunately, for any practical setting, the actual performance of these methods should be evaluated numerically, since nothing about their theoretical approximation or running time complexity bounds is known so far.

At last, the third approach is concentrated on design of polynomial time approximation algorithms and schemes with theoretical bounds. The main advantage of such algorithms is that, for any instance of the considered problem, their performance (with respect to both the accuracy and running time) is guaranteed[1].

It is known, that the CVRP is NP-hard even in the Euclidean plane [26]. Although, the metric CVRP is APX-hard (even for any fixed capacity $q > 2$), its geometric settings admit quasi-polynomial and even polynomial time approximation schemes. The first approximation scheme for the CVRP in the Euclidean plane was introduced by Haimovich and Rinnooy Kan. In their celebrated paper [12], they proved that their algorithm is a Polynomial Time Approximation Scheme (PTAS), when the capacity $q = o(\log \log n)$. Later, Asano et al. [3] extended their approach and proposed the PTAS for more week restriction $q = O(\log n / \log \log n)$. Further, Das and Mathieu [8,9] proposed a Quasi-Polynomial Time Approximation Scheme (QPTAS) for the general case of the planar Euclidean CVRP extending the famous Arora's result for the Euclidean TSP [2]. Employing their result as a "black box", Adamaszek et al. [1] proposed PTAS for the case, when $q \leq 2^{\log^\delta n}$, where $\delta = \delta(\varepsilon)$. Also, there were approximation schemes for several extensions of the planar CVRP, among them are Efficient Polynomial Time Approximation Schemes (EPTAS) for the CVRP in the Euclidean spaces of any fixed dimension and multiple depots [16,21,22], QPTAS [27,28] and EPTASs [17,18,20] for the CVRP with Time Windows.

Although, the aforementioned approximation results appear to be promising, all of them address the simplest case of the Capacitated Vehicle Routing Problem with uniform demand. Without loss of generality, in this case, one can assume that the demand of any customer is unit. Evidently, such an assumption is far from the real-life applications, where demand is non-uniform as a rule.

Recently, the PTAS for the Euclidean CVRP with Time Windows and splittable non-uniform demand was proposed [19]. In this paper, to the best of our knowledge, we propose the first approximation scheme for the restriction of the Euclidean CVRP with non-splittable demand, where each customer demand is a divisor of the capacity bound. We show that the scheme is PTAS if capacity and the number of distinct demands are $o(\log \log n)$ and EPTAS for any constant values of these parameters. Hereinafter, we employ the usual notation Efficient Polynomial Time Approximation Scheme (EPTAS) for a PTAS with running time bound $f(1/\varepsilon) \cdot n^{O(1)}$ for an arbitrary computable function f (see, e.g. [10]).

The rest of the paper is organized as follows. Section 2 describes the basic points of our proposed scheme and the main theorem. In turn, Sect. 3 contains the auxiliary lemmas and time complexity description needed for the proof of our main result. At last, Sect. 4 summarize our work and annotate the future work.

---

[1] Although, in some cases, they can perform even better.

2   Main Idea

Our approach extends the famous framework introduced by Haimovich and Rinnooy Kan in their seminal paper [12]. In a nutshell, our approximation scheme consists of several consecutive stages, which can be specified as follows.

(i) Relabel the customer set $\{x_1, \ldots, x_n\}$ by descending their distances $r(x_i) = w(x_i, y)$ from the depot y.

(ii) Given an accuracy $\varepsilon > 0$, find the smallest number $k = k(\varepsilon, q, J, \rho)$ independent on the number of customers, for which

$$q(2k + (\rho + 1)J)\frac{r_k}{\sum_{x \in X} d(x)r(x)} + 2\sqrt{\pi}\rho q J \sqrt{\frac{r_k}{\sum_{x \in X} d(x)r(x)}} < \varepsilon, \quad (1)$$

where q is the given capacity bound, $J = |D|$, and $\rho \geq 1$ is an approximation factor of the algorithm used for the the auxiliary instance of the metric Traveling Salesman Problem (TSP).

(iii) If $n < k$, the instance is assumed to be "small" and solved to optimality by the dynamic programming [14]. Otherwise, we split the customer set X into subsets $X(k) = \{x_1, \ldots, x_{k-1}\}$ and $X'(k) = \{x_k, \ldots, x_n\}$ of *outer* and *inner* customers respectively. After that, the subset $X'(k)$ is partitioned into subsets $X'_1 \cup \ldots \cup X'_J$, where X'_j comprises inner customers having the same demand d^j. Such a way, we decompose the initial instance CVRP-NSD(X) into one outer subinstance CVRP-NSD($X(k)$) and (at most) J inner subinstances CVRP-NSD(X'_j), each of them can be solved in parallel.

(iv) We output the union of the following partial solutions $S_0^{DP}, S_1^{ITP}, \ldots, S_J^{ITP}$, where S_0^{DP} is an optimal solution of the subinstance CVRP-NSD($X(k)$) obtained by the dynamic programming [14] and S_j^{ITP} is an approximate solution of the auxiliary instance of the classic unit-demand CVRP equivalent to the subinstance CVRP-NSD(X'_j) constructed by the famous Iterative Tour Partition technique [12].

Our main result is claimed in the following theorem.

Theorem 1. *For any $\varepsilon > 0$ an $(1 + \varepsilon)$-approximate solution for the Euclidean CVRP-NSD can be obtained in time*

$$O\left(qk^3 2^k\right) + TIME(TSP, \rho, n) + O(n^2),$$

where

$$k = k(\varepsilon, q, J, \rho) = O\left((\rho + 1) \cdot J \cdot \exp\left(\frac{2q}{\varepsilon}\right) \cdot \left(\exp\left(\sqrt{\frac{qJ}{\varepsilon}}\right)\right)^{\sqrt{\pi\rho}}\right)$$

and $TIME(TSP, \rho, n)$ is the time needed to approximate the metric TSP within the approximation factor ρ.

Remark 1. The proposed scheme is PTAS for $q = o(\log \log n)$ and $J = o(\log \log n)$ and becomes EPTAS for any fixed q and J.

3   Sketch of the Proof

To prove our main result we need several technical lemmas. Lemma 1 extends the known result for the classic CVRP [12] and provides a lower bound for the optimum value of the metric CVRP-NSD.

Lemma 1. *For any instance CVRP-NSD(X) of the metric CVRP-NSD the following equation*

$$\text{CVRP-NSD}^*(X) \geq \max \left\{ 2 \cdot r_1, \text{TSP}^*(X), \frac{2}{q} \sum_{x \in X} d(x)r(x) \right\}$$

is valid.

Proof. Since the first two inequalities follow from the triangle inequality straightforwardly, we prove the third one. Let $S = \{R_1, \ldots, R_l\}$ be an arbitrary optimal solution of the instance CVRP-NSD(X). By definition, its weight $w(S) = \sum_{j=1}^{l} w(R_j)$. To any route R_j, assign the subset $X_j \in X$ of customers visited by the route R_j. By the triangle inequality, for every $x_i \in X_j$, we have $w(R_j) \geq 2w(y, x_i) = r_i$. Therefore,

$$w(R_j) \geq 2 \frac{\sum_{x_i \in X_j} d(x_i)r_i}{\sum_{x_i \in X_j} d(x_i)} \geq \frac{2}{q} \sum_{x_i \in X_j} d(x_i)r_i,$$

since $\sum_{x_i \in X_j} d(x_i) \leq q$. Thus,

$$\text{CVRP-NSD}^* = w(S) = \sum_{j=1}^{l} w(R_j) \geq \frac{2}{q} \sum_{j=1}^{l} \sum_{x_i \in X_j} d(x_i)r_i = \frac{2}{q} \sum_{i=1}^{n} d(x_i)r_i,$$

since every customer is serviced by a single route. Lemma is proved.

For our further constructions we need the well-known upper bound [12] for the cost of approximate solutions provided by the Iterative Tour Partition (ITP) heuristic for the classic (not necessarily metric) CVRP with unit demands.

Lemma 2. *Let S^{ITP} be the approximate solution for the instance $CVRP(X)$ of the classic unit-demand CVRP induced by a Hamiltonian cycle H in the subgraph $G\langle X \rangle$. Then, for the cost $w(S^{\text{ITP}})$ the following bound*

$$w(S^{\text{ITP}}) \leq 2\frac{\lceil n/q \rceil}{n} \sum_{x \in X} r(x) + w(H) \tag{2}$$

holds.

For the sake of brevity, we skip both the proof of Lemma 2 and even the pseudo-code of the ITP, which can be found, e.g. in [12,22].

Relying on Lemma 2, we obtain the similar upper bound for the metric CVRP-NSD. Indeed, let an arbitrary instance of CVRP-NSD(X) be given. Partition the customer set X into non-empty subsets $X_1, \ldots, X_J$, such that $X_j = \{x \in X \colon d(x) = d^j\}$. Then, given by an arbitrary Hamiltonian cycle H for the set X, we shortcut it subcycles H_j, each of them visits the customers of X_j only. Further, we spawn exactly J auxiliary subinstances of the classic unit-demand CVRP specified by graphs $G\langle X_j \cup \{y\}\rangle$ and adjusted capacity bounds $q^j = \frac{q}{d^j}$ (by condition, any $d^j \in D$ is a divisor of the initial capacity q). Finally, for any j-th subinstance, we construct an approximate partial solution S_j^{ITP} applying the classic ITP heuristic. After that, we output their union $S^{\mathrm{ITP}} = S_1^{\mathrm{ITP}} \cup \ldots \cup S_J^{\mathrm{ITP}}$ as a desired approximate solution.

Lemma 3. *The cost $w(S^{\mathrm{ITP}})$ of the obtained approximate solution S^{ITP} satisfies the following upper bound*

$$w(S^{ITP}) \leq \frac{2}{q} \sum_{x \in X} d(x)r(x) + 2Jr_{max} + Jw(H).$$

Here $r_{max} = \max_{x \in X} r(x)$, and H is the given Hamiltonian cycle spanning the set X.

Proof.

$$w(S^{ITP}) = \sum_{j=1}^{J} w(S_j^{ITP}) \leq 2 \sum_{j=1}^{J} \left\lceil \frac{|X_j|}{q^j} \right\rceil \frac{\sum_{x \in X_j} r(x)}{|X_j|} + \sum_{j=1}^{J} w(H_j)$$

$$\leq 2 \sum_{j=1}^{J} \left(\frac{|X_j| d^j}{q} + 1 \right) \frac{\sum_{x \in X_j} r(x)}{|X_j|} + Jw(H)$$

$$= \frac{2}{q} \sum_{j=1}^{J} d^j \sum_{x \in X_j} r(x) + 2 \sum_{j=1}^{J} \frac{\sum_{x \in X_j} r(x)}{|X_j|} + Jw(H)$$

$$\leq \frac{2}{q} \sum_{x \in X} d(x)r(x) + 2Jr_{max} + Jw(H).$$

Lemma is proved.

Remark 2. As can be easily seen from Lemmas 1 and 3, for any fixed parameter J, the ITP heuristic is the polynomial time approximation algorithm for the metric CVRP-NSD with the constant factor $(1 + (\rho + 1)J)$.

The main point of our scheme is based on decomposition of the customer set X into subsets $X(k) = \{x_1, \ldots, x_{k-1}\}$ and $X'(k) = X \setminus X(k)$ driven by some number $k = k(\varepsilon, q, J, \rho)$, whose choice will be explained later. The following lemma claims that any time the subinstances obtained can be solved in parallel.

Lemma 4.

$$CVRP\text{-}NSD^*(X(k)) + CVRP\text{-}NSD^*(X'(k)) \leq CVRP\text{-}NSD^*(X) + 4(k-1)r(x_k).$$

As for many modifications of the metric CVRP (see, e.g. [18,20,21]), the proof of Lemma 4 can be obtained by the route transformation technique proposed in [12]. We skip it for the brevity.

To prove the correctness of our scheme, we estimate the relative approximation error of the combined approximate solution $S = S_0^{DP} \cup S^{ITP}(X'(k))$, where $S^{ITP}(X'(k)) = S_1^{ITP} \cup \ldots \cup S_J^{ITP}$, as follows

$$
\begin{aligned}
e(k) &= \frac{w(S) - CVRP\text{-}NSD^*(X)}{CVRP\text{-}NSD^*(X)} \\
&= \frac{CVRP\text{-}NSD^*(X(k)) + w(S^{ITP}(X'(k))) - CVRP\text{-}NSD^*(X)}{CVRP\text{-}NSD^*(X)}.
\end{aligned} \tag{3}
$$

Lemma 5. *For any $k \leq n$,*

$$e(k) \leq q\,(2k + (\rho+1)J)\,\frac{r(x_k)}{\sum_{i=1}^{n} d(x_i)r(x_i)} + 2\sqrt{\pi}\rho q J \sqrt{\frac{r(x_k)}{\sum_{i=1}^{n} d(x_i)r(x_i)}}. \tag{4}$$

Proof. Using the simple transformation, represent Eq. (3) in the form

$$
\begin{aligned}
e(k) &= \frac{CVRP\text{-}NSD^*(X(k)) + CVRP\text{-}NSD^*(X'(k)) - CVRP\text{-}NSD^*(X)}{CVRP\text{-}NSD^*(X)} \\
&\quad + \frac{w(S^{ITP}(X'(k))) - CVRP\text{-}NSD^*(X'(k))}{CVRP\text{-}NSD^*(X)} \\
&\leq \frac{4(k-1)r(x_k) + w(S^{ITP}(X'(k))) - CVRP\text{-}NSD^*(X'(k))}{CVRP\text{-}NSD^*(X)},
\end{aligned}
$$

by Lemma 4. Then, by Lemma 3, for an arbitrary Hamiltonian cycle H in the graph $G\langle X \rangle$, we have

$$
\begin{aligned}
w(S^{ITP}(X'(k))) &\leq \frac{2}{q} \sum_{x \in X'(k)} d(x)r(x) + 2Jr(x_k) + Jw(H) \\
&\leq \frac{2}{q} \sum_{x \in X'(k)} d(x)r(x) + 2Jr(x_k) + \rho J\,TSP^*(X'(k)),
\end{aligned}
$$

since, in our scheme, the cycle H is found by a ρ-approximation algorithm. Finally, applying Lemma 1 and the known upper bound (see, e.g. Lemma 4 in [18]) for optimum value of the Euclidean TSP, we obtain

$$
\begin{aligned}
e(k) &\leq \frac{4(k-1)r(x_k) + 2Jr(x_k) + \rho J\left(2r(x_k) + 4\sqrt{\pi r(x_k) \sum_{x \in X} d(x)r(x)}\right)}{\frac{2}{q} \sum_{x \in X} d(x)r(x)} \\
&\leq q\,(2k + (\rho+1)J)\,\frac{r(x_k)}{\sum_{x \in X} d(x)r(x)} + 2\sqrt{\pi}\rho q J \sqrt{\frac{r(x_k)}{\sum_{x \in X} d(x)r(x)}}.
\end{aligned}
$$

Lemma is proved.

To complete the accuracy bound of our scheme, we show (in Lemma 7) that, for any $\varepsilon > 0$ and sufficiently large n, there exists such a value k that $e(k) < \varepsilon$. In turn, the proof of Lemma 7 is based on the following technical lemma.

Lemma 6. *If $A, B, C > 0$, and $s_1, s_2 \ldots, s_{\tilde{K}} > 0$, and $\sum_{k=1}^{\tilde{K}} s_k^2 \leq \alpha \leq 1$ satisfy the condition*

$$s_k^2 + \frac{2B}{2k + A} s_k - \frac{C}{2k + A} \geq 0 \tag{5}$$

for every $k, 1 \leq k \leq \tilde{K}$, then for $\tilde{K}$ the following bound is valid

$$\tilde{K} \leq \left(1 + \frac{A}{2}\right) \cdot \exp\left(\frac{2\alpha}{C} + \frac{4B}{\sqrt{AC}}\right).$$

Proof. Fix a $k \in [1, \tilde{K}]$. An arbitrary positive solution of inequality (5) satisfies the lower bound

$$s_k \geq -\frac{B}{2k + A} + \sqrt{\frac{B^2}{(2k + A)^2} + \frac{C}{2k + A}} \geq -\frac{B}{2k + A} + \sqrt{\frac{C}{2k + A}}.$$

Since $\alpha \geq \sum_{k=1}^{\tilde{K}} s_k^2$,

$$\alpha \geq \sum_{k=1}^{\tilde{K}} \left(\frac{B^2}{(2k + A)^2} - \frac{2B\sqrt{C}}{(2k + A)^{3/2}} + \frac{C}{2k + A}\right) \geq \sum_{k=1}^{\tilde{K}} \left(\frac{C}{2k + A} - \frac{2B\sqrt{C}}{(2k + A)^{3/2}}\right)$$

$$= \sum_{k=1}^{\tilde{K}} \frac{C}{2k + A} - \sum_{k=1}^{\tilde{K}} \frac{2B\sqrt{C}}{(2k + A)^{3/2}} \geq \int_1^{\tilde{K}} \frac{C \, dx}{2x + A} - \int_0^{\tilde{K}} \frac{2B\sqrt{C} \, dx}{(2x + A)^{3/2}}$$

$$= \frac{C}{2} \ln \frac{2\tilde{K} + A}{2 + A} + \frac{2B\sqrt{C}}{\sqrt{2\tilde{K} + A}} - \frac{2B\sqrt{C}}{\sqrt{A}} \geq \frac{C}{2} \ln \frac{2\tilde{K} + A}{2 + A} - \frac{2B\sqrt{C}}{\sqrt{A}}.$$

Therefore,

$$\tilde{K} \leq \left(1 + \frac{A}{2}\right) \cdot \exp\left(\frac{2\alpha}{C} + \frac{4B}{\sqrt{AC}}\right).$$

Lemma is proved.

Lemma 7. *For arbitrary $\varepsilon > 0, J \geq 1, \rho \geq 1$ and $q > 1$ there exists*

$$\tilde{K} = \tilde{K}(\varepsilon, J, \rho, q) = O\left((\rho + 1) J \cdot \exp\left(\frac{2\alpha q}{\varepsilon}\right) \cdot \left(\exp\left(\sqrt{\frac{qJ}{\varepsilon}}\right)\right)^{\sqrt{\pi\rho}}\right)$$

such that the inequality

$$q\left(2k + (\rho + 1)J\right) \frac{r(x_k)}{\sum_{i=1}^n d(x_i)r(x_i)} + 2\sqrt{\pi}\rho qJ \sqrt{\frac{r(x_k)}{\sum_{i=1}^n d(x_i)r(x_i)}} < \varepsilon \tag{6}$$

holds at least for one $1 \leq k \leq \tilde{K}$. Here $\alpha^{-1} = \min_{x \in X} d(x)$.

Proof. Apply Lemma 6 with substitution as follows

$$s_k = \sqrt{\frac{r(x_k)}{\sum_{x \in X} d(x) r(x)}}, \ A = J(\rho + 1), \ B = \frac{\sqrt{\pi} J \rho}{4}, \text{ and } C = \frac{\varepsilon}{q}.$$

Lemma is proved.

Proof of Theorem 1. As it follows from Lemma 7, the proposed scheme do provide a $(1 + \varepsilon)$-approximate solution of CVRP-NSD for any given $\varepsilon > 0$. To obtain its complexity bound just remind that

(i) an optimal solution of the subinstance CVRP-NSD$(X(k))$ can be found by the dynamic programming in time $O(qk^3 2^k)$
(ii) a 3/2-approximate solution of the auxiliary metric TSP instance can be found by Christofides-Serdukov algorithm in time $O(n^3)$
(iii) time complexity of the ITP heuristic is $O(n^2)$.

Theorem is proved.

4   Conclusion

We consider an extension of the Capacitated Vehicle Routing Problem, where customer demand is non-uniform and non-splittable. Although this problem appears to be highly relevant having numerous applications in practice, its approximability in the class of algorithms with theoretical bounds still remains weak investigated.

To the best of our knowledge, in this paper, we proposed the first approximation scheme for the Euclidean setting of this problem in the case when each customer demand is constrained to be a divisor of the capacity bound q. The proposed scheme is PTAS if the capacity q and the number of distinct demands J are $o(\log \log n)$ and becomes EPTAS, when q and J are fixed.

In the forthcoming paper, we plan to extend the proposed scheme to the general case of the Euclidean Capacitated Vehicle Routing Problem with Non-splittable Demand (CVRP-NSD).

Acknowledgments. This research was supported by the Russian Foundation for Basic Research, grants no. 17-08-01385 and 19-07-01243.

References

1. Adamaszek, A., Czumaj, A., Lingas, A.: PTAS for k-tour cover problem on the plane for moderately large values of k. Int. J. Found. Comput. Sci. **21**(06), 893–904 (2010)
2. Arora, S.: Polynomial time approximation schemes for Euclidean traveling salesman and other geometric problems. J. ACM **45**, 753–782 (1998)

3. Asano, T., Katoh, N., Tamaki, H., Tokuyama, T.: Covering points in the plane by k-tours: towards a polynomial time approximation scheme for general k. In: Proceedings of the Twenty-ninth Annual ACM Symposium on Theory of Computing, STOC 1997, pp. 275–283. ACM, New York (1997)
4. Blocho, M., Czech, Z.: A parallel memetic algorithm for the vehicle routing problem with time windows. In: 2013 Eighth International Conference on P2P, Parallel, Grid, Cloud and Internet Computing, pp. 144–151 (2013)
5. Borčinova, Z.: Two models of the capacitated vehicle routing problem. Croatian Oper. Res. Rev. 8, 463–469 (2017)
6. Bula, G.A., Gonzalez, F.A., Prodhon, C., Afsar, H.M., Velasco, N.M.: Mixed integer linear programming model for vehicle routing problem for hazardous materials transportation**universidad nacional de colombia. universite de technologie de troyes. IFAC-PapersOnLine 49(12), 538–543 (2016). http://www.sciencedirect.com/science/article/pii/S2405896316309673. 8th IFAC Conference on Manufacturing Modelling, Management and Control, MIM 2016
7. Dantzig, G., Ramser, J.: The truck dispatching problem. Manag. Sci. 6, 80–91 (1959)
8. Das, A., Mathieu, C.: A quasi-polynomial time approximation scheme for Euclidean capacitated vehicle routing. In: Proceedings of the Twenty-first Annual ACM-SIAM Symposium on Discrete Algorithms, SODA 2010, pp. 390–403. Society for Industrial and Applied Mathematics, Philadelphia (2010)
9. Das, A., Mathieu, C.: A quasipolynomial time approximation scheme for Euclidean capacitated vehicle routing. Algorithmica 73, 115–142 (2015)
10. Fomin, F.V., Lokshtanov, D., Raman, V., Saurabh, S.: Bidimensionality and EPTAS. In: Proceedings of the Twenty-Second Annual ACM-SIAM Symposium on Discrete Algorithms, SODA 2011, pp. 748–759. Society for Industrial and Applied Mathematics, Philadelphia (2011). http://dl.acm.org/citation.cfm?id=2133036.2133095
11. González, O., Segura, C., Valdez Peña, S.: A parallel memetic algorithm to solve the capacitated vehicle routing problem with time windows. Int. J. Comb. Optim. Probl. Inform. 9(1), 35–45 (2018). https://ijcopi.org/index.php/ojs/article/view/77
12. Haimovich, M., Rinnooy Kan, A.H.G.: Bounds and heuristics for capacitated routing problems. Math. Oper. Res. 10(4), 527–542 (1985)
13. Hashimoto, H., Yagiura, M.: A path relinking approach with an adaptive mechanism to control parameters for the vehicle routing problem with time windows. In: van Hemert, J., Cotta, C. (eds.) EvoCOP 2008. LNCS, vol. 4972, pp. 254–265. Springer, Heidelberg (2008). https://doi.org/10.1007/978-3-540-78604-7_22
14. van Hoorn, J.J.: A note on the worst case complexity for the capacitated vehicle routing problem. In: Research Memorandum, vol. 5. Faculteit der Economische Wetenschappen en Bedrijfskunde (2010)
15. Kara, I.: Arc based integer programming formulations for the distance constrained vehicle routing problem. In: Proceedings of LINDI 2011–3rd IEEE International Symposium on Logistics and Industrial Informatics (2011)
16. Khachai, M.Y., Dubinin, R.D.: Approximability of the vehicle routing problem in finite-dimensional Euclidean spaces. Proc. Steklov Inst. Math. 297(1), 117–128 (2017). https://doi.org/10.1007/978-3-319-44914-2_16
17. Khachai, M., Ogorodnikov, Y.: Polynomial time approximation scheme for the capacitated vehicle routing problem with time windows. Trudy instituta matematiki i mekhaniki UrO RAN 24(3), 233–246 (2018). https://doi.org/10.21538/0134-4889-2018-24-3-233-246

18. Khachay, M., Ogorodnikov, Y.: Efficient PTAS for the Euclidean CVRP with time windows. In: van der Aalst, W.M.P., et al. (eds.) AIST 2018. LNCS, vol. 11179, pp. 318–328. Springer, Cham (2018). https://doi.org/10.1007/978-3-030-11027-7_30

19. Khachay, M., Ogorodnikov, Y.: Approximation scheme for the capacitated vehicle routing problem with time windows and non-uniform demand. In: Khachay, M., Kochetov, Y., Pardalos, P. (eds.) MOTOR 2019. LNCS, vol. 11548, pp. 309–327. Springer, Cham (2019). https://doi.org/10.1007/978-3-030-22629-9_22

20. Khachay, M., Ogorodnikov, Y.: Improved polynomial time approximation scheme for capacitated vehicle routing problem with time windows. In: Evtushenko, Y., Jaćimović, M., Khachay, M., Kochetov, Y., Malkova, V., Posypkin, M. (eds.) OPTIMA 2018. CCIS, vol. 974, pp. 155–169. Springer, Cham (2019). https://doi.org/10.1007/978-3-030-10934-9_12

21. Khachay, M., Dubinin, R.: PTAS for the Euclidean capacitated vehicle routing problem in R^d. In: Kochetov, Y., Khachay, M., Beresnev, V., Nurminski, E., Pardalos, P. (eds.) DOOR 2016. LNCS, vol. 9869, pp. 193–205. Springer, Cham (2016). https://doi.org/10.1007/978-3-319-44914-2_16

22. Khachay, M., Zaytseva, H.: Polynomial time approximation scheme for single-depot Euclidean capacitated vehicle routing problem. In: Lu, Z., Kim, D., Wu, W., Li, W., Du, D.-Z. (eds.) COCOA 2015. LNCS, vol. 9486, pp. 178–190. Springer, Cham (2015). https://doi.org/10.1007/978-3-319-26626-8_14

23. Nalepa, J., Blocho, M.: Adaptive memetic algorithm for minimizing distance in the vehicle routing problem with time windows. Soft Comput. **20**(6), 2309–2327 (2016)

24. Necula, R., Breaban, M., Raschip, M.: Tackling dynamic vehicle routing problem with time windows by means of ant colony system. In: 2017 IEEE Congress on Evolutionary Computation (CEC), pp. 2480–2487 (2017)

25. Ng, K., Lee, C., Zhang, S., Wu, K., Ho, W.: A multiple colonies artificial bee colony algorithm for a capacitated vehicle routing problem and re-routing strategies under time-dependent traffic congestion **109**, 151–168 (2017). http://www.sciencedirect.com/science/article/pii/S0360835217301948

26. Papadimitriou, C.: Euclidean TSP is NP-complete. Theoret. Comput. Sci. **4**, 237–244 (1977)

27. Song, L., Huang, H.: The Euclidean vehicle routing problem with multiple depots and time windows. In: Gao, X., Du, H., Han, M. (eds.) COCOA 2017. LNCS, vol. 10628, pp. 449–456. Springer, Cham (2017). https://doi.org/10.1007/978-3-319-71147-8_31

28. Song, L., Huang, H., Du, H.: Approximation schemes for Euclidean vehicle routing problems with time windows. J. Comb. Optim. **32**(4), 1217–1231 (2016)

29. Toth, P., Vigo, D.: Vehicle Routing: Problems, Methods, and Applications. MOS-Siam Series on Optimization, 2nd edn. SIAM, Philadelphia (2014)

30. Vidal, T., Crainic, T.G., Gendreau, M., Prins, C.: A hybrid genetic algorithm with adaptive diversity management for a large class of vehicle routing problems with time-windows. Comput. Oper. Res. **40**(1), 475–489 (2013)

Analysis of Dynamic Behavior Through Event Data

Detection of Anomalies in the Criminal Proceedings Based on the Analysis of Event Logs

Alexandra A. Kolosova$^{(\boxtimes)}$ (ID) and Irina A. Lomazova (ID)

National Research University Higher School of Economics, Moscow, Russia
aakolosova@edu.hse.ru, ilomazova@hse.ru

Abstract. Process mining makes it possible to solve a task of finding and analyzing deviations in the process. System event logs record information about real process behavior. Weaknesses and errors of a workflow can be found during the analysis of logs. This is especially important in areas associated with significant responsibility and risk.

In this paper the focus is on the criminal procedure analysis via process mining methods. A model of this process allows for flexibility only in a strictly regulated framework. However, in practice undesired deviations appear and, therefore, need to be detected and prevented.

We adopted conformance checking techniques to determine the anomaly of the trace, taking into account the specifics of the process. We also did clustering of anomaly cases to reveal behavior patterns. They will be helpful for identification of potential causes of such anomalies.

Keywords: Clustering · Conformance checking · Criminal proceedings · Event logs · Process mining

1 Introduction

Process mining (PM) is a research discipline that provides techniques for business process analysis. PM allows for discovering and improving artificial process models, and inspecting system behavior through analysis of event logs as well [1]. Organizations in various spheres benefit from application of these techniques to maintain their processes. There are several studies on the use of PM in medicine, economy, industrial systems, government systems. For example, Driessen [9] demonstrates the usability of PM in improvement of the Netherlands Ministry of Defense processes. Mapikou and Etoundi in their work adopt α-algorithm to facilitate structure of administrative processes [11]. Rojas, Munoz-Gama, Sepúlveda and Capurro [15] cover 74 papers on the usage of PM in healthcare, in particular, in the medical treatment process and the organization process. Improvements in this area impact on the quality and life expectancy of patients.

This work is supported by the Basic Research Program at the National Research University Higher School of Economics.

© Springer Nature Switzerland AG 2019
W. M. P. van der Aalst et al. (Eds.) AIST 2019, LNCS 11832, pp. 401–410, 2019.
https://doi.org/10.1007/978-3-030-37334-4_36

An important perspective of process analysis focus is conformance checking—comparing an existing process model with the real behavior [2]. Conformance between a process and a model is extremely important in the law enforcement sphere. There are already several papers which aim at simplifying and improving monitoring of activities and data in organizations involved in criminal procedure. Van Dijk, Kalidien and Choenni [7] developed a data space system in order to gain an insight into the success and possible bottlenecks in the Dutch criminal justice system. Chang, Lu and Jen introduced the Integrated Criminal Justice Data Base System based on the supply chain management model [6]. However, all these studies aim at workflow and data management, but do not monitor the insight process as it is. As for the process itself, we found a case study concerning the control flow of employees in law-enforcement agencies, which is based on quantitative analysis of statistics published by the Government [12]. Also worth noting is a text mining application for extracting data from police reports. Several works were made using this approach. As a result, reports classification was improved and the process of extracting significant information has been revised [13].

In this paper we concentrate on the criminal procedures. The criminal process is the normative one and is described in the criminal procedure code. It contains various restrictions on ordering of activities, limits on duration and repetitions of activities. Any undesired process deviations must be detected and prevented. So, this is particularly the main goal of our study: apply process mining technique to detect anomalies in the process of criminal proceedings. In the course of the study, we will analyze event logs that contain records about "as-is" criminal case processing. This study addresses the following research questions:

Question 1. How to properly consider the specifics of the process? What requirements should be taken into account?

Question 2. How to model criminal procedures, which notations and formalisms to use?

Question 3. How to check the conformance between the actual process behavior, recorded in the event log, and the process model taking into account the specifics of the process?

Question 4. How to use this conformance checking to detect and classify process deviations?

Question 5. How to visualize the discovered deviations and make them transparent for analysis?

2 Specific of the Criminal Prosecutions

The subject of the study is the rigid process with predefined activities, deadlines and undesirable outcomes for participants. To obtain a process model and properly compare it with the given event logs, we first need to determine the requirements that influence execution analysis.

2.1   The Law Nature of the Criminal Proceedings

We should take into account that this procedure was established by the law, and, therefore any deviation from the prescribed behavior is a signal that a trace contains anomalies. Moreover, we cannot rely on automatic model discovery. Instead we should create a model of the process manually with respect to the Criminal Procedure Code.

2.2   Uniqueness of a Criminal Case

A trace in a log representing one criminal case contains records about processing one instance of case documents. Physically this case cannot be divided. This means that parallelism in processing is not allowed.

2.3   Time Limits Between Activities

The Criminal Procedure Code fixes time limits on the stages of the trajectory of the criminal case. Performing actions that go beyond the time frame will be considered as a violation.

2.4   Undesired Outcomes in Case Processing

Some operations and decisions in a case may be caused by insufficient investigation, reluctance to work a case or procedural errors. We need to catch and assess such activities while conducting trace analysis.

2.5   Looping Activities

One of an undesirable behaviors, observed in processing criminal cases, is looping of operations performed within a case. For example, multiple termination and reopening of the case. The number of performed activities is one of the crucial factors in process correctness evaluation.

3   Model of the Criminal Proceedings

To compare process event logs with the expected behavior, we first need to obtain a model of the process. The model should correspond to the flow of activities described by the law, which will be considered as the standard behavior. Non-conformance between the model and the logs should not be caused by the imperfection of the model.

In our work we use Petri Nets as a basic notation for formal modeling of the criminal proceedings. A criminal case documents are represented as tokens and performance of activities are matched to transitions. Places in the model indicate intermediate states of the case runs.

Modeling of the criminal procedure is addressed by two major issues: (1) process restrictions and (2) a large number of different activities. The first issue is resolved by adding attributes to elements of a Petri Net. The second problem requires transformation of low-level logs into logs over more abstract activities [4].

3.1 Attributed Petri Net

There are extensions of Petri nets which take into account conditions imposed on transition firing and places states. The most commonly used are Petri Net with time and cost [3,5]. Among existing time constraints patterns are *restricting execution times* and *recurrent process elements* [10] which are significant for our case study. We store information about time execution of transitions in the model. Moreover, we need to associate activities with *weight* and *count* attributes for further calculation of transition cost. Therefore, we use an attributed Petri Net to model the process of criminal procedure.

Definition 1. *An Attributed Petri Net N is a tuple (P, T, F, I, C, W) where*

- *P is a finite set of places,*
- *T is a finite set of transitions,*
- *$F \subseteq (P \times T) \cup (T \times P)$ is a flow relation,*
- *$I : T \to \tau$ associates with each transition a firing interval $(0, \tau)$,*
- *$C : T \to c$ associates with each transition a maximal number of firings allowed in a case execution,*
- *$W : T \to w$ associates with each transition a weight (or a firing cost).*

We also define transition firing rules taking into account execution time bounds. Conditions and results for transition firings are defined as for Time Petri nets [14], where transition $t \in T$ is assigned a time interval $(0, I(t))$.

The transition weight determines the maximal number of firings for this transition in one Petri net run. It does not affect the transition enabling. However, it is taken into account in the estimation of the overall trace conformance quality.

3.2 Transformation of Low-Level Activities

There are over 100 different events logged in the criminal procedure. To reduce the complexity of the process model, we manually mapped events to more abstract transitions in the model relative to the actual process control flow. As a result, we get a model with 14 places (including one source place and one sink place) and 34 transitions (some of them are hidden and were created artificially). Places in the model are for intermediate states and resources, while activities in a log correspond to transitions. Thus, transitions should be considered as the main perspective of the analysis.

3.3   Model Visualization

The model of an investigated process was developed in the proposed notation. It was created based on the information from Trajectory of a Criminal Case [12] and Criminal Procedure Code of the Russian Federation. In order to visualize the process for the concerned parties, we made an illustration using standard graphical elements of Petri net notation along with some extra symbols denoting Petri net attributes. This attributes comply with the constraints imposed on the process.

The model of the process being studied is shown in Fig. 1. In Table 1 we give a description of the symbols used in the model.

Table 1. Model symbols description

Symbol	Description
⬤	Initial place in the model
◯	Regular place
◎	Finish place in the model
◯	Places in the model, which can also be final. Introduced to simplify the model.
▯	Regular transition
▮	Hidden transition
▮	Transition with cost attribute
🕐	Transition time limit attribute
📖	Transition firing quantity attribute

4   Checking the Conformance of the Criminal Proceedings

Conformance checking allows us to find discrepancies between required case prosecution and real process behavior [16]. In general, the concept of conformance checking is an assessment of the distance between the event log and the process model. We use the token-replay based algorithm to calculate conformance values for criminal cases instances. This algorithm uses the *fitness* metric. It refers to the percentage of traces in the log allowed by the model [4]. Decision to use this particular metric instead of computing alignments was made based on the specifics of the process in question. Variety of possible traces is large, therefore by concentrating on how the event log fits the model, we can make our calculation process not too complicated. In addition, to study event logs of the criminal proceeding requires not only the perspective of control flow, but also the weights of transitions, time stamps and the number of firings. All these attributes reflect possible anomalies.

Note that the restrictions described above require an extended fitness metric. Each transition must be taken into account depending on its weight and time constraints. The average weight is used as a cost function for fitness of a trace σ:

$$weight\_fitness(\sigma, N) = \left(\frac{1}{2} \left(1 - \frac{m}{c} \right) + \frac{1}{2} \left(1 - \frac{r}{p} \right) \right) * \frac{\sum weight}{fired\_trans}. \qquad (1)$$

In (1), p and c denote tokens, respectively, produced and consumed by transitions, m—the number of missing tokens, r—the number of tokens remaining after completion.

The weight of a transition firing consists of compliance with time bounds, the number of firings and the cost of this transition. All these components can be considered as signals of various anomalies.

5   Detection of Anomalies in the Process

We adopted existing notations and metrics to apply them in the process of criminal proceedings. With weighted token-replay we can calculate the value of conformance for each trace in event logs. However, the resulting value will only identify the presence of deviations. For now, we cannot establish an unambiguous, well-interpreted scale of case anomalies. Instead, we use this values in addition to aspects anomalies for further clustering.

While clustering, we need to consider criminal procedure workflow in addition to the results of conformance checking. Information about sequence of performed actions is selected from high-level event logs. It represents most of the volume of clustered data. As a sequence of characters, the workflow can be clustered using Levenshtein distance [8]. As for conformance testing results, one numeric value does not be help when comparing strings annotated with set of transition labels. The anomaly signals, detected during the conformance check must be added into workflow. We decided to insert special anomaly symbols into the transitions sequence to highlight weaknesses and aggregate all information about a specific deviation. Table 2 gives an example of such transitions sequence extension for two possible anomaly cases.

Table 2. Symbols insertion

	Initial set of transitions	Set of transitions with inserted symbols
Time limit	$\ldots t_{4.2} t_{5.3} t_{6.3} \ldots$	$\ldots t_{4.2} t_{5.3} \tau t_{6.3} \ldots$
Activities loop	$\ldots t_{5.3} t_{6.2} t_{7.1} t_{6.2} t_{7.1} t_{6.2} t_{7.1}$ $t_{6.2} t_{7.1} t_{8.1} \ldots$	$\ldots t_{5.3} @ t_{6.2} t_{7.1} 4 t_{8.1} \ldots$

Now we can compare sets of events in anomalous traces taking into account the revealed deviations. It is important to compare traces with highlighted deviation locations, since then these locations can be easily matched with e.g. relevant departments. Through clustering, we get various anomaly patterns that can be used to further manage the processes.

6   Conclusion

This study was designed to diagnose the discrepancies between the normative criminal procedure and the real process of criminal proceedings in Russia. To achieve this goal, we developed a process model in the notation of attributed Petri Nets, adopted a fitness metric to capture the peculiarities of the process and revealed patterns of anomalies using the results of conformance checking for clustering of traces.

During the study, multiple event logs were assessed using the approach introduced in the paper. The results revealed that the most common problem in handling criminal cases is cyclical activity. Possible reasons for the observed behavior may be the lack of necessary information, unwillingness to work with a specific case. Resources that are responsible for the frequent re-execution of work redoing should be taken under control.

A further direction of this work is the automatic construction of a process model with special characters introduced. Since these symbols do not relate to any activity, we cannot use standard process discovery techniques. Therefore, we plan to adapt the α-algorithm so that the transition and the special symbol are treated as one unit. With this approach, the same transitions with and without symbols will be considered different. This will allow us to distinguish between different anomaly traces in the scope of one cluster.

We also need to visualize the workflow with highlighted anomalies in a convenient way in order to uniquely identify the shortcomings. In this form, the tool can be used in the analysis of criminal proceedings by researchers of the institute for the rule of law. It will significantly simplify the process of monitoring litigation.

Appendix

See Table 3.

Table 3. Transitions used in the criminal procedure model

Transition label	Transition name
t1.1	Accident report to other agencies
t1.2	Accident report to police
t2.1	Stop accident report consideration
t2.2	Crime report registration
t3.1	Preliminary inquiry completed
t3.2	Send for preliminary inquiry to another agency
t4.0	Extension of preliminary inquiry
t4.1	Transfer a crime report
t4.2	Refuse to open a criminal case
t4.3	Open a criminal case
t5.0	Criminal case entry
t5.1	Reverse a refusal to pen the case
t5.1.1	Convert to a crime report
t5.1.2	Convert to a criminal case
t5.2	Reversal of opened criminal case
t5.3	Start investigation
t6.0	Add case file materials
t6.1	Suspend a case
t6.1.1.	Reopen pending case
t6.2	Close a case
t6.2.1	Reopen close case
t6.3	Send criminal case files to prosecutor
t6.3.1	Send a case to another prosecutor
t6.4	Transfer criminal case files
t6.5	The case was sent to the court by investigation agencies
t7.0	Private prosecution case received
t7.1	The case was sent to the court by the prosecutor
t7.2	Prosecutor returns criminal case files
t8.1	Case is considered by the court
t9.1	Decision of the court
t9.2	The court returns criminal case files

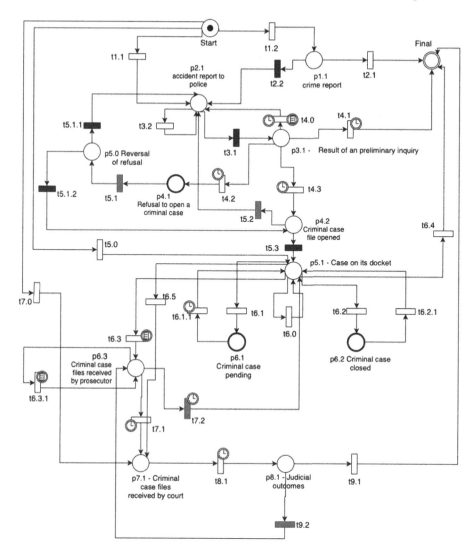

Fig. 1. Model of the criminal proceedings

References

1. van der Aalst, W.M.P.: Process Mining: Discovery, Conformance and Enhancement of Business Processes. Springer, Heidelberg (2011). https://doi.org/10.1007/978-3-642-19345-3
2. Carmona, J., van Dongen, B., Solti, A., Weidlich, M.: Conformance Checking. Springer, Cham (2018). https://doi.org/10.1007/978-3-319-99414-7
3. Abdulla, P.A., Mayr, R.: Petri nets with time and cost. In: Atig, M.F., Rezine, A. (eds.) 14th International Workshop on Verification of Infinite-State Systems, pp. 9–24 (2013)

4. Begicheva, A., Lomazova, I.: Checking conformance of high-level business process models to event logs. In: 8th Spring/Summer Young Researchers' Colloquium on Software Engineering, Saint Petersburg, pp. 77–82 (2014)
5. Bérard, B., Cassez, F., Haddad, S., Lime, D., Roux, O.H.: Comparison of different semantics for time Petri nets. In: Peled, D.A., Tsay, Y.-K. (eds.) ATVA 2005. LNCS, vol. 3707, pp. 293–307. Springer, Heidelberg (2005). https://doi.org/10.1007/11562948_23
6. Chang, W., Lu, C.C., Jen, W.Y.: A study of integrated criminal justice data base system. In: 2008 IEEE International Conference on Intelligence and Security Informatics, pp. 290–291. IEEE (2008)
7. van Dijk, J., Kalidien, S., Choenni, S.: Smart monitoring of the criminal justice system. Gov. Inf. Q. 35(4), S24–S32 (2016)
8. Doan, A., Halevy, A., Ives, Z.: Principles of Data Integration. Morgan Kaufmann, Burlington (2012)
9. Driessen, R.: The usability of the Process Mining analysis method to improve processes of the Netherlands Ministry of Defence. Delft University of Technology (2017)
10. Lanz, A., Weber, B., Reichert, M.: Time patterns for process-aware information systems. Requirements Eng. 2(19), 113–141 (2014)
11. Mapikou, G.L.M., Etoundi, R.A.: A process mining oriented approach to improve process models analysis in developing countries. In: 2016 IEEE/ACS 13th International Conference of Computer Systems and Applications (AICCSA), pp. 1–8. IEEE (2016)
12. Paneyakh, E., Titaev, K., Shklyaruk, M.: Trajectory of a Criminal Case: Institutional Analysis of Russian Criminal Justice System. EUSP Press, Saint Petersburg (2018)
13. Poelmans, J., Elzinga, P., Viaene, S., Dedene, G.: Formally analysing the concepts of domestic violence. Expert Syst. Appl. 38(4), 3116–3130 (2011)
14. Popova-Zeugmann, L.: Time and Petri Nets. Springer, Heidelberg (2013). https://doi.org/10.1007/978-3-642-41115-1
15. Rojas, E., Munoz-Gama, J., Sepúlveda, M., Capurro, D.: Process mining in healthcare: a literature review. J. Biomed. Inform. 61, 224–236 (2016)
16. Rozinat, A.: Process mining: conformance and extension. Technische Universiteit Eindhoven (2010)

A Method to Improve Workflow Net Decomposition for Process Model Repair

Semyon E. Tikhonov$^{(\boxtimes)}$ and Alexey A. Mitsyuk

National Research University Higher School of Economics,
20 Myasnitskaya St., 101000 Moscow, Russia
setikhonov@edu.hse.ru, amitsyuk@hse.ru
https://pais.hse.ru/en/

Abstract. Creating a process model (PM) is a convenient means to depict the behavior of a particular information system. However, user behavior is not static and tends to change over time. In order for them to sustain relevant, PMs have to be adjusted to ever-changing behavior. Sometimes the existing PM may be of high value (e.g. it is well-structured, or has been developed continuously by experts to later work with), which makes the approach to create a brand-new model using discovery algorithms less preferable. In this case, a different and better suitable approach to adjust PM to new behavior is to work with an existing model through repairing only such PM fragments that do not fit the actual behavior stated in sub-log. This article is to present a method for efficient decomposition of PMs for their future repair. It aims to improve the accuracy of model repair. Unlike the ones introduced earlier, this algorithm suggests finding the minimum spanning tree of undirected graph's vertices subset. It helps to reduce the size of a fragment to be repaired in a model and enhances the quality of a repaired model according to various conformance metrics.

Keywords: Process model repair · Workflow nets · Event logs

1 Introduction

Process modeling is a common technique applied when designing modern information systems. Formal models are convenient to depict the structure and the behavior of a system. Keeping models relevant is a complex task in the continuously changing environment. *Process mining* algorithms can help to cope with that problem, when a system records its behavior in event logs [19].

The key characteristic of a process model is its *recall* (or *fitness*) to an event log with behavior of the system [6]. This characteristic shows if the model fully reflects the real-life behavior of the system. Unfitting models do not reflect the reality, have inconsistencies, and are less applicable.

This work is supported by the Basic Research Program at the National Research University Higher School of Economics.

© Springer Nature Switzerland AG 2019
W. M. P. van der Aalst et al. (Eds.) AIST 2019, LNCS 11832, pp. 411–423, 2019.
https://doi.org/10.1007/978-3-030-37334-4_37

It is possible to discover a new up-to-date model of the system using process discovery techniques [19]. However, if the model is of high value (e.g. it is well-structured, or has been developed continuously by experts to later work with), it is better to preserve this model. Instead of re-discovering a completely new model, the existing process model can be repaired. Many repair techniques have been proposed in the field of process mining [3,4,7–9,12–14,16].

One of the recent techniques to repair process models w.r.t. model recall is called *modular process model repair* [14]. This technique is based on decomposed conformance checking and process model discovery [5]. The technique repairs workflow models through decomposing them, and then replacing the unfitting sub-nets. The smaller the number of small sub-nets has been replaced by the repair, the better the original structure of the model is preserved. Authors have proposed techniques for local [1,14] and non-local [13] model repair.

The greedy technique for the non-local repairs [13] has a disadvantage. Usually, it selects the larger sub-net-to-replace than it is actually needed. The technique's greedy nature is the main reason of this disadvantage. In this paper, we show how this greedy technique can be improved to select the smaller sub-nets-to-replace. Indeed, we present the algorithm to find the smaller sub-nets which should be replaced to repair the model.

The paper is organized as follows. It begins with this Introduction. Basic notions are presented in Sect. 2. Section 3 describes process model repair, and how the greedy repair technique works. Besides, this section discusses why the greedy techniques needs to be improved. We present the new algorithm to select the sub-nets-to-replace in Sect. 4. This algorithm has been implemented as a part of the prototype repair tool *Iskra*. The implementation issues and results of prototype evaluation are described in Sect. 5. Finally, Sect. 6 concludes the paper.

2   Preliminaries

By $\mathbb{N}$ we denote the set of natural numbers including zero. Let S be a set. A mapping $B : S \rightarrow \mathbb{N}$ is a multiset over S. It maps each element of S to the number of its occurrences. By $\mathcal{B}(S)$ we denote the set of all multisets over S. A finite multiset can be specified by enumerating all its elements and indicating their multiplicity. For example, $B = [e, e, e, c, d, d] = [e^3, c, d^2]$ denotes the multiset with three occurrences of e, one c, and two d. Besides, we extend the standard set operations to multisets in the usual way.

For a function $f : X \rightarrow Y$, $dom(f)$ denotes its domain, and $f : X \nrightarrow Y$ denotes a partial function. A function $f \upharpoonright_Q : Q \nrightarrow Y$ is a projection of a (partial) function f onto a set $Q \subseteq X$ such that $\forall x \in Q : f \upharpoonright_Q (x) = f(x)$. This notation can be extended to multi-sets, e.g. $[e^3, c, d^2] \upharpoonright_{\{c,d\}} = [c, d^2]$.

By X^* we denote the set of all finite sequences over a set X, and we use triangle brackets for sequences, e.g. $\sigma = \langle x_1, x_2, ..., x_n \rangle$ is a sequence of length n. By $\sigma_1 \cdot \sigma_2$ we denote the concatenation of two sequences, $\sigma \upharpoonright_Q$ is the projection of the sequence σ onto the set Q.

Definition 1 (Event log). *Let $A \subseteq \mathcal{U}_A$ be a set of process activities. A trace σ is a finite sequence of activities from A, i.e. $\sigma \in A^*$. An* event log *(or simply* log*) L is a finite multi-set of traces, i.e. $L \in \mathcal{B}(A^*)$.*

A projection of an event log $L = [\sigma_1, \sigma_2, \dots, \sigma_n]$ onto a set of activities B is a log $L{\restriction}_B = [\sigma_1 {\restriction}_B, \sigma_2 {\restriction}_B, \dots, \sigma_n {\restriction}_B]$, where $\sigma_i {\restriction}_B$ is a projection of the trace σ_i onto B. We call such log a *sub-log* of L.

A *Petri net* is a triple (P, T, F), where P and T are disjoint sets of *places* and *transitions*, and $F : (P \times T) \cup (T \times P) \to \mathbb{N}$ is a flow relation. For a transition $t \in T$ a preset $\bullet t$ and a postset $t \bullet$ are defined as the subsets of P such that $\bullet t = \{p \mid F(p, t) \neq 0\}$ and $t \bullet = \{p \mid F(t, p) \neq 0\}$.

A labeled Petri net is a tuple (P, T, F, l), where $l : T \to \mathcal{U}_A \cup \{\tau\}$ is a labeling function, which maps transitions to activity labels from $\mathcal{U}_A$. A transition t is called invisible, if $l(t) = \tau$, otherwise it is called visible.

A *marking* $M : P \to \mathbb{N}$ specifies a current state of a Petri net. A transition $t \in T$ is *enabled* in a marking M iff $\forall p \in P, M(p) \geq F(p, t)$. An enabled transition t may *fire* yielding a new marking M', such that $M'(p) = M(p) - F(p, t) + F(t, p)$ for each $p \in P$ (denoted $M \xrightarrow{t} M'$).

Definition 2 (Workflow net). *A* workflow net *is a labeled Petri net $N = (P, T, F, l)$ with distinguished initial M_i and final M_o markings such that*

- *$\exists! \ i \in P$ (source place) such that $\bullet i = \emptyset$, and $M_i = [i]$,*
- *$\exists! \ o \in P$ (sink place) such that $o \bullet = \emptyset$, and $M_o = [o]$,*
- *every node $n \in P \cup T$ is on a path from i to o.*

Figure 1 shows the example of a workflow net with transitions $t_1, \dots, t_8$ and places $c_1, \dots, c_4$, *source*, *sink*. Transitions of this net have labels a, b, c, d, e, f, g, and h. Its initial marking shown with the black token in the place *source*.

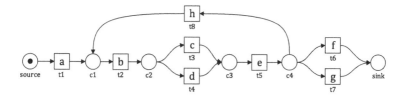

Fig. 1. Workflow net

By $T_v(N)$ we denote the set of all visible transitions in N. A transition label is called *unique* if it labels exactly one transition in the net. $T_v^u(N)$ denotes the set of all visible transitions with unique labels in N. The union of two WF-nets is a WF-net $N^U = N^1 \cup N^2$, which is built by the union of sets of places, transitions, and flows: $N^1 \cup N^2 = (P^1 \cup P^2, T^1 \cup T^2, F^1 \cup F^2, l^u)$, where $l^u \in (T^1 \cup T^2) \nrightarrow \mathcal{U}_A$ is a union of l^1 and l^2, $dom(l^u) = dom(l^1) \cup dom(l^2)$, $l^u(t) = l^1(t)$ if $t \in dom(l^1)$, and $l^u(t) = l^2(t)$ if $t \in dom(l^2) \setminus dom(l^1)$. We assume further that all labeled transitions have unique labels.

3   Decomposed Process Model Repair

Process model repair aims at improving the model quality w.r.t. specified criteria such that this model preserves its characteristics as far as possible. This differs the repair from discovery.

As it has already been stated in the introduction, *recall* (or *fitness*) is the main characteristic of a process model w.r.t. to an event log. It shows to what extent the behavior from the event log (i.e. its traces) can be replayed by the process model. Let L denotes a set of behavior variants (traces) recorded in the log, and let M be a set of all possible model runs. $L \cap M$ denotes the behavior seen in the log L which can be replayed by the model M. Then, the *recall* or *fitness* is (see Carmona et al. [6] for details):

$$\text{recall} = \frac{|L \cap M|}{|L|}.$$

Adriansyah [2] proposed the method to calculate the recall of a Petri net based on so called *alignments* between log traces and model runs. The method is commonly used in process mining practice [6,19]. In this work, we also will employ this method.

In this paper, the following model repair problem is considered. Let $N = (P, T, F, l)$ be a workflow net, and L be an event log. Note that $L \in \mathcal{B}(A^*)$, and $l\colon T \to A \cup \{\tau\}$. Less technically speaking, transitions labeled either with activities seen in the log, or with τ. Model N does not perfectly fit L, i.e. recall$(L, N) < 1$. In other words, there are traces in L which can not be replayed by N. A repair algorithm $N^r = \text{repair}(N, L)$ needs to construct a Petri net N^r (*repaired model*) such that (1) recall$(L, N^r) = 1$, i.e. N^r can replay all traces of L; (2) N and N^r have a *similar* [22] graph structure.

So called *modular technique* for process model repair has been proposed in [14]. This technique decomposes a workflow net using one of *valid* [18] decompositions. Then, it finds the erroneous sub-nets which are the reason of a model recall reduction. Finally, the algorithm replaces these sub-nets with correct workflow sub-nets con-

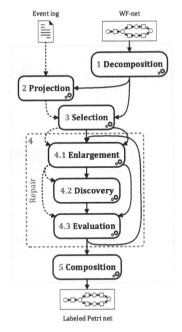

Fig. 2. The greedy model repair [13]

structed using a *process discovery* algorithm. Authors show [14] how two discovery algorithms can be used as they guarantee the discovery of models with recall equal to 1. These algorithms are *Inductive Miner* [11] and *ILP Miner* [20].

An inconsistency is called local when it is isolated within a single sub-net [14]. The methods described in [1,14] are able to repair local inconsistencies while preserving model precision [6]. However, they reduce model precision and tend to change the model significantly when repairing non-local inconsistencies. The greedy technique suitable for these cases has been proposed in [13].

Figure 2 shows the scheme of the greedy repair from [13]. The five steps are performed to repair the model: (1) Decomposition, (2) Projection, (3) Selection, (4) Repair, and (5) Composition. The fourth step is iterative and consists of three sub-procedures: (4.1) Enlargement, (4.2) Discovery, and (4.3) Evaluation. The algorithm begins with the decomposing a model into the set of sub-nets using a valid decomposition. Then, the iterative process begins from an arbitrary erroneous sub-net which recall w.r.t. corresponding sub-log is less than 1.

Figure 3 shows an example of the greedy algorithm execution. In this particular example, the decomposition of the workflow net contains two erroneous sub-nets: N^1 and N^2. The error is that labels of two transitions (the first one from N^1 and the second from N^2) have been swapped. The greedy algorithm started from N^1.

Each step of the iterative process is the enlargement of the current sub-net. This enlargement is done by joining the sub-net with all its neighbor sub-nets. In Fig. 3, after the first step of the iterative procedure N^1 will be joined with two its neighbors marked with 1.

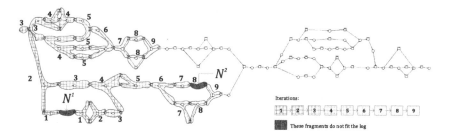

Fig. 3. The greedy algorithm selecting a fragment-to-repair [13]: sub-nets are labeled with iteration numbers

After the enlargement, the algorithm re-discovers a sub-net from the sub-log that corresponds to the enlarged fragment. Then, it tests if the re-discovered sub-net actually repairs the model. The iteration stops if the recall of the model with the re-discovered sub-net is 1. This condition is satisfied when all erroneous sub-nets are joined to the sub-nets that is re-discovered. In Fig. 3, 8 iterative steps are needed. The algorithm from [13] makes one additional step to preserve the border between the re-discovered sub-nets and the remaining model.

The greedy algorithm works well and produces models that fit corresponding event logs perfectly. Moreover, it preserves a part of the model that contains no erroneous sub-nets.

However, the algorithm joins unnecessary sub-nets to the sub-net-to-repair. One can easily see in Fig. 3 that the whole *upper* part of the model will be re-discovered whereas significantly smaller number of sub-nets are important for the repair.

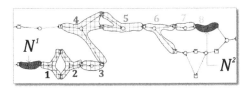

Fig. 4. The necessary sub-nets selected

Figure 4 shows the sub-net consisting of the sub-nets needed to connect N^1 and N^2. This disadvantage of the algorithm comes with its greedy nature. At each iterative step, all neighbors are joined with the sub-net whereas most of them we can leave untouched. Thus, it is possible to reduce the size of the model fragment being changed during repair. Since the size of this fragment affects conformance metrics such as generalization, precision and simplicity of the repaired model, it is more beneficial to re-discover a smaller sub-net, especially when the existing model is of high quality. The following section describes the algorithm to select the smaller fragment-to-repair.

4   Algorithm for Constructing Model Decomposition

We propose the following approach to construct the sub-net that will be re-discovered. Firstly, the model is decomposed using the *maximal decomposition* [18]. Secondly, the so called *decomposition graph* is constructed. Thirdly, the algorithm, that is described in the remainder, is executed to find the minimum spanning tree in this graph. We need to find the specific tree which will include all nodes corresponding to the erroneous sub-nets. Let us consider the proposed approach in detail.

The remainder of this section presents an algorithm for selecting the minimal sub-net that will include all initial model sub-nets to be repaired. The algorithm works with the decomposition graph of a workflow net.

Definition 3 (Decomposition graph). *For a given Petri net $N = (P, T, F)$ a decomposition graph G is a pair (V, E), where $V = P$ is the set of all places in a Petri net, and $E = \{(p_1, p_2) \,|\, p_1, p_2 \in P, \exists t \in T : (F(p_1, t) \neq 0) \wedge (F(t, p_2) \neq 0) \wedge (l(t) \neq \tau)\}$ is a set of edges.*

Decomposition graph is a way to depict a relation between two particular places in a Petri net, regardless of the number of transitions between those places. An edge in this graph shows that places p_1 and p_2 are connected by some transition, but it does not specify how many parallel transitions may connect these places.

In general, a decomposition graph is directed one. However, this graph can be considered as undirected one for the purpose of finding the minimal sub-net that includes sub-nets to be repaired. Thus, arcs are considered as undirected edges in the remainder.

Figure 5(b) shows an example of a decomposition graph for a workflow net which maximal decomposition (Fig. 5(a)) contains 63 sub-nets.

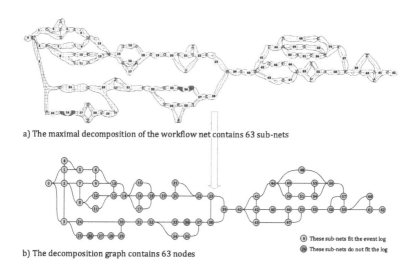

a) The maximal decomposition of the workflow net contains 63 sub-nets

b) The decomposition graph contains 63 nodes

○ These sub-nets fit the event log
● These sub-nets do not fit the log

Fig. 5. Decomposition graph (Color figure online)

The developed algorithm also works with the notion of dominator. We define dominator as follows:

Definition 4 (Graph dominator). *Let $G = (V, E)$ be an undirected graph. Let $r \in V$ be the (arbitrary selected) root (or source vertex). A dominator is the relation between two vertices. A vertex $v \in V$ is dominated by another vertex $w \in V$ if every path in the graph from the root r to v have to go through w.*

Let $G = (V, E)$ be the undirected decomposition graph, where V is the set of vertices and E is the set of edges. Our goal is to find the minimum spanning tree of a subset of vertices in G that will include all vertices corresponding to the sub-nets with recall smaller than 1. Further this subset of vertices will be referred to as R $(R \subseteq V)$.

Let C be all vertices in G that are not included in R, i.e. $C = \{v \mid v \in V \setminus R\}$. In Fig. 5(b) vertices from the set R are colored with red while vertices from the set C are colored with green. Our goal is to find a subgraph of G that will span all red vertices operating the least possible amount of green vertices.

We have developed the Algorithm 1 for finding the minimum spanning tree of undirected graph's vertex subset. Let us consider this algorithm in detail.

Its input contains a decomposition graph $G = (V, E)$ in which all vertices are either marked C (which means that they correspond to the sub-nets without errors w.r.t. event log) or R (which correspond to erroneous sub-nets).

The output contains two graphs—the first one ($G_{unfitting}$) representing the desired spanning tree, and the second one ($G_{fitting}$) containing all vertices and edges from the input (original) decomposition graph which were not included in $G_{unfitting}$. Note that $G_{fitting}$ might be a disconnected graph whereas $G_{unfitting}$ needs to be connected.

4.1   GetAllShortestPaths Procedure

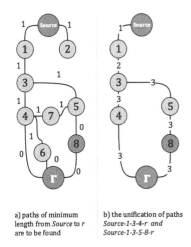

a) paths of minimum length from *Source* to r are to be found

b) the unification of paths *Source-1-3-4-r* and *Source-1-3-5-8-r*

The procedure `GetAllShortestPaths` has been developed for this algorithm. This procedure finds the set *Paths* of all paths of minimum length from *Source* to vertex r in a given graph G using the standard depth-first search algorithm. However, when searching for each path of minimum length we consider the cost of an edge ending with a vertex $r \in R$ be 0, and the cost of an edge ending with a vertex $b \in C$ be 1.

Each path is represented as a set of edges $Edges_{path}$. Let set E_r be the unification of all $e_r \in Edges_{path}$. The function returns sub-graph G_r of G as a set of edges E_r. An example is depicted in Fig. 6.

Fig. 6. Example of paths

4.2   Algorithm Steps

The Algorithm 1 steps are as follows.

(1) The algorithm randomly chooses a *Source* vertex among all vertices in R.

(2) It creates an empty set *Doms* that will contain graph dominators discovered in step (3.b).

(3) Then, a number of operations is performed for each vertex $r \in R$. (3.a) The algorithm executes $GetAllShortestPaths(G, Source, r)$ procedure that finds a graph G_r consisting of all shortest paths in G from *Source* to r. (3.b) It finds all vertices that dominate r in G_r. (3.c) Finally, all discovered dominators are added to *Doms*.

Then, the algorithm constructs the set $Edges_{unfitting}$. **(4)** For each vertices d_1, $d_2 \in Doms$ such that there exist edge $(d_1, d_2) \in E$ the algorithm **(4.a)** adds d_1 and d_2 to the set $Vertices_{unfitting}$, and **(4.b)** adds edge (d_1, d_2) to the set $Edges_{unfitting}$.

Algorithm 1: Finding the spanning tree of the decomposition graph

Input : $G = (V, E)$ is a decomposition graph, where V is a pair (C, R)
Output: $G_{unfitting}, G_{fitting}$

$Source \leftarrow$ GetRandomElement(R);
$Doms \leftarrow \emptyset$;
foreach $r \in R$ **do**
   | $G_r \leftarrow$ GetAllShortestPaths$(G, Source, r)$;
   | $dominators_r \leftarrow$ FindDominators$(G_r, Source, r)$;
   | $Doms \leftarrow Doms \cup dominators_r$;
end
$Vertices_{unfitting} \leftarrow \emptyset$;
$Edges_{unfitting} \leftarrow \emptyset$;
foreach $d_1, d_2 \in Doms$ **do**
   | **if** $(d_1, d_2) \in E$ **then**
      | $Vertices_{unfitting} \leftarrow Vertices_{unfitting} \cup d_1 \cup d_2$;
      | $Edges_{unfitting} \leftarrow Edges_{unfitting} \cup (d_1, d_2)$;
   | **end**
end
foreach $r \in R$ **do**
   | **if** $r \notin Vertices_{unfitting}$ **then**
      | $(V_{Path}, E_{Path}) \leftarrow$ GetPathToTheNearestVertex$(G, r, Vertices_{unfitting})$;
      | $Vertices_{unfitting} \leftarrow Vertices_{unfitting} \cup V_{Path}$;
      | $Edges_{unfitting} \leftarrow Edges_{unfitting} \cup E_{Path}$;
   | **end**
end
$Vertices_{fitting} \leftarrow \{v | v \in V \setminus Vertices_{unfittng}\}$;
$Edges_{fitting} \leftarrow \{e | e \in E \setminus Edges_{unfitting}\}$;
$G_{fitting} \leftarrow (Vertices_{fitting}, Edges_{fitting})$;
$G_{unfitting} \leftarrow (Vertices_{unfitting}, Edges_{unfitting})$;
return $(G_{fitting}, G_{unfitting})$;

Then, the last of three iterative procedures is executed. **(5)** For each vertex r such that $r \in R$ and $r \notin Vertices_{unfitting}$ the algorithm **(5.a)** finds $P = (V_{Path}, E_{Path})$—a path from r to the nearest vertex in $Vertices_{unfittng}$ (a vertex that corresponds to an erroneous sub-net), **(5.b)** adds all vertices on this path $v_p \in V_{Path}$ to $Vertices_{unfitting}$, and **(5.c)** adds all edges of this path $e_p \in E_{Path}$ to $Edges_{unfitting}$.

Finally, the output sub-graphs $G_{fitting} = (Vertices_{fitting}, Edges_{fitting})$ and $G_{unfitting} = (Vertices_{unfitting}, Edges_{unfitting})$ are constructed. Here $Vertices_{fitting} = \{v|v \in V \setminus Vertices_{unfittng}\}$ is the set of vertices in the spanning tree and $Edges_{fitting} = \{e|e \in E \setminus Edges_{unfitting}\}$ is the set of edges of this tree.

The number of steps performed by the algorithm is bounded by $O((|V|^2 + |E|) \cdot |V|)$ where $|V|$ and $|E|$ are the number of vertices and the number of edges in the decomposition graph. However, the time complexity may alter depending on the time complexity of an algorithm used to find graph dominators in step **(3.b)**. In this case, we were using an approach which is quadratic in the number of graph nodes.

This algorithm can be directly used to reconstruct the building block (4) Repair of the approach described in Sect. 3. Indeed, the greedy iterative procedure can be replaced by this improved procedure which selects the sub-net-to-repair in a more precise manner. Then, the selected sub-net is re-discovered as in the greedy approach.

5   Implementation and Evaluation

We implemented a tool prototype in Java based on ProM 6.7 Framework [21] to evaluate the developed algorithm. Source code is freely available at the *Iskra* project page: https://pais.hse.ru/research/projects/iskra.

The Algorithm 1 has been implemented in `SmartDecomposerImpl` class alongside with other Iskra decomposers.

We have performed a number of experiments on various models of different size. Experimental data is available at the Iskra project page as well. Our main goal was to compare the efficiency of the developed algorithm versus the efficiency of the other existing approaches to select the fragment-to-replace. Experiments were performed on two models: M-1 (consisting of 19 places and 21 transitions) and M-2 (consisting of 63 places and 70 transitions).

At first we were breaking our initial models by swapping labels of two random transitions in model M-1 (which makes ≈9.5% of all transitions) and four random transitions in model M-2 (≈5.7% of all transitions). Then we were running two repair algorithms for each broken model with a perfectly-fitting event log. These steps were repeated 10 times for each model using each one of the algorithms (40 experiments were performed overall). We used *Inductive miner* to re-discover the erroneous sub-nets.

The results suggest that the algorithm presented in this paper decomposes the initial model in such a way that a significantly smaller model fragment is touched by the repair procedure. However, it appears that the difference between the basic greedy approach and the presented algorithm is only apparent when large models are repaired. In larger models based on M-2, the greedy approach was more likely to produce poorly-generalized models compared to the developed algorithm. Even though there were cases when the developed approach produced models with a greater number of silent transitions (compared to the greedy one), those models still tended to be better-generalized. The advantage that the greedy approach has over the developed one is that it takes significantly less time to repair large models.

Table 1. Comparison of the developed and the greedy algorithms on model M-2

	Size of Wrapper		Iterations		Size of Rep. Model		Time (ms)	
	Smart	Greedy	Smart	Greedy	Smart	Greedy	Smart	Greedy
1	47	113	1	5	142	170	242862	2692
2	55	126	1	12	152	171	225070	1336
3	50	125	1	11	135	167	6262	1248
4	80	133	1	15	174	166	135474	1058
5	60	123	1	8	175	168	162256	932
6	48	120	1	9	135	168	4892	1254
7	31	45	1	1	142	149	16264	452
8	53	117	1	9	144	167	13368	906
9	38	61	1	4	143	154	69310	456
10	62	123	1	9	170	170	578948	1084
Average	52,4	108,6	1	8,3	151,2	165	145470,6	1141,8
Median	51,5	121,5	1	9	143,5	167,5	102392	1071
max	80	133	1	15	175	171	578948	2692
min	31	45	1	1	135	149	4892	452

The detailed comparison of two approaches performed on model M-2 is shown in Table 1. Values in `Size of Wrapper` column depict the total number of places and transitions in the fragment-to-be-repaired. The `Iterations` column shows the number of iterations that the greedy algorithm was performing. Values in the `Size of Rep. Model` column depict the total number of places and transitions in the repaired model.

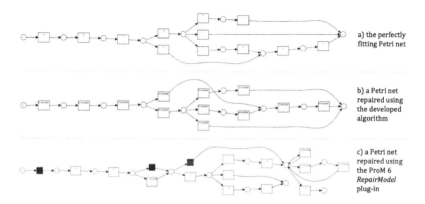

a) the perfectly fitting Petri net

b) a Petri net repaired using the developed algorithm

c) a Petri net repaired using the ProM 6 *RepairModel* plug-in

Fig. 7. Process model repair using different methods

Let us illustrate how the algorithm works with a single example. Figure 7 shows the comparison of results of two different approaches for model repair: the presented algorithm and the method from [8] as it is implemented in ProM 6. One can note that the developed algorithm (see Fig. 7b) creates a more precise model rather than the other method (see Fig. 7c).

6   Conclusions

This paper present the algorithm to find smaller sub-nets-to-repair which can improve the modular repair technique from [14]. The algorithm has been implemented as a prototype and evaluated. The evaluation shows that the repair approach touches a smaller part of the model using the *smart* decomposition algorithm presented in the paper than using the greedy algorithm. Thus, such an improvement can be considered useful.

Unfortunately, our procedure to find the spanning tree covering a sub-set of vertices is time consuming. The related problem is called *minimum Steiner tree problem* and is known to be $\mathcal{NP}$-hard. However, there are algorithms which are efficient in specific cases [15, 17]. The approximate solution can also be found using algorithms [10] for minimal spanning tree finding. One of future work directions will be to investigate existing efficient algorithms and to integrate the most suitable one into our approach. This should significantly improve performance of the whole model repair algorithm.

References

1. Mitsyuk, A.A., Lomazova, I.A., Shugurov, I.S., van der Aalst, W.M.P.: Process model repair by detecting unfitting fragments. In: Supplementary Proceedings of the 6th International Conference on Analysis of Images, Social Networks and Texts (AIST-SUP 2017), Moscow, Russia, 27–29 July 2017. CEUR-WS.org, vol. 1975, pp. 301–313. CEUR-WS.org (2017)
2. Adriansyah, A.: Aligning observed and modeled behavior. Ph.D. thesis, Technische Universiteit Eindhoven (2014)
3. Armas Cervantes, A., van Beest, N.R.T.P., La Rosa, M., Dumas, M., García-Bañuelos, L.: Interactive and incremental business process model repair. In: Panetto, H., et al. (eds.) OTM 2017. Lecture Notes in Computer Science, vol. 10573, pp. 53–74. Springer, Cham (2017). https://doi.org/10.1007/978-3-319-69462-7_5
4. Buijs, J.C.A.M., La Rosa, M., Reijers, H.A., van Dongen, B.F., van der Aalst, W.M.P.: Improving business process models using observed behavior. In: Cudre-Mauroux, P., Ceravolo, P., Gašević, D. (eds.) SIMPDA 2012. LNBIP, vol. 162, pp. 44–59. Springer, Heidelberg (2013). https://doi.org/10.1007/978-3-642-40919-6_3
5. Carmona, J.: Decomposed process discovery and conformance checking. In: Sakr, S., Zomaya, A. (eds.) Encyclopedia of Big Data Technologies. Springer, Cham (2018). https://doi.org/10.1007/978-3-319-63962-8_95-1
6. Carmona, J., van Dongen, B., Solti, A., Weidlich, M.: Conformance Checking: Relating Processes and Models. Springer, Cham (2018). https://doi.org/10.1007/978-3-319-99414-7

7. Fahland, D., van der Aalst, W.M.P.: Repairing process models to reflect reality. In: Barros, A., Gal, A., Kindler, E. (eds.) BPM 2012. LNCS, vol. 7481, pp. 229–245. Springer, Heidelberg (2012). https://doi.org/10.1007/978-3-642-32885-5_19

8. Fahland, D., van der Aalst, W.M.P.: Model repair - aligning process models to reality. Inf. Syst. **47**, 220–243 (2015)

9. Gambini, M., La Rosa, M., Migliorini, S., Ter Hofstede, A.H.M.: Automated error correction of business process models. In: Rinderle-Ma, S., Toumani, F., Wolf, K. (eds.) BPM 2011. LNCS, vol. 6896, pp. 148–165. Springer, Heidelberg (2011). https://doi.org/10.1007/978-3-642-23059-2_14

10. Karger, D.R., Klein, P.N., Tarjan, R.E.: A randomized linear-time algorithm to find minimum spanning trees. J. ACM **42**(2), 321–328 (1995)

11. Leemans, S.J.J., Fahland, D., van der Aalst, W.M.P.: Discovering block-structured process models from incomplete event logs. In: Ciardo, G., Kindler, E. (eds.) PETRI NETS 2014. LNCS, vol. 8489, pp. 91–110. Springer, Cham (2014). https://doi.org/10.1007/978-3-319-07734-5_6

12. Maggi, F.M., Bose, R.P.J.C., van der Aalst, W.M.P.: A knowledge-based integrated approach for discovering and repairing declare maps. In: Salinesi, C., Norrie, M.C., Pastor, Ó. (eds.) CAiSE 2013. LNCS, vol. 7908, pp. 433–448. Springer, Heidelberg (2013). https://doi.org/10.1007/978-3-642-38709-8_28

13. Mitsyuk, A.A.: Non-local correction of process models using event logs. In: Proceedings of the 2017 Ivannikov ISPRAS Open Conference, pp. 6–11. IEEE Computer Society, Los Alamitos (2018)

14. Mitsyuk, A.A., Lomazova, I.A., van der Aalst, W.M.P.: Using event logs for local correction of process models. Autom. Control Comput. Sci. **51**(7), 709–723 (2017)

15. Pajor, T., Uchoa, E., Werneck, R.F.: A robust and scalable algorithm for the Steiner problem in graphs. Math. Program. Comput. **10**(1), 69–118 (2018)

16. Polyvyanyy, A., van der Aalst, W.M.P., ter Hofstede, A.H.M., Wynn, M.T.: Impact-driven process model repair. ACM Trans. Softw. Eng. Methodol. (TOSEM) (2016)

17. Polzin, T., Vahdati-Daneshmand, S.: Approaches to the Steiner problem in networks. In: Lerner, J., Wagner, D., Zweig, K.A. (eds.) Algorithmics of Large and Complex Networks. LNCS, vol. 5515, pp. 81–103. Springer, Heidelberg (2009). https://doi.org/10.1007/978-3-642-02094-0_5

18. van der Aalst, W.M.P.: Decomposing Petri nets for process mining: a generic approach. Distrib. Parallel Databases **31**(4), 471–507 (2013)

19. van der Aalst, W.: Process Mining: Data Science in Action, 2nd edn. Springer, Heidelberg (2016). https://doi.org/10.1007/978-3-662-49851-4

20. van der Werf, J.M.E.M., van Dongen, B.F., Hurkens, C.A.J., Serebrenik, A.: Process discovery using integer linear programming. Fundam. Inform. **94**(3–4), 387–412 (2009)

21. Verbeek, H.M.W., Buijs, J.C.A.M., van Dongen, B.F., van der Aalst, W.M.P.: ProM 6: the process mining toolkit. In: La Rosa, M. (ed.) Proceedings of BPM Demonstration Track 2010. CEUR Workshop Proceedings, vol. 615, pp. 34–39 (2010)

22. Voorhoeve, M.: Structural Petri net equivalence. Technical report, Eindhoven University of Technology, Department of Mathematics and Computing Science (1996)

Author Index